TOWNSHIP

PETER LANG
New York • Washington, D.C./Baltimore • Bern
Frankfurt am Main • Berlin • Brussels • Vienna • Oxford

Michael D. Sublett

TOWNSHIP

Diffusion and Persistence of Grassroots Government in Illinois, 1850–2000

PETER LANG
New York • Washington, D.C./Baltimore • Bern
Frankfurt am Main • Berlin • Brussels • Vienna • Oxford

Library of Congress Cataloging-in-Publication Data

Sublett, Michael D.
Township: diffusion and persistence of grassroots government in Illinois, 1850–2000 / Michael D. Sublett.
p. cm.
Includes bibliographical references and index.
1. Local government—Illinois—History. I. Title.
JS451.I35S83 320.8'3—dc22 2003025252
ISBN 0-8204-7055-4

Bibliographic information published by **Die Deutsche Bibliothek**.
Die Deutsche Bibliothek lists this publication in the "Deutsche Nationalbibliografie"; detailed bibliographic data is available on the Internet at http://dnb.ddb.de/.

Cover design by Lisa Barfield
Cover concept by Michael D. Sublett

The paper in this book meets the guidelines for permanence and durability of the Committee on Production Guidelines for Book Longevity of the Council of Library Resources.

275 Seventh Avenue, 28th Floor, New York, NY 10001
www.peterlangusa.com

Printed in Germany

For Patricia Ann and Jennifer Ann

Table of Contents

Illustrations

Tables

Preface

THE YEAR WAS 1984. Colleague William D. Walters, Jr., and I were returning to central Illinois after a conference in Cape Girardeau, Missouri. As we motored northward along Interstate 55 between St. Louis and Springfield, I observed that, according to my calculations, we were in the zone where a hundred years or so before a struggle had occurred between forces that wanted to retain a form of local government that had come early into Illinois from the Upland South and those that coveted access to a system they had witnessed in New England or elsewhere out east. Bill and I speculated about how the confrontation might have materialized and whether it moved steadily across the state or if it had jumped around from county to county, following no particular spatial pattern. The outcome of the struggle I already knew from having examined, years before, a map depicting the 17 small, lightly populated, generally peripheral counties that do not subscribe to the replacement form of local government, the one favoring named townships and an assortment of township officers (figure 1). What sorts of arguments had proponents of township government used to sign up such an overwhelming majority of the 102 Illinois counties? How had the 17 been able to resist? In what other states had townships invaded and upset the prevailing political landscape? Our conversation drifted on to other topics (probably campus politics or Cardinal/Cub baseball), but I now mark that brief interchange as the beginning of the research agenda that would lead me to this book.

Anyone looking at the map of the 17 commission counties, which has not changed since 1933, ought to pause and wonder what is and what was at work here to create this pattern. Clearly some force, some trend, some confrontation worked to make the map appear this way. Could the pattern be simply a function of the distance (however measured) from downtown Chicago? Could there have been partisan politics involved? Did some legislative, judicial, or constitutional quirk require or encourage these counties to desist from joining their 85 township-forming brethren? Had these 17 counties been holdouts from the outset or had one or more dipped a toe into the township trough only to withdraw it later? Did the 17 pop up as a package of resisters or did the adopters increase gradually in number to reach their plateau of 85? An academic research topic seemed present here, but how far would it take me?

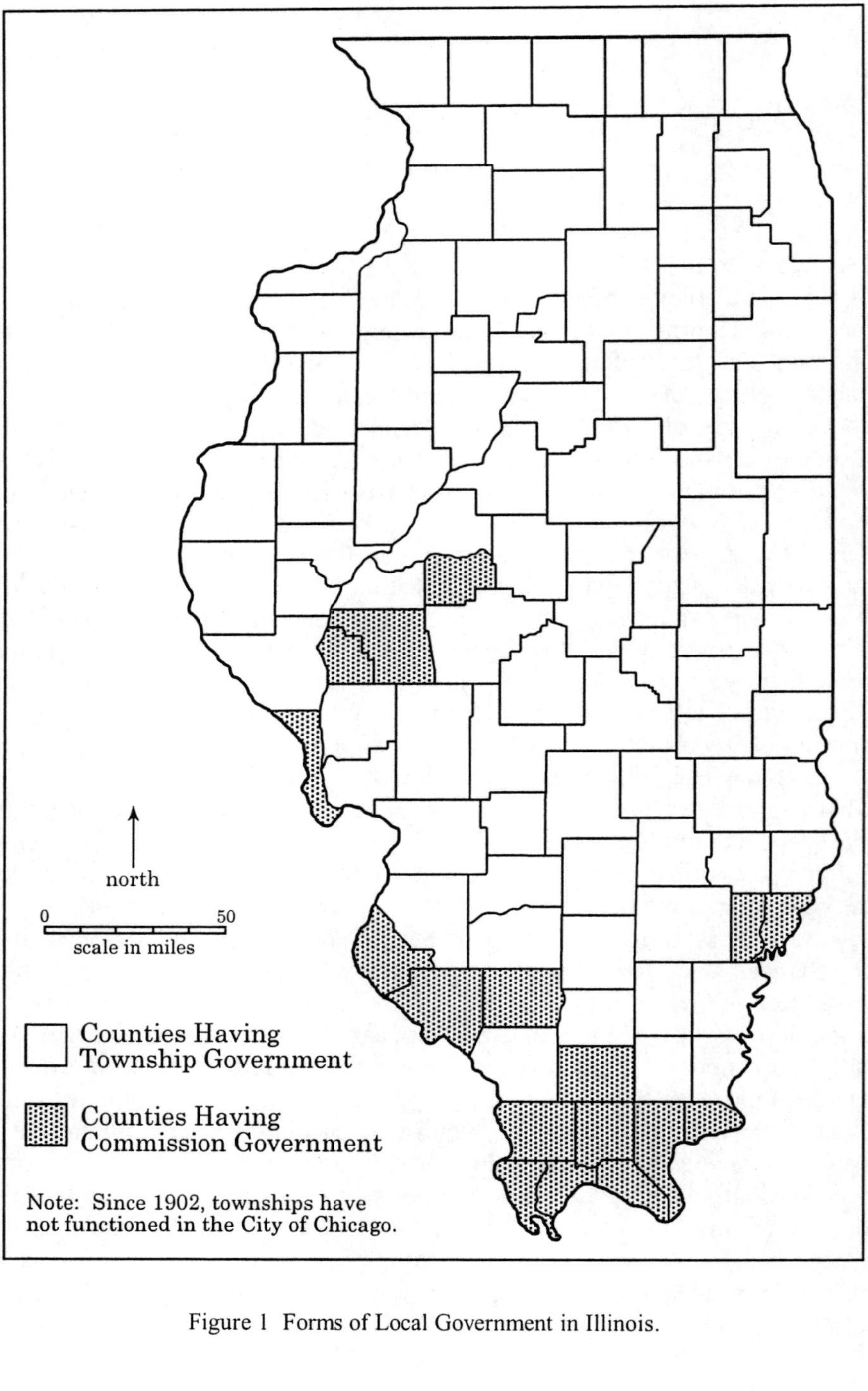

Figure 1 Forms of Local Government in Illinois.

My intention in 1990, when I finally got a chance to bring that map of the 17 versus 85 to the forefront, was to examine a sample of nineteenth century newspapers, beginning with 1850 because I knew townships such as Illinois now has did not exist before the ratification of a revised constitution in 1848 and the subsequent legislation and referenda of 1849. I did not know then if contemporary newspapers even bothered to cover township organization as an election issue. At that juncture, I thought I would be investigating a topic that would have no modern legs, a topic that I could safely observe from the vantage point of a century or at least several decades removed from the action. I had no idea then that during the summer of 1990, while I was rolling microfilm of old newspapers through a reader in Springfield, two determined men and their meager band of township critics were circulating petitions door to door and at parades, picnics, and plays in the park, trying to collect several thousand signatures in the hope of bringing the abolition of townships before the electorate that November in Rock Island County.

This project did not commence with a book in mind. Instead, I just found the map of 17 and 85 intriguing; and it was especially intriguing for one who had used the governmental township in several states as a convenient research vessel and someone who likes to focus on political boundaries as a landscape element. Somewhere, somehow the topic took on vigor, a character, and a life. At some point I recognized that maybe there was more to the topic than the answers to some questions tossed back and forth in a car on a highway south of the state capital in 1984.

Township focuses on township government during its first 15 decades in Illinois. I have chosen to highlight diffusion because the pattern of 17 and 85 is the result of a clear case of an innovation moving through space and through time across a territory. I emphasize persistence because, despite dozens of ballot challenges, township government has almost always survived and lives on into the new century. Grassroots is in the title due to the frequency of its appearance in the debate over whether this form of government, so close to the people, ought to continue to levy taxes and provide services in an era far removed from the middle of the nineteenth century when settlers brought it in from the East.

A portion of this monograph appeared in an article in the spring 1997 issue of the *Bulletin of the Illinois Geographical Society.* I am grateful to the editor for permission to use here parts of my "Efforts to Abolish Illinois Townships: The 1990s."

Years ago a geographer told his class at the University of Missouri-Columbia that an investigator could not claim to have contributed to the geographical literature if he or she did not utilize, truly utilize, at least one map in presenting the results. I nodded my head then and concur today. Maps enhance this study, and I express my gratitude to my cartographic colleague, Jill Freund Thomas, for her efforts over the years to illustrate the numerous iterations of my township research and for her contributions to the maps and graphs that I feature herein.

Many others have played significant roles as well. Kathaleen Caisse, then an undergraduate honors student majoring in political science at Illinois State, provided early research assistance related to published county histories. Principal 1990s activists Michael D. Richardson and Robert G. (Bob) Anderson offered me access to the inner circle of their township abolition movement and kept me updated, through materials mailed and periodic conversations over the telephone and in person. Other activists, like Randy Popp, Gail Dunham, Rosie Fitzpatrick, Janice Schnobrich, Carl Edwards, David Partney, Bill Bethel, Charles Horton, and Judy Anderson (spouse of Bob Anderson), were helpful and courteous. Township officials were forthcoming, when I requested their assistance. Former Township Officials of Illinois executive director George H. Miller and his successor (and former associate director), Bryan E. Smith, answered my questions and sent useful printed material originating with their organization. Bryan Smith allowed me to consult elusive back issues of the organization's magazine and gave me permission to publish photographs he provided of Troy Kost, George Miller, and himself with George Miller. A number of township officials at the local level, like my township's supervisor during the 1990s, Mary Caisley, were generous with time and insights. I also spoke with several county officials about the differences between commission-run counties and those with townships, and therefore a different type of county board. Keith Sculle, historian (and sometime geographer) at the Illinois Historic Preservation Agency, offered vital encouragement, helped check a reference now and then for my citations, and served as the first critical reader of the manuscript. Former colleague, now emeritus professor of geography at Illinois State University, Paul Mattingly read and commented on the early chapters, with a particular eye to one of his research specialties—spatial diffusion. Management at the *Rock Island Argus* granted permission to publish a photograph of Michael Richardson and Randy Popp.

Financial assistance came from two principal sources. Illinois State University provided funds to support research travel and related expenses, like photocopying; granted me a semester-long sabbatical leave, in the fall of 2000, plus regular reductions in teaching assignments to make available more time for research; and secured the services of Mary T. Van der Hoven to help with technical aspects of editing the text. The Illinois Humanities Council funded preparations for, delivery of, and evaluation of a series of four public lectures that I gave in four counties where the idea of township government had to struggle mightily for acceptance.

Annotated Abbreviations

APA	Auditor of Public Accounts. State officer charged with keeping track of the state's townships.
BCCGG	Bipartisan Committee to Continue Good Government. Citizen group, which in 1966 stood against countywide abolition of townships in Saline County.
CAC	Citizen Advocacy Center. Not-for-profit organization established in 1993 to assist, especially with legal services, Chicagoland residents attempting to participate in community affairs.
CBTCG	Committee for a Better Type of County Government. Citizen group, which in 1966 sought to abolish all township government in Saline County.
CCC	County Commission Committee. Group of Saline County citizens that attempted to abolish townships there in 1996.
CLG	Committee on Local Government. Official name of the committee of the 1970 Constitutional Convention that dealt with township government.
CMT	Citizens for McHenry Township. Group that formed in 1998 to counter a threat of abolition against McHenry Township (McHenry County).
CTG	Citizens for Township Government. McHenry County group that citizens and township officials organized in 1994 to combat an effort to abolish all townships in that county.
CTIG	Concerned Taxpayers for Integrity in Government. McHenry County organization formed in 1993 to challenge actions of local officials and, if necessary, seek the abolition of governmental units.
FTG	Friends of Township Government. Franklin County citizen group, composed largely of township officials, which worked to save township government there during 2000 abolition effort.
IATO	Illinois Association of Township Officials. Predecessor group to the Township Officials of Illinois, during the period 1936-1945.
ICTE	SIllinois Commission on Taxation and Expenditures. Creation of the Fifty-eighth General Assembly tasked to find ways to reduce the cost of government during the mid-1930s.
ICTO	*Illinois County and Township Official*. Magazine that served as main means of communication between TOI and its member townships and their officials prior to the name change, in 1999, to *Township Perspective.*
ISAHCTC	Illinois State Association of Highway Commissioners and Town Clerks. Predecessor group to the Illinois Association of Township Officials and the Township Officials of Illinois, during the period 1907-1936.

JLC	Joint Legislative Committee. Creation of the Forty-seventh General Assembly tasked to study and make recommendations about counties, townships, and road districts during 1912 and 1913.
P-TAC	Property Tax Awareness Committee. Franklin County citizen group that collected signatures in 2000 in an attempt to abolish all of that county's civil-political townships.
RT	Retrenchment Commission. Creation of the Fifty-eighth General Assembly tasked to study and recommend consolidation or abolition of governmental units during the mid-1930s.
SCOTS	Some Citizens Opposed to Township Spending. Group that fielded candidates in 1972 pledged, if victorious, to dismantle the governments of Niles and Palatine townships (Cook County).
SCTA	Saline County Taxpayers Association. Group that worked during the 1970s to take Saline County back to county commissioner government.
STC	State Tax Commission. State agency that between 1919 and 1943 conducted research and made recommendations about Illinois property tax assessment and other local tax matters.
STOP	Send Township Officials Packing. Multi-county citizen group founded in 1995 to harass and abolish townships one by one or on a countywide basis.
TFTG	Taxpayers for Township Government. Saline County group that opposed abolition of townships there in 1976.
TGPG	Township Government Problems Commission. Township-friendly creation of the Seventieth General Assembly designated, in the late 1950s, to examine township government and make recommendations.
TIP	Township Improvement Party. Niles Township (Cook County) group that ran candidates pledged to continuation of township government, in 1972.
TOCC	Township Officials of Cook County. First of many county organizations of townships officials in Illinois dedicated to the improvement and continuation of township government.
TOI	Township Officials of Illinois. Statewide organization, tracing its lineage back to 1907, that serves as the chief advocate for the state's civil-political townships and their officials.
TOIPAC	Township Officials of Illinois Political Action Committee. A 1990 creation of the Township Officials of Illinois that has the ability to collect funds, endorse legislative candidates, and make campaign contributions.
TOIRMA	Township Officials of Illinois Risk Management Association. The self-insurance component of the Township Officials of Illinois. Dates from 1986.
TSG	Township Study Group. Rock Island County organization that challenged township government there during the early 1990s.

1 Introduction

TOWNSHIP GOVERNMENT PERSISTS in Illinois more than a century and a half after its inception. Persistence occurs in part because townships offer citizens reasonably priced services at a scale that citizens connect with and appreciate. On the other hand, it also persists because of constant vigilance by local officials and their statewide umbrella organization, the Township Officials of Illinois (TOI), in the face of those who believe the system is wasteful and outdated. In fact, a protracted struggle has characterized the confrontation between this product of an invasion from the northeastern part of the United States and the commissioner form of local government, which had been in place from the time of statehood in 1818. Gradually, county by county, the township innovation pushed south and westward from its hearth near the shores of Lake Michigan into the strongholds of the Upland South, into counties generally south of Springfield where immigrants from Kentucky, Tennessee, Virginia, and the Carolinas dominated politics for much of the nineteenth century. Township government, featuring its signature, citizen-friendly, annual town meeting, was only one aspect of what Timothy C. Frazer called a "Yankee conquest" of the inland empire, a conquest that "was to be peaceful" though the "military spirit was common." Township government seems to have been one of this peaceful conquest's "tangible victories."[1]

Today local township officials and TOI are still on alert, still responding immediately to even the most innocuously negative or misinformed newspaper editorial or letter to the editor. Announcement of a petition drive to place the issue of discontinuance on a countywide or single-township ballot for the public to consider brings out the big guns from TOI for serious engagement. Persistence of the township has not been and will not be without its price on the part of its defenders, although now the confrontation is less between townships and the commission format and more about whether townships or county or municipality can deliver services better and cheaper.

The reader should not confuse the townships upon which I am focusing here with those townships that resulted from the grid that federal surveyors laid down across most of Illinois. Generally six miles square, or nearly

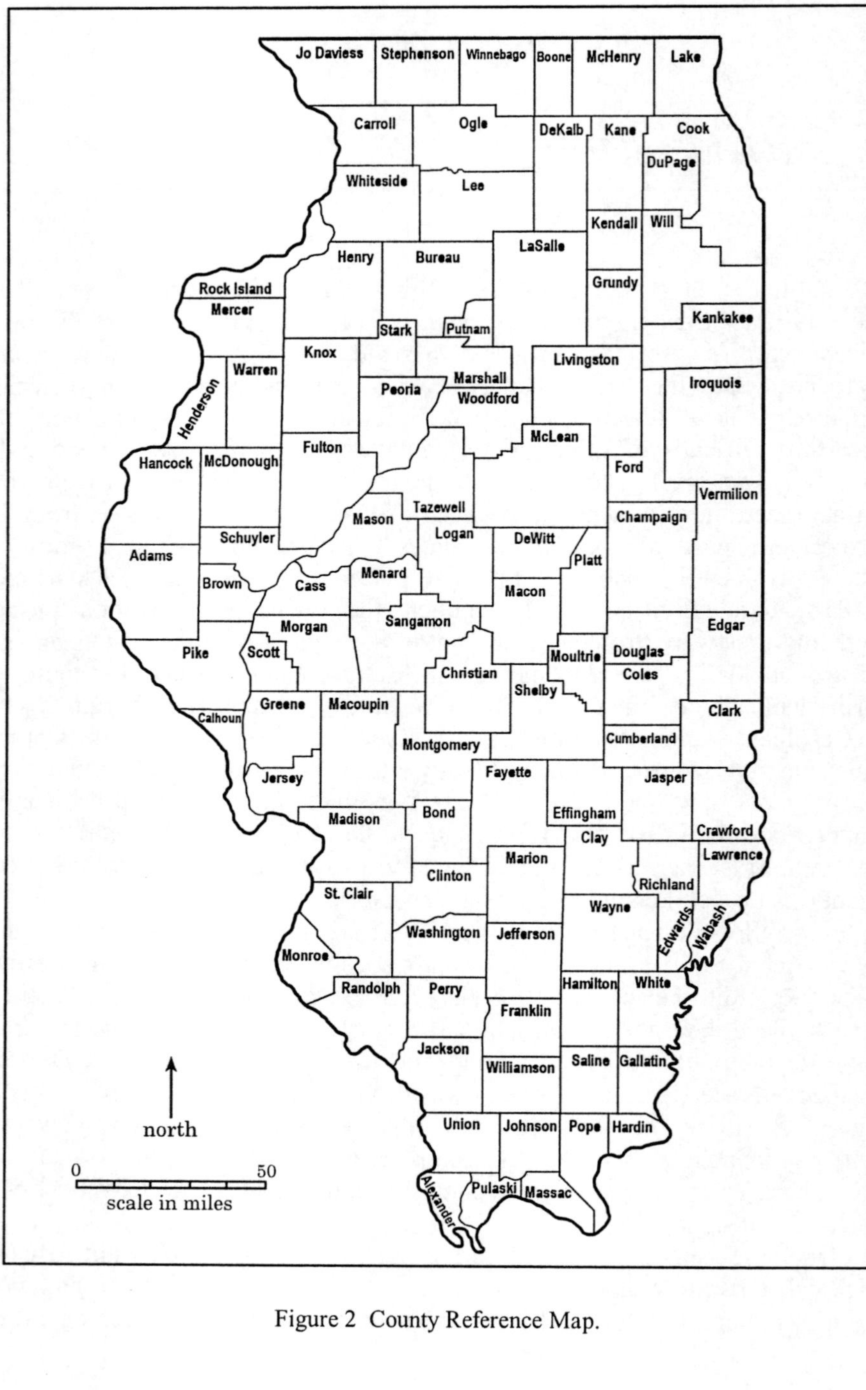

Figure 2 County Reference Map.

square given the convergence of the meridians along their eastern and western margins and surveyor error, federal townships go by such names as survey, congressional, land, and geographical; and they occur in many states that lack township government.[2]

Do not confuse my focus here with those townships that also served early Illinoisans as the framework for administrating the state's common (public) schools. Elementary school districts, based on federal townships, came into existence immediately after statehood because of action by the First General Assembly: "An Act Relating to the Lands Reserved for the Use of Schools."[3] Trustees of elementary districts utilized proceeds from their federally granted land (typically Section 16 [one square mile] of the 36 sections in the standard federal township) to support building and staffing one-room school buildings scattered about one per 4–6 square miles. In the common school district, for instance, that occupied Township 1 North, Range 2 East of the Third Principal Meridian (now Raccoon Township in Marion County [see figure 2]), trustees sold the granted land, loaned the profits to local residents, and used the interest earned to run their schools.[4]

School consolidation, largely occurring right after World War II, ended many of these township-size elementary districts. High school districts based on federal township dimensions date from the years after the Civil War in Illinois when the general assembly provided for Princeton Township High School in 1867 and passed general legislation in the 1870s making such districts possible elsewhere in the state. Numerous township high school districts came on line in subsequent years, and many township high schools remain active.[5]

In Illinois, the federal townships appeared between 1800 and 1840, so they were waiting when modern township government became a reality in 1850. These latter townships, the real focus for our conversation here, have a variety of prefatory descriptors: civil, political, municipal, and governmental. Because the federal townships preceded them, were about the right size for a nineteenth century minor civil division, and were having such a profound effect on layout of counties and rural roads, the general assembly encouraged the men (they were always men) who were to divide adopting counties into civil townships, to use the existing survey township boundaries if feasible. Many times counties did make the two coterminous. In McLean County, for example, the original correspondence between the two was striking, but not total. Twenty were coterminous, nine civil townships were larger than the surveyed townships, and two were smaller.[6] Later the City of Bloomington Township came along to disrupt some of this historic correspondence of boundaries in McLean County (figure 3). Henceforth, unless I specify otherwise, all township references here will be to these civil-political-municipal-governmental units.

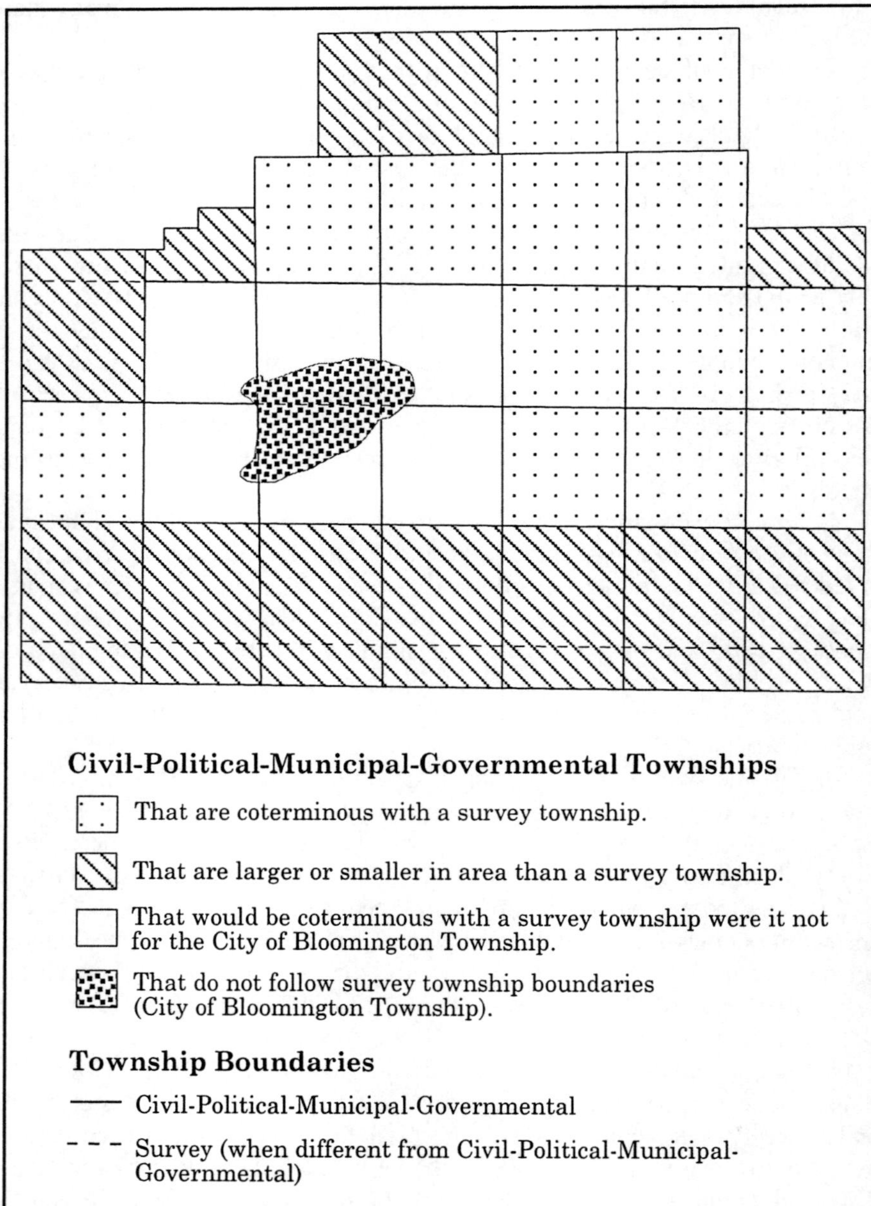

Figure 3 Townships in McLean County, Illinois.

Not Just an Illinois Phenomenon

"It has been amply demonstrated," wrote Paul W. Wager in 1957, "that in most areas there is no longer a need for township government." Of Iowa townships, in 1952, Richard C. Spencer said, "Most students of local government agree that under modern conditions the township is not and cannot be an effective governmental unit." Clyde F. Snider concluded his ominously titled 1952 piece, "Twilight of the Township," with these elegiac words:

> From whatever angle considered, the conclusion seems inescapable that the sunset of township government even now is past. Year by year its twilight deepens. As an instrument of government, the civil township has outlived its usefulness and those interested in improving our local institutions will do well to foster, in every legitimate way, its ultimate demise.[7]

Yet, today, a half-century later, townships are thriving in states as far east as New York (as far east as Maine, if we stretch the definition to include the somewhat different New England town) and as far west as the Dakotas.

One simple indicator of how well the township is doing in a particular state is the existence of a state organization to lobby its cause. Joining TOI are similar organizations in Connecticut, Indiana, Michigan, Minnesota, New York, North Dakota, Ohio, Pennsylvania, South Dakota, and Wisconsin, as well as the National Association of Towns and Townships. Many of the state groups maintain Websites and publish materials ranging from monthly magazines to manuals for the officials in the trenches. Other indicators of township wellness include the authority to levy taxes, provide mandated functions, and offer additional services. See chapter 3 of Monica Dwyer Abress's *Quietly at Work: Township Government in America* for a state-by-state compilation of what townships do in most of the places where they operate.[8]

Scope and Purpose of Study

Township is a study in historical/political geography. Critical to the book are the 438 county-specific, countywide referenda that I have documented. These referenda, occurring in every decade from the 1840s through the 1990s, provided county officials with citizen authorization to replace the county commissioner form of government with the township setup or, once townships were in place, to abolish townships and revert to the three commissioners elected at large for staggered terms of first one, then three, then four, and now six years. My main purposes here are to track the spread of the township across Illinois through time and space, and then to follow efforts to reverse the process by removing townships one by one, county by county, or statewide.

Although the referenda provide raw material for the book, they tell us nothing about what the voters knew, felt, or believed regarding township organization, the official name in Illinois for this governmental system. So, whenever possible, I provide the local flavor and prevailing sentiments. I attempt to portray the problems that townships faced, especially during periods of stress, when abolition talk and action were commonplace. Overall, *Township* offers a balanced and careful commentary on a subject that has generated considerable interest, both lay and academic, over the years since a determined group of Yankees brought the township to the Midwest. I do not attempt here, however, to link local decision making necessarily with the broad currents of national and state history, unless these larger arenas were clearly relevant to township diffusion and persistence. The Civil War and Great Depression did play notable roles in the township tale, whereas the Panic of 1873 and World War I did not. The Progressive movement and its backwash in the 1910s and 1920s mattered some; Prohibition and its repeal seemed not to matter at all. Furthermore, I leave to the political scientists and the citizenry the task of deciding whether or not townships really belong. My intention was never to provide a cost/benefit analysis of the Illinois township or to come down on the side of sustainment or abolition. My guess is that townships and their elected officials will face challenges, make adjustments, and (above all) persist.

Referenda Sources

Data collection has taken me from the humidity-controlled, pencils-only manuscript room of the Illinois State Historical Library to the dark and filthy, stoop-over, watch-your-head basement of an ancient courthouse on the National Road to the hot, stuffy, and dusty attic of a small courthouse in the Illinois Military Tract. In between was an assortment of library reading rooms, county clerk offices, the seven Illinois Regional Archives Depositories, the State Archives, the Illinois State Library, and a whole lot of time in the twilight of the Illinois State Historical Library's microfilm-viewing quarters. For the most part, and to my satisfaction, collection guardians were courteous and helpful, even the county clerk/recorder in that courthouse on the National Road who led me through the catacombs of his domain to find and retrieve the minutes of the county board of supervisors that I was seeking. Here is a rundown of the sources I used to document the 438 countywide referenda.

At about the time I was seriously beginning to investigate townships, Sheila Kelly, an employee of the Illinois State Archives, authored the *County and Township Gazetteer.*[9] She listed for each county the date that county adopted, if it had adopted, township organization. In some cases, when the counties adopted, disadopted, and adopted again, she provided multiple dates. Approximately one-fifth of the referenda I documented came from Kelly. Almost all of these 81, however, are the readily available dates on which voters initially approved townships, many of which I validated later in other references. Nevertheless, the *Gazetteer* was and is a solid source of valuable township-related information, organized in alphabetical order by county. She also provided a useful index, listing all fourteen hundred plus modern-day Illinois townships.

If county clerks had submitted the results of every referendum on township organization to the secretary of state, my task of assembling the referenda roster would have been much simpler. Unfortunately, they were not always forthcoming with results, especially in the nineteenth century and especially if township organization failed. At the State Archives, first in bound volumes, by year and containing election results from diverse counties, then, after 1882, continuing with files by county through the years are the secretary of state's Election Returns collection. From that archive, I was able to extract 146 referenda results or 33 percent of the total. Entries therein give votes tallied for and against the issue, and one can obtain an idea of the total vote cast at the election by looking for the race or issue that generated the most votes for and against. Total vote was a key number because the legislation making townships possible once required, for adoption, a favorable majority of all votes cast. I was able frequently to use the Election Returns as a cross check on other sources, so, in actuality, I could have deferred to Election Returns for nearly half of the 438.

Another useful collection at the State Archives comes under the auditor of public accounts (APA). In the APA's "Township Organization" file, by county, went correspondence, abstracts of votes taken, maps, and other items pertinent to these referenda and the subsequent division of the counties into townships. In fact, the Township Organization Act of 1849 ordered the county clerk to transmit to the APA, within a month, materials relevant to the division into townships. APA files yielded several otherwise undocumented referenda, a total of 23 or 5 percent of the 438. A major reason the auditor was to receive the organization materials was to avoid duplication of township names; but several duplications did occur, such as Ash Grove Township in Iroquois County and Ash Grove Township in Shelby County.[10]

Also at the State Archives is the product of the Illinois Historical Records Survey of the Works Projects Administration, which before 1939 was the Work Progress Administration. Part of the Federal Writers' Project, the Illinois Historical Records Survey sent unemployed or underemployed but literate men and women into courthouses looking for any existing county documents. Some items they transcribed; others they inventoried. Many writers prepared historical sketches of their assigned county or counties, basing their sketches on a questionnaire the Historical Records Survey furnished, in which local government was a significant issue. Their transcripts of county board/county commission minutes, either cursive or typed, exist in the State Archives for all but Brown, Carroll, and Cook counties—although coverage through time, for many counties, is incomplete. Often key periods are missing, periods for which I expected to find reference to the presentation of petitions and subsequent ballot authorization of the county board/commission. In Coles County, for instance, even though the index to the transcripts tells us there is coverage from 1832 to 1887, a key period, 1863–1868, is missing, a period just after the adoption of township organization in Coles in 1859. Attempts to discontinue townships often followed adoption in the next few years. Anyway, transcripts and historical sketches were useful in the documentation of often otherwise unavailable information: 28 (6 percent) from transcripts and 11 (3 percent) from sketches.

Actual county board/commission minutes, not transcripts, do exist for some counties and for periods significant to this project. Unfortunately, county officials have shown much less interest in preserving minutes than they have their land records, which are the key to a major source of county revenue through the property tax. So, while deeds, mortgages, and grantee/grantor indexes hold a place of prominence in courthouse vaults, clean and accessible to abstract companies, lawyers, county employees, and the occasional genealogist, one often finds minutes stacked above the reach of mere mortals, stuffed under cabinets, buried in the basement, hidden in the attic, stored offsite, and unfamiliar to most county clerical help.[11] With notable exceptions, like Jersey and Randolph counties, I had little luck even finding the bound volumes of minutes, for the years I needed, in the couple dozen courthouses I visited. On the other hand, some counties have voluntarily taken the State Archives up on its offer to preserve actual county and township documents, or to film and store copies, in the depositories of the Illinois Regional Archives (IRAD). I visited, at least once, all seven of these facilities, all on college campuses; and the IRAD sites provided 14 otherwise unaccounted for township referenda, or 3 percent of the total.

While at the IRAD sites, I also looked at their collections of election papers, which consist of tally sheets and poll books giving the names of the voters by precinct. Election papers yielded five new referenda, or 1 percent of the 438. Election papers simply did not exist in the courthouses I visited. They are apparently even less worthy of preservation than board/commission minutes.

Published county histories, despite their reputation as subscription-driven mugbooks, did provide me useful information, especially if the writer seemed to know the area well and had lived through some of the conflict over township organization. Overall, I used county histories to identify 17 referenda (4 percent) and to provide backdrop in some cases.

A lot of effort went into the deliberate search for referenda. Serendipitous discoveries did occur, however. While perusing every available copy of the TOI magazine, under its various titles, I discovered a pair of referenda in the period right after World War II, which I was not expecting. Among the last of the countywide referenda I unearthed was the one that occurred in Brown County in 1869. During the February 1870 debate among constitutional convention delegates over the issue of township government, William L. Vandeventer admitted that he had joined township opponents there the previous spring and had "succeeded in getting a vote on it," only to find "there was one thousand two hundred majority in Brown County for it."[12]

The other 110 referenda, the other 25 percent, came from contemporary newspaper accounts. Most of my newspaper research, on papers published prior the 1990s, took place at the Illinois State Historical Library in their microfilm reading area. Editors generally covered township referenda, but sometimes they did so only in the after-action matrix of votes tallied. Other editors took the space to explain the differences between the two approaches to county government, and often used their editorial prerogative to urge voters in one direction or another. A great asset was the editor who looked beyond the home county, at elections in other counties. Editors seemed especially likely to look outward if their county had been struggling with the issue of adopting township organization or returning to the predecessor system.[13] A reoccurring problem for me, especially for the 1850s and 1860s, was editors so tied up with partisan politics that they showed very little interest in local referenda.[14] Even worse, however, is the fact that key newspapers have not survived to inform us what was happening and how people felt 100–150 years ago. Newspaper accessibility is much better for the twentieth century; but, to assure myself of having immediate access, in the 1990s, I subscribed for several months each to the *Moline Dispatch* and the *Crystal Lake Northwest Herald* during periods when local abolition controversies were at their height.

All sources for referenda considered, I believe I have unearthed most efforts to alter the fundamental form of county government in Illinois. My confidence level is extremely high in the northern half of the state because of early adoption of township organization and good newspaper preservation. The farther south one travels the murkier the picture becomes because of multiple referenda per county and spotty survival of newspapers and other event records. I know some referenda escaped scrutiny, perhaps another 20 to 25. Such is the risk when one attempts to document precisely and for every county what happened so many years ago.[15]

Additional Sources

My quest for township organization referenda took early precedence in this project. That quest did not dampen, however, my desire to uncover relevant contemporary and subsequent reflective commentary on the process of supplanting a traditional system of local government, with township organization, or a reversion to the commission form. I sought out such additional sources vigorously, and I accepted them gratefully when principals in recent confrontations over abolition of townships offered them to me.

Newspapers again proved invaluable. I collected over three thousand newspaper items, dating from 1849 to the present, mostly with an eye toward local issues such as an upcoming or recently decided referendum. A quick sort of this collection revealed that about eleven hundred are articles, presenting facts and no opinion; eight hundred plus are editorials, which provide the editor's and therefore the newspaper's perspective; and five hundred or so are letters to the editor, almost always expressing an opinion about the issue of township government or some personalities involved. The other newspaper items run the gamut from advertisements, election result tables, and correspondent columns to poetry (one item) and music (one set of lyrics).

Books and journal articles make up another big portion of my additional materials. A few of the hundreds of county histories I consulted belong here because of authors who seemed to know their county well and perhaps had lived through the conflict over township organization. Good examples of the most valuable county histories are one for Madison County and another of Hancock.[16] Substantial studies of township government itself do exist. Mason H. Newell, in 1904, and Frederick Mercer Van Sickle, in 1982, focused on Illinois; and Monica Dwyer Abress, in 2000, looked at the national scene.[17] None emphasized the referenda and the confrontations they inspired. Township critics, scholars, and activists, produced a number of journal articles and freestanding booklets in the twentieth century, often citing the argument that townships are a throwback to a simpler age and in need of modification or elimination.

Magazines and newsletters constitute a slightly less durable source but one nonetheless that served me well. I found several hundred potentially relevant items in the TOI publication, *Township Perspective,* and its predecessors, *Illinois County and Township Official* and *Illinois Supervisor and County Commissioner.* Among them were editorials, legislative reports, articles about lobbying efforts with the general assembly, and rules for contests designed to encourage and reward good communication initiatives among member townships. *The Illinois Voter,* and its forerunner, *Bulletin of the Illinois League of Women Voters,* fall somewhere between a magazine and a newsletter. In any case, it helped me discern the various views that the League held about townships, including the periods when the League considered whether to call for abandonment of this layer of local government. Newsletters emanate from many Illinois townships, as supervisors attempt to communicate better with their citizenry. Such newsletters as *The Town Crier* (Schaumburg Township in Cook County) provided me with up-to-date evidence of the wide variety of services many townships today offer. A short-lived and not-so-complimentary newsletter that I consulted extensively was the *TSG news,* which highlighted scandals and other controversies surrounding various townships in the early 1990s and promoted abolition efforts in several counties.

Propaganda came in a variety of packages. Flyers went out through the mail and moved hand to hand at citizens' homes and in public places. Bumper stickers, yard signs, t-shirts, radio spots, and videotapes found favor, too. Recent abolitionists made good use of news releases to keep the local media invested and the cause of abolition before the public.

Legal matters occupied more of my attention than I had anticipated. Court documents—briefs and opinions—helped me keep track of the several lawsuits in progress during the 1990s. Records of the 1847, 1862, 1869–1870, 1920–1922, and 1969–1970 constitutional conventions play a role in several chapters. Journals, bills, and laws from the general assembly are a key component for laying out the sequence of events correctly.

Events, of course, played out since 1990 on my watch. Therefore, I am able here to utilize insights that I gained from personal correspondence and interviews with the principals, as well as from attendance at meetings and court sessions, so that I do not for 1990–2000 have to depend nearly so much on second-hand accounts as I did for the years before 1990. Abolitionists Michael Richardson and Bob Anderson were especially helpful in keeping me apprised of what was happening and in providing me with copies of correspondence, newspaper items, and confrontational literature from both sides.

A Chronological Approach

Diffusion occurs across time and space, but time seemed here the preferable organizing element. Therefore, I have chosen to examine diffusion and persistence in a series of chronological chapters beginning in chapter 2 with a focus on the events leading up to the adoption of township organization at the November 1849 general election by 24 counties. Chapter 3 covers the period 1850–1864, a time when uncertainty about township enabling legislation and subsequent court decisions in the early 1850s gave way to a rush to adopt in the northern half of the state toward the latter part of the decade and then to lack of interest statewide during the Civil War years. Chapter 4 (1865–1899) brings in the post-war surge of later adopters, the confrontation that township organization faced south of Springfield, and the first county to return permanently from a flirtation with township government to the commission format. The last three adopters, the last county to discontinue, and township critics are the focus of chapter 5, as I take the chronology of referenda from 1900 to 1939, picking up in November 1923 the most recent adoption and in the 1930s one of township government's most stressful decades. Chapter 6 covers a half-century of considerable activity, 1940–1989, which included condemnation, threats, and referenda (to adopt and discontinue) but no changes in the map of township versus commission counties. Chapters 7 and 8 take the chronology from 1990 through 1999, a period when abolitionists shook township officials severely and pushed them to do a better job, and a better job promoting their good works and protecting their turf. Diffusion of township government in Illinois may be complete, but persistence in the 85 township counties continues to be an issue. The last chapter, chapter 9, therefore, is not the end of the story but rather my attempt at closing the first century and a half of a conflict between those who believe in government at the grassroots and those who believe Illinois, like so many other states, could do just as well without this layer of government.

Notes

1. Timothy C. Frazer, "The Language of Yankee Cultural Imperialism: Pioneer Ideology and 'General American,'" in *Heartland English: Variation and Transition in the American Midwest,* ed. by Timothy C. Frazer (Tuscaloosa: University of Alabama Press, 1993), 62.
2. Confusion among residents of northwestern Illinois prompted the editor of the *Warren Independent* to explain for readers how government-surveyed townships, municipalities called towns, and political townships differed. "Town and Township," *Warren Independent,* 8 April 1859, 2.
3. Illinois, General Assembly, *Laws Passed by the First General Assembly of the State of Illinois at Their Second Session* (Kaskaskia: 1819), 107–9. For a thorough review of legislation and other actions affecting the public schools in Illinois over the better part of two centuries, see Charles W. Clabaugh, *Township School Trustees and Treasurers in Illinois: 1819–1980* (n.p.: Township School Trustees of Cook County, 1980).
4. Marion County. Township 1 North, Range 2 East. School Trustees' Minutes. 1838–1875; and School Fund Loan Record. 1839–1867. Originals are at the Illinois State Archives, Springfield.
5. Horace A. Hollister, *The Township High School in Illinois,* Bulletin, no. 8 (Urbana-Champaign: School of Education, University of Illinois, 1912).
6. In a large volume at the State Archives is a county-by-county, year-by-year roster of names chosen, names changed, and government-survey descriptions of the political townships as they became a part of the Illinois landscape. For instance, see, Illinois, Auditor of Public Accounts, Township Organization Records, Record of Township Boundaries, McLean County, 156–58, State Archives, Springfield.
7. Paul W. Wager, "Townships on Way Out," *National Municipal Review* 46 (October 1957): 475; Richard C. Spencer, "Iowa Townships Still Here?" *National Municipal Review* 41 (September 1952): 397; Clyde F. Snider, "The Twilight of the Township," *National Municipal Review* 41 (September 1952): 396.
8. Monica Dwyer Abress, *Quietly at Work: Township Government in America* (North Branch, Minnesota: Specialty Press, 2000).
9. Sheila Kelly, *County and Township Gazetteer* (Springfield: Illinois State Archives, 1988).
10. Kelly, *County and Township Gazetteer,* "Index to Illinois Townships."
11. Karl Moore, supervisor of the Illinois Regional Archives Depositories, pointed out to me that county highway officials do sometimes utilize board/commission minutes to gather information about characteristics of early roads. Karl Moore, interview by author, 4 October 2000.
12. Illinois, Constitutional Convention, Debates and Proceedings of the Constitutional Convention of the State of Illinois, Convened at the City of Springfield, Tuesday, December 13, 1869 (Springfield: E. L. Merritt & Brother, Printers to the Convention, 1870): 877.

13. For instance, see "Election in Calhoun County," *Carrollton Gazette,* 20 November 1869, 2, in which editor G. B. Price commented on a letter received from S. G. Lewis, an attorney living in Hardin, the seat of Calhoun County. Price wrote, "A vote was taken on Township organization, which resulted in its defeat by a large majority, the people preferring to let well enough alone."
14. Here is an example of an editor who took pains to illuminate the local situation. "We have seldom an election," he began, "which called upon the people to decide on matters of greater importance to them than the approaching one in November. While we may trouble ourselves a great deal, and spend our time and money in regard to the Presidential contest, yet the points of first importance to us are local." "Township Organization," *McLeansboro Golden Era,* 19 September 1872, 2.
15. On several occasions, I opted not to include a referendum for which I had found inconclusive documentation. Here are some examples of such exclusions. (1) Piatt County in 1849 showed a total of 46 votes for township organization and none against the change. The vote total should have been greater; and the outcome, given the location of Piatt, should not have given such a strong affirmation to the innovation in 1849. Illinois, Secretary of State, Abstract of Election Returns November 6th for and Against Township Organization: 1849, Volume 1, 1818–1850, 607, State Archives, Springfield. (2) DeWitt County's Long Point Precinct, in 1850, showed one vote for township organization and 14 against. Other precinct totals in the same record did not show a vote on township organization, so I excluded. DeWitt County. Election Papers. 5 November 1850. Original at the Illinois Regional Archives Depository, Illinois State University, Normal. (3) Morgan County poll books for two precincts, but neither from Jacksonville (the county seat and largest municipality), showed over a hundred votes each for township organization and not a single vote against in the 5 November 1850 election. Two other poll books in the record for that election did not list this referendum as an issue. Since I knew there were several precincts unaccounted for and because I did not have a tally from Jacksonville, I chose not to include. If I had included this referendum, Morgan County would have tied Williamson County for the lead in total number of township referenda per county. Morgan County. Election Returns. 1832–1888. On microfilm at the Illinois State Historical Library, Springfield.
16. *History of Madison County, Illinois* (Edwardsville: W. R. Brink Co., 1882); and Newton Bateman and Paul Selby, gen. eds., *Historical Encyclopedia of Illinois and History of Hancock County,* ed. Charles J. Scofield (Chicago: Munsell Publishing Company, 1921).
17. Mason H. Newell, "Township Government in Illinois," *Transactions of the Illinois State Historical Society* 9 (1904): 467–504; Frederick Mercer Van Sickle, *Grassroots Democracy: A Celebration of the Township Officials of Illinois 75th Anniversary, 1907–1982* (Astoria, Illinois: Township Officials of Illinois, 1982); and Abress, *Quietly at Work.*

2 A Yankee Idea Arrives

On the sixth of November in 1849, 59 Illinois counties asked voters to choose "For Township Organization" or "Against Township Organization." All 99 counties were to have placed the issue on the ballot, according to a law enacted by the Sixteenth General Assembly entitled "AN ACT to provide for township and county organization, under which any county may organize whenever a majority of the voters of such county, at any general election, shall so determine."[1] That 40 counties failed to vote on the issue is testimony to the confusion and resistance that surrounded township organization (figure 4).

Townships before Township Organization

In addition to federal townships and school district townships (chapter 1), there were also scattered administrative areas in Illinois called townships prior to the 1849 referendum. As early as 1790, when Governor Arthur St. Clair of the Northwest Territory created St. Clair County, which encompassed much of what is now Illinois, local judges had the responsibility of dividing counties into administrative townships. Duties of township officials, to be sure, failed to reach the level of true self-government; but such townships did assist the poor, supply viewers to settle fence squabbles between neighboring farmers, and provide a means of livestock identification, among other minor tasks.[2] The oldest evidence that I have uncovered of such an administrative township, in what is now Illinois, is the Marks and Brands Book of Eagle Township in St. Clair County for part of 1796.[3] Townships continued as the Northwest Territory yielded to Indiana Territory in 1800 and to Illinois Territory in 1809. Territorial legislators, according to Clyde F. Snider, seemed to consider townships "in no true sense of the word a unit of local self-government, but rather a mere administrative area for the purpose of carrying on certain local functions under county supervision."[4] One of the duties of oversight officials was to subdivide townships from time to time, as the situation and good of the population warranted. In St. Clair County, commissioners created Wood River Township out of the northern reaches of Goshen Township in June 1808 by drawing a line "beginning on the Mississippi River above Bet Whitesides [house?], & running with the Surveyors line a due east course...."[5] By May of 1816, St. Clair County, which then

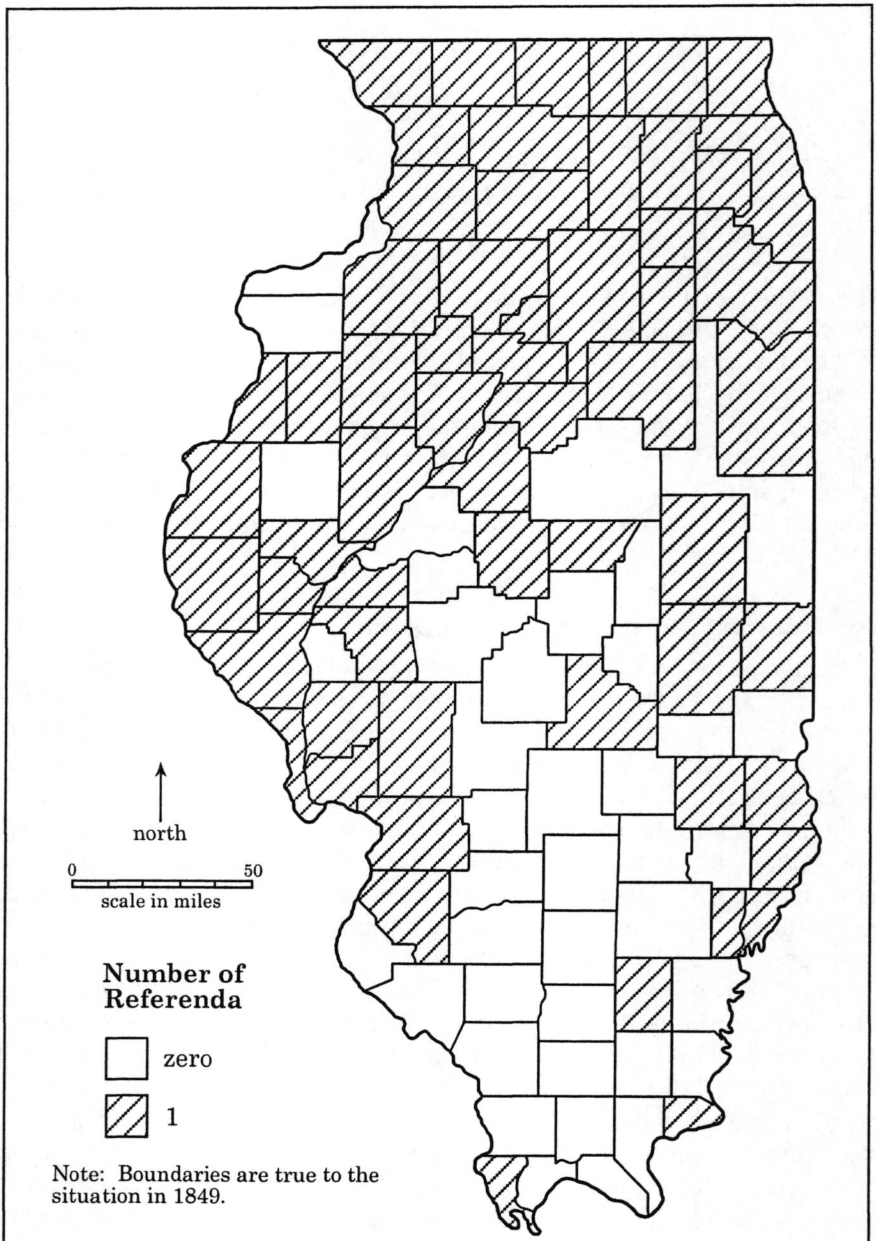

Figure 4 Countywide Referenda on Township Organization, by County: 1849. (Abstract of Election Returns, State Archives, Springfield)

included, in addition to its modern territory, all of what is now Clinton and Washington counties, plus the eastern tip of Monroe, had nine such administrative townships: Cahokia, Ogle, Turkey Hill, Clinton, Spring, Scott, Chambers, Silver Creek, and Sugar Creek.[6] Figure 5 illustrates how Edwards County would have looked administratively, after February 1817, when its county commissioners revised the township structure by laying off Coffee, Palmyra, Prairie, and Embarrass townships, all stretching westward from the Wabash River to the Third Principal Meridian.[7]

Administrative townships and county commissioners' courts carried over into statehood for Illinois, in 1818. The First General Assembly, in March of 1819, passed one law defining the county commissioners' court as the authoritative body for conducting county affairs and another law that specified how corporate bodies like counties might sue and suffer suits against them. In part that second law read, "That each of the several townships which now exist, or which may hereafter be established in this state, shall be a body corporate and politic; and all suits which shall hereafter be brought by or against the several townships in this state, shall be brought by or against the trustees of the township of K_____."[8] In other words, townships were to carry proper names.

References to administrative townships persisted into the first few years of statehood. Greene County's original townships, at the time of Greene's creation, numbered six, two for what is now Jersey County, two for Greene, and two for Morgan and Scott counties, since Greene either encompassed or administered this sizable territory prior to the creation of the named adjacent counties. The commissioners of Lawrence County, in March 1824, partitioned their domain (which included modern Lawrence and eastern Richland) into three townships to aid in the conduct of business by the overseers of the poor. Pike County's court, on 28 January 1828, divided that area into three large townships "for convenience in voting, making roads, etc."[9]

Gradually, however, other administrative units took the place of the township. The election precinct, justice precinct, polling district, militia district, and road district prevailed in commissioners' minutes through the 1830s, 1840s, and beyond, while the township became primarily a survey-related or school-related term. When Whiteside County's commissioners met for the first time, 15 May 1839, they laid off a number of road districts, including "No. 4 to comprise all that part of Elkhorn Precinct north of [the] township line running east and west between [survey] townships 21 and 22." In Kane County, the commissioners rearranged all the county's precincts in 1843 "to contain a single congressional township each." When Albert Stone and John Lawrence reported to the county court of Boone County on 4 March 1850 regarding certain paupers, they did so as overseers of the poor for Deerfield Precinct.[10]

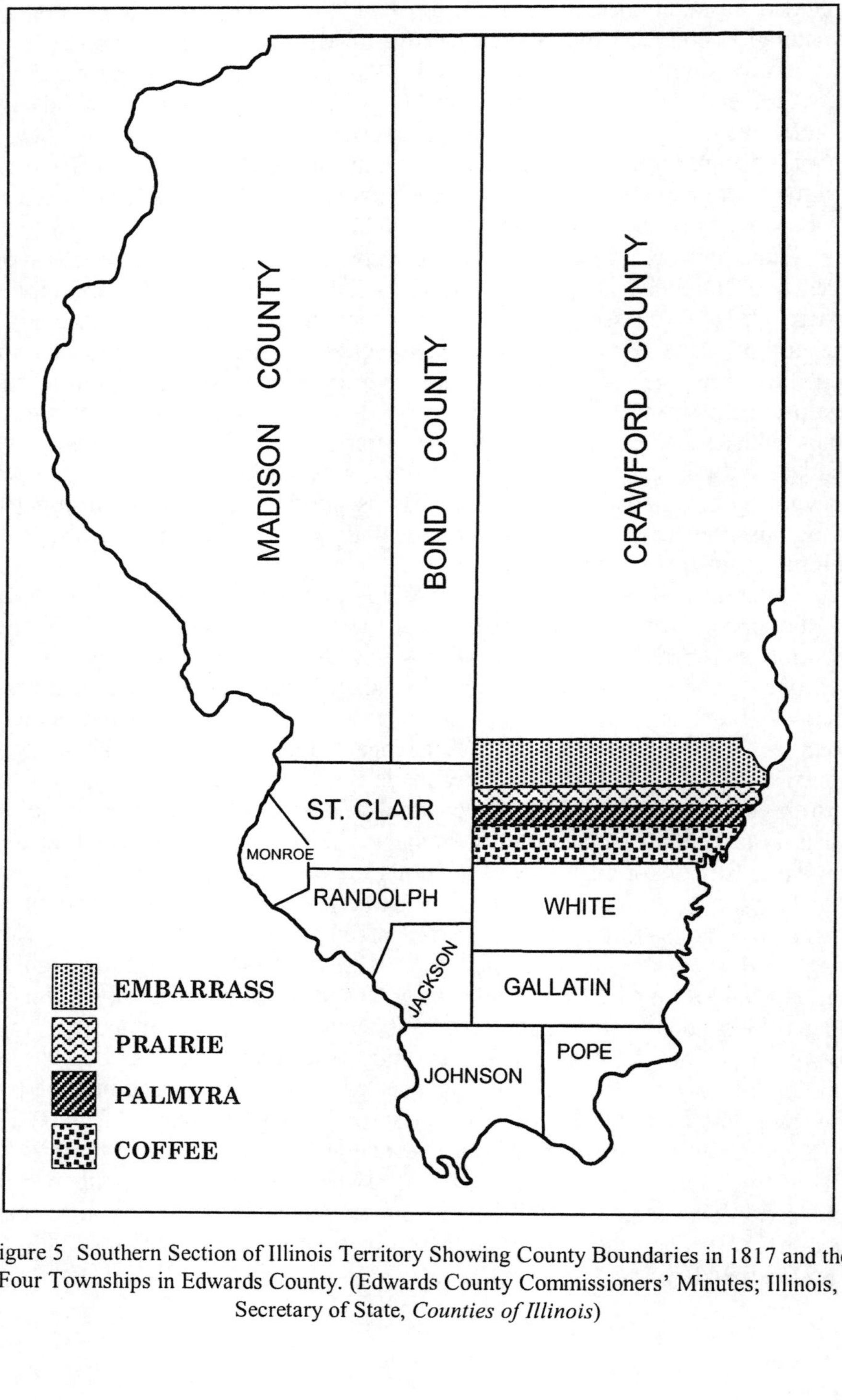

Figure 5 Southern Section of Illinois Territory Showing County Boundaries in 1817 and the Four Townships in Edwards County. (Edwards County Commissioners' Minutes; Illinois, Secretary of State, *Counties of Illinois*)

Cultures Collide

The Upland South had a head start, dominating Illinois politics and life before the middle decades of the nineteenth century. Douglas K. Meyer, however, in his *Making the Heartland Quilt,* showed how that southern element had lost its dominance by 1850, when, according to the Census of Population, southerners comprised just 35 percent of native-born immigrants. New Yorkers, by 1850, had replaced Kentuckians as the largest adult male group in Illinois.[11] By mid-century, what had once been a distant province of Virginia—later settled by Virginians, Kentuckians, Tennesseans, and Carolinians, with a smattering of French and others—was a state facing cultural conflict over a whole array of institutions and ways of life. Partly as a result of that growing presence of invading Yankees and other native-born immigrants came a call for a new constitution in the 1840s.[12] Under the influence of a fervor to replace, with "enterprise and improvements," a way of life that Frazer said "they regarded as inferior,"[13] this northern contingent insisted on a delegate distribution based on the (more favorable to them) 1845 Census instead of the decennial Census of 1840.[14] Among those improvements had to be a better way, in their opinion, of conducting the affairs of the government then closest to the people—the county. The Yankees and their northern brethren were not content to leave county government in the hands of three commissioners, elected at large, perhaps from the county seat or some other populated point of the county, "to the detriment," as Ben C. Allensworth put it, "of the sparsely settled localities."[15] Some of those recent immigrants no doubt shared Newell's view that the southern or Virginia form of county administration reminded them too much of the "landed aristocracy of England" and later the Tidewater, which lived apart from the masses, believing it knew best what the people needed in the way of services and infrastructure.[16]

The 1848 Constitution

TOI began its three-year sesquicentennial celebration in 1998, in recognition of the impact the 1848 Constitution has had on the lives of the 11,000 plus township officials and the 8 million or so citizens who live in the 1,433 Illinois townships.[17] A host of laws and a couple of subsequent constitutions have come since, but the nativity point for this book and for Illinois's modern governmental township consists of just 88 words, those words in article VII laying the groundwork for township organization as an alternative governmental model for Illinois.[18]

Delegates to the convention numbered 162, of which only 7 had drawn first breath in Illinois. There also assembled 26 born in New England, 38 from New York and the other middle coastal states, 35 from Virginia and coastal states southward, 41 from Kentucky and Tennessee, 10 from Ohio and Indiana, and 5 from Europe. Farming was the predominant occupation of the delegates, and Democrats outnumbered Whigs 91-71 when the convention opened on 7 June 1847.[19]

Township organization did not magically appear at the end of the convention process. One or more of the 162 had to step forward and champion its cause. But who was this champion, bringing such a fundamental innovation to Illinois? Among the dozen plus men that Arthur Charles Cole singled out as especially effective and active delegates at the convention[20] was attorney Hezekiah Morse Wead, a resident of Fulton County, which abuts Peoria County on the south along the Illinois River and, despite being a long way from the apparent hearth of township support in northeastern Illinois, was among the original adopters of the township model in 1849.

Born in 1810 in Vermont, Wead worked as a teacher and at other jobs before beginning his law career in St. Albans, Vermont; but he had the urge to move farther west. Traveling through Chicago, he pressed on in 1840 to Lewistown, the seat of Fulton County, where he set up a law practice and soon married the daughter of a local Methodist minister. Wead left a wonderful legacy of his convention service in the form of a daily journal. Unfortunately, he not once mentioned therein his considerable role in the fate of township government.[21]

Wead stepped into the township debate, at least for the record, on the eighteenth of June, when he spoke at length about the importance of protecting large existing counties from division into smaller counties—as lobbyists repeatedly urged general assemblies of that era to do.[22] Near the end of his address to the convention, Wead said that new, smaller counties (each needing a crew of officers and a courthouse) were not the answer to complaints about taxes but rather that such "high county taxes and burthens arise from our defective system of county government, and the people can obtain relief only by abolishing the county commissioners' court."[23] This condemnation of the southern-born commissioners' court came two weeks into the convention, after the issue of county politics had bounced from convention floor to committee and back for redirection. First, the delegates had refused a motion by Onslow Peters, representing Peoria and Fulton counties and formerly of Massachusetts, on 10 June to add "to the standing committees of this convention a committee on Townships, with instructions to report whether it is expedient to amend the constitution as to provide for the incorporation of the several townships in this state for municipal and other purposes."[24] Second, later that day, another transplanted Vermont-born delegate, Linus Worcester, then a resident of Greene County, successfully proposed the township/county government issue go the "committee on Counties" (really the Committee on the Division of the State into Counties and their Organization).[25] Third, Horace Butler, a native of New Hampshire representing Lake County, had adopted on 14 June his motion to instruct the Judiciary Committee to "inquire into the expediency of abolishing the county commissioners' court and substituting a precinct or township organization instead thereof."[26] Then, three days later, the chair of the Judiciary Committee, William Bennett Scates, representing Jefferson, Marion, and Franklin counties, came back asking relief from the county court and township referral because his committee felt "the subject matter of the resolution properly belonged to another committee." Which committee, asked Delegate Charles Henry Constable of Wabash County? "The committee on County Organizations," said Scates.[27] For a brief time, it appeared that the local government reform, so desirable to the Yankees, might fade from view; but on the twenty-fourth of June, Wead rose to propose "that a special committee of eleven persons be appointed to inquire into the expediency of abolishing the county commissioners' court and providing for the organization of townships, which townships shall have the general superintendance of their fiscal affairs; and also to report a plan for better administration of county affairs." The convention agreed to the motion, appointing Wead, Peters, and nine others to that special committee.[28] Finally, township organization had not only a champion but also a dedicated subset of delegates, plus still the chance of having a hearing in the Committee on Counties. Coming to the convention, by the way, were also petitions from northern counties urging delegates to scrap the commissioners in favor of townships.[29]

The Committee on the Division of the State into Counties and their Organization reported back both majority and minority reports regarding the proposed article on county affairs. Among the provisions of the majority report were replacement of the at-large county commissioners with three county judges elected from population-based districts within the county and the option of the county court determining "districts or townships" with appointed constables and justices of the peace. The committee's minority report called for election of all township officers, not appointment, and for several additional posts such as town clerk, supervisor, and overseer of the poor.[30] So the members of the committee wanted to retain the three-person court but had at least considered expanding the role of the minor civil divisions, which might carry the name "township," a scenario not greatly different from the administrative townships of territorial and early statehood times.

Ten days later, on Friday, the sixteenth of July, Wead brought forth a different proposal. His committee believed the general assembly should provide a law making township government possible, fixing the powers and duties of officers. Officers should stand for re-election every year. The general assembly needed to provide by law "for the creation of a board of supervisors...composed of one or more officers from each township." And the county commissioners' court would cease to function when the voters decided to create townships and a board of supervisors. After hearing Wead's presentation, without apparent debate, the convention tabled the proposal and ordered two hundred copies printed for distribution to delegates and others.[31]

The township champion was getting closer to success, but a month passed before the issue would reach the convention floor again. On the sixteenth of August, just two weeks prior to adjournment, while the convention debated judiciary matters, Delegate Stephen A. Hurlburt, representing Boone County, on the Wisconsin line, had adopted his motion to amend the proposed judiciary article mandating the general assembly "pass a general law authorizing township organization in all counties in which a majority of legal voters, at any general election, vote for such township organization" and subsequent cessation of the county court. By a substantial majority, the delegates approved the Hurlburt motion; but it failed to appear in the final judiciary article.[32] Apparently Hurlburt and other delegates still thought of the county court as a judicial rather than administrative court.[33] Or perhaps on that matter they had succumbed to a malaise that Wead was observing about that time in his convention journal: "There is a very general disposition manifested to hurry over the business without discussion or much examination. The members are anxious to get home." He further opined that "the business is likely to be imperfectly...done, and the haste of the members may produce great injury to the State."[34]

The next afternoon the delegates reconsidered his special committee's report and promptly adopted sections 1 and 2 without debate. On the following day, 21 August, delegates revisited section 1, struck it out, and substituted this: "The legislature shall provide by law, that the legal voters of any county in this state may adopt a township form of government within each county, by a majority of votes cast at any general election within said county." That substitution had come from Lincoln B. Knowlton, a Peoria attorney, born in Massachusetts. The delegates voted 87 to 52 in favor of this amendment, which, after change offline, came to be the 88 words that appear as section 6, article VII of the 1848 Constitution.[35]

Even though Wead's special committee's words failed to be the exact words on which the townships of Illinois hang, the township champion had prevailed. Figure 6, showing the distribution of delegate votes for and against the Knowlton amendment, serves as a surrogate for a map depicting support of and opposition to township organization in the state as of the summer of 1847. Township advocates dominated the northern quarter of the map, with only Daniel Hilton Whitney of Boone County casting a negative vote from a county north of what is now Interstate 80.[36] In the Military Tract, west of the Illinois, townships found numerous supporters, with the principal opposition being toward the Illinois-Mississippi confluence. Central Illinois, east of the Illinois, featured a mixed review of the township section, no doubt anticipating the uncertainty that would occur there regarding townships in the next few years. Except for pockets of support, the southern third of the state stood steadfast against townships and the board of supervisors in 1847. Of the 11 on the Wead committee, by the way, 7 voted in favor of the Knowlton resolution (including Wead and Peters); and the other 4 were among the 23

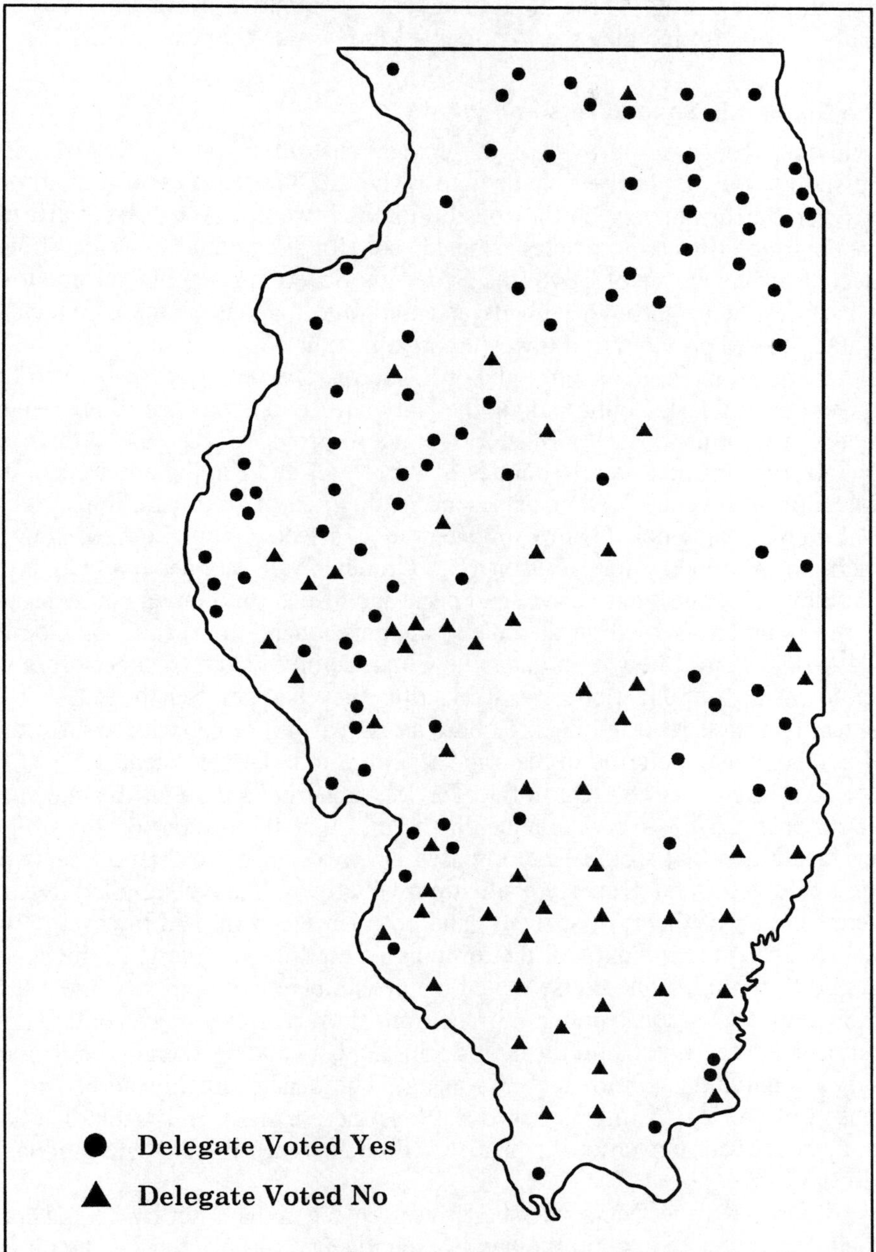

Figure 6 Delegate Vote, at the 1847 Constitutional Convention, on the Matter of Authorizing the General Assembly to Provide for Township Organization. (Illinois, Constitutional Convention, *Journal of the Convention*)

delegates not voting. Three of the nonvoting committee members were from southern Illinois, and the other represented Jo Daviess County.

The Sixteenth General Assembly Acts

Following delegate approval of the entire constitution on 31 August 1847, the state's voters had their chance on 6 March 1848. By nearly four to one, they voted affirmatively on the constitution as a whole and by lesser affirmative margins on two articles singled out for individual consideration.[37] Elected in November of 1848, the Sixteenth General Assembly set about the task of following through with its constitutional mandates, one of which, of course, was to provide for a township organization.

Legislation came up through the house of representatives, where Joshua P. Cooper of Clark County, along the National Road in eastern Illinois, introduced on 5 January 1849 a bill for "An act to provide for a general township organization in the several counties in this state, and for the assessment and collection of revenue." After its second reading, on motion from Rep. Francis C. Sherman of Cook County, it went to a select committee consisting of Sherman, Cooper, and Wilson H. Crandall of Winnebago County.[38] Sherman, a Connecticut native, had been one of the 162 convention delegates in 1847 and had served on the county commissioners' court in Cook County, 1840–1845. Later he would chair the Cook County board of supervisors under township organization.[39] A week after they had formed the select committee, Crandall reported back Cooper's original bill along with a substitute, which the house preferred to the original, after modest amendments.

Rep. Cyrenius B. Denio of Jo Daviess must have thought the substitute bill's content a bit overwhelming. In the midst of the discussion, he offered an amendment that would have imposed a tax of not more than 50 cents per hundred dollars real property evaluation, which would have "applied towards the education of the most sprightly and promising boy of said township" who was to "interpret and explain the meaning of the foregoing law, to the people of said township." The house tabled his amendment and approved the bill 57 to 8, with opposition coming entirely from those representing counties lying south of an east-west line through Springfield, including committee member and original bill sponsor Rep. Cooper.[40] The senate quickly agreed to the house bill, voting 15 to 6 to approve. Nays there came mainly from southern counties as well. Governor Augustus C. French signed the bill into law on 12 February 1849.[41]

Representative Denio may have been on the right track by asking for a bright youngster to explain the law to the citizenry. As printed for the public, the township law's 24 articles cover more that 30 pages and probably contain as many words as the entire 1848 Constitution. The law ordered all counties to allow citizens to vote on township organization in their respective counties "at the next general election." In order to replace the county court, which was about to replace the county commissioners' court, township organization needed to receive "a majority of all the votes cast for or against township organization," a simple majority. The law specified the procedures for dividing the county and naming the new townships. Division commissioners were to use the federal township lines (running east-west) and range lines (running north-south) if possible.[42] Subsequent articles of the law dealt with rights and powers of townships, town meetings, town officers, officer duties, and officer compensation.[43] In article 16, the lawmakers finally got to the second (after townships themselves) key element of the law—the board of supervisors, a diminutive general assembly. The supervisors were to meet at least annually to pay the county's bills and make other decisions as needed. Among the designated duties of the board and of the supervisors in each township was the charge "to take care of the poor."[44] Later articles of the law covered taxes, including the need for proper attention to property assessment and tax collection by township officers; and the last 12 pages of the law, the last three articles, focused on road and bridge responsibilities of the townships.[45] From this 1849 law, therefore, modern townships and their coterminous road districts derive their three mandated duties: taking care of the poor, assessing property for taxation, and providing township roads and bridges.[46]

Anticipating the November Referendum

Newspaper editors attempted to prepare readers for the referendum mandated in the township law of 12 February 1849 by summarizing as much of the law as possible. Only one of the several papers in which I discovered discussion of the issue, the *Warsaw Signal,* actually suggested readers vote against townships: "On this question we have but little to say. In a more thickly populated county, it might be well for townships to organize—but in Hancock, where many of them are very sparsely settled, we think nothing can be gained by organization."[47] Another paper, the *Pittsfield Union* out of Pike County, bordering both the big rivers, ran a long story containing a copy of sections 1 and 2 of the law's first article and a summary reprinted from the *Springfield Illinois Journal.* The *Union* editor added at the end that in his opinion "the counties hereabouts are not in a situation to adopt this new system," but no matter whether readers voted for or against, he would be "satisfied."[48]

Support predominated in papers published farther north and east. The *Chicago Daily Democrat* told voters on election day, "Do not forget to-day to vote for township organization." Ottawa's *Free Trader* came out favorably in late September, telling readers that township organization

> is so much more convenient than the present plan of having to post off to the county seat for every little thing, that the people are much more capable of attending to at home; and then every body will have a more direct voice, and therefore take a deeper interest in the management of many affairs of a public nature, which

> are now left to a few individuals at the county seat, who frequently do business but half way and that bunglingly.

Complaints about lack of solid information as to the particulars of the law marked an October editorial from the *Aurora Beacon:*

> If properly understood, we are confident that the law would be generally approved, but owing to the miserable arrangements of our law-makers in providing for its publication, it appears that nine-tenths of the people have no means of informing themselves on the subject, and hence its probable defeat in a majority of the counties.

This prediction came true, but the cause was likely a combination of ignorance and true opposition in 1849. The *Beacon* editor said he was in favor of township organization and would "go for it blind" if need be.[49]

Strong support also came out of northwestern Illinois. The editor of the *Freeport Journal* expressed surprise at how little argument the topic had fostered and remarked how familiar were terms like "town meeting, the supervisor, [and] the town clerk" from his previous life. He felt the people of Stephenson County were probably, like himself, in favor of the change. Among the most fervent affirmative editorial commentaries to have survived was one from Galena. The editor of the *North-Western Gazette* wrote, "Wherever it has been adopted, we never heard a word of complaint against it, and to abolish it by the free will of those who are immediately affected by it would be an impossibility." He went on to praise its impact on self-government, roads and bridges, and schools (because he hoped this could be the first step in creating a "closely organized system of Common Schools"). Among the best prophets offering commentary on the referendum was the editor of the *Rockford Forum,* who after stating strong agreement with the concept of local self-government inherent in township organization, wrote, "In the southern part of the State, we expect to see the old plan adhered to for a while longer. In the North, most of the counties, perhaps all, will go into the town organization, the people having long desired to get rid of the burden, inconvenience and expence [*sic*] of the old system."[50]

Referendum Results

Exactly two dozen Illinois counties adopted township organization on 6 November and followed through with all aspects of the process (figure 7). A handful of other counties took some of the necessary steps, including apparent voter approval, but ultimately pulled back because of confusion over the outcome.

A nearly solid swath of adopting/organizing counties stretched from Lake Michigan to Fulton County. Support for townships proved strongest in the counties closest to its birthplace—New England and New York—counties where immigrants from those areas in 1849 were predominant segments of the population, according to Meyer.[51] In many counties, the support

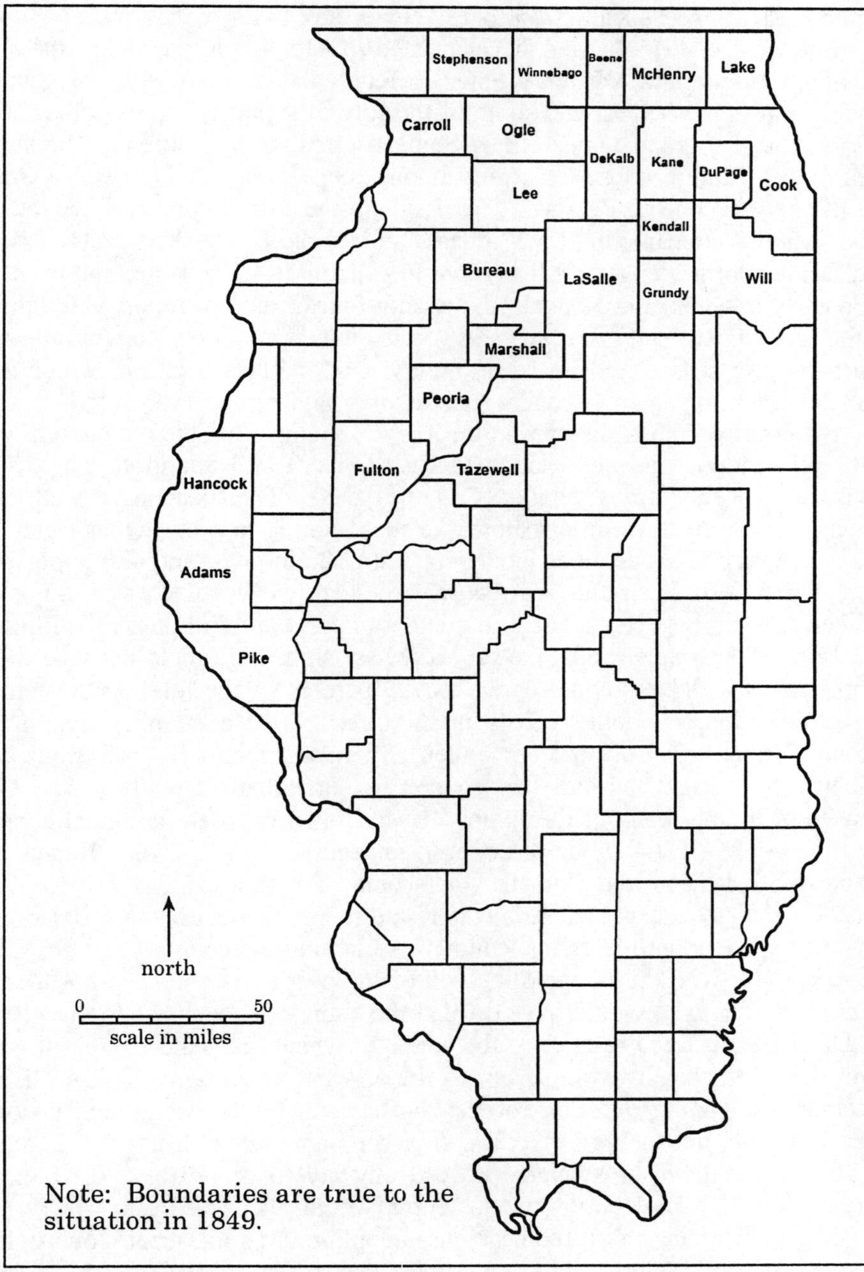

Figure 7 Counties Adopting Township Organization: 1849. (Abstract of Election Returns, State Archives, Springfield; Newell)

was overwhelming: Cook, 3,376 to 27; Lake, 1,692 to 3; DuPage, 773 to 1; Grundy, 310 to 4; DeKalb, 755 to 1; Winnebago, 1624 to 12; Will, 1436 to 29; Boone, 1,134 to 0; and McHenry, 1,923 to 0. Not a single man in neighbors Boone and McHenry voted to retain the southern style of county government. Farther west and south of the core of support, success ratios still ran 10:1 or better, in Ogle,[52] Lee, Stephenson, LaSalle, Bureau, Marshall, Peoria, and Fulton. Closest margins among the 24 were in Tazewell (across the Illinois River from Peoria) at 1.5:1; along the Mississippi in the Military Tract where Pike came in at 5:1, Adams at 4:1, and Hancock at nearly 2.5:1; and farther north along the Mississippi in Carroll at 4:1.[53] Kane, the twenty-third early adopter, and Kendall, the twenty-fourth, did not report vote tallies to the secretary of state; but analysis of the dates that county commissioners' court minutes ceased and board of supervisors' minutes began does confirm the 1849 adoption and subsequent successful organization for both.[54]

As court opinions directed in the 1850s, having a simple majority in the 1849 referendum was not enough to ensure successful adoption. The 1848 Constitution called for a "majority of the voters of such county," and that phrase turned out to mean a majority of those voting in a particular election. So, in Military Tract counties like Knox, with 728 in favor and 420 opposing; Schuyler (673 to 205); and Warren (471 to 220), which together would have bridged the gap between successful adopters Peoria and Fulton on the Illinois and Hancock, Adams, and Pike on the Mississippi, fell short because their approval votes did not equal or exceed 50 percent of the total votes at that general election.[55] County courts in all three of these counties appointed commissioners to lay off and name their new townships, only to find this action was for naught, as did their counterpart in Putnam County (152-129), situated at the big bend of the Illinois River and surrounded by actual adopters in 1849.[56] In Jo Daviess County, adjacent in northwestern Illinois to adopters Stephenson and Carroll, voters opted for organization 930 to 596; and the *North-Western Gazette*'s editor said it was "an event of considerable importance". A month later, the editor was blaming the county court for failing to comply with the law and the will of the people by choosing not to appoint three men who would have divided the county. In response to that item, JO DAVIESS offered a letter to the editor in which the writer sought to unravel the confusion by going back to the 1848 Constitution. JO DAVIESS concluded it was "a question for the *Supreme Court* to decide and no one else," certainly not the *county court.* In a companion commentary, the editor suggested that townships might proceed anyway to organize; and, if there were a "defect[,]...no doubt the Legislature [would] rectify it."[57]

Most surprising of all the near-miss adoptions was in Jasper County, far removed in southeastern Illinois by almost a hundred miles from the nearest adopter. Voters there in 1849 gave a 142 to 132 edge to township organization; and the county court had it divided as the law specified, only to find their action premature because 142 was less than half of the voters.

J. H. Battle contended, "A feeling had arisen as early as 1848, that the county-seat exercised an undue influence upon the affairs of the various precincts, and the people of Saint [*sic*] Marie, especially, felt that the county-seat was thriving at the expense of its more commercial rival."[58] Furthermore, Meyer reported that Jasper County had then a sizable minority of German-speaking French immigrants living in and around Ste. Marie, several miles from Newton, the county seat.[59] Perhaps foreigner treatment at the hands of the Upland South majority in Jasper County could have helped trigger this premature plurality so far from the adopting counties of northern and western Illinois. Whatever the causes, confusion, such as that in Jasper, took several more years and many more referenda to untangle.

The Transitions Begins

County courts met for the first time (after succeeding county commissioners' courts) throughout the state in December of 1849 at the regularly scheduled fourth quarter session, but change was in the air. The county courts in many counties knew their tenure would be short. A main order of business in the 24 adopters and several pseudo-adopters was to "appoint three commissioners, residents of the county, to divide the county into towns or townships"; name the townships "in accordance with the expressed wish of the inhabitants of the town"; "make a written report of their proceedings"; and present the "report to the clerk of the county court on or before the first day of March in the year 1850."[60] For those counties, like Knox, that were wavering between approval and status quo, the courts would have to intervene. For counties like Henderson, which could muster only 82 votes in favor to counter 337 against, the process would drag on for years. At Oquawka, the Mississippi port that served Henderson County as its seat, the editor of the *Spectator* observed after the November 1849 election that even though many counties seemed to have approved, "Of its advantages, over the old system, we are not prepared to speak—never having resided in any state where the system was in operation. We shall endeavor to note its progress in the adjoining Counties, and communicate results hereafter to our readers."[61] So shall I.

Notes

1. Illinois, General Assembly, *Laws of the State of Illinois Passed at the First Session of the Sixteenth General Assembly* (Springfield, 1853), 190–224.
2. Clyde F. Snider, "Indiana Counties and Townships," *Indiana Magazine of History* 33 (June 1937): 120–26; and Newell, "Township Government in Illinois," 475–76.
3. St. Clair County (Northwest Territory). Eagle Township. Marks and Brands Book. 15 February 1796–16 July 1796. Original at Illinois State Archives, Springfield.
4. Snider, "Indiana Counties and Townships," 125.

5. St. Clair County (Indiana Territory). Record of Orphan Court. June Term, Special Session. 1808. Original at Illinois State Archives, Springfield.
6. St. Clair County (Illinois Territory). Record of Orphan Court. May Term, 1816, 159–61. Original at Illinois State Archives, Springfield.
7. Edwards County (Illinois Territory). County Commissioners' Minutes. 1815–1841. 17 February 1817. Original at the Illinois Regional Archives Depository, Charleston.
8. Illinois, General Assembly, *Laws Passed by the First General Assembly of the State of Illinois at Their Second Session* (Kaskaskia: 1819), 175 and 184.
9. *History of Greene County, Illinois: Its Past and Present* (Chicago: Donnelley, Gassette & Loyd, Publishers, 1879), 292; *Combined History of Edwards, Lawrence and Wabash Counties, Illinois* (Philadelphia: J. L. McDonough & Co., 1883), 104; History of Pike County, Illinois (Chicago: C. C. Chapman & Co., 1880), 264.
10. Charles Bent (ed.), *History of Whiteside County, Illinois* (Morrison, Illinois: Whiteside Sentinel, 1877), 59; John A. Gustafson, *Historic Batavia, Illinois* (Batavia, Illinois: Batavia Historical Society, 1962), 58; Boone County. Transcript of County Court Record. 4 March 1850. Transcript is in State Archives, Springfield.
11. Douglas K. Meyer, *Making the Heartland Quilt: A Geographical History of Settlement and Migration in Early-Nineteenth-Century Illinois* (Carbondale and Edwardsville: Southern Illinois University Press, 2000), 141 and 173.
12. For a thorough review of the issues prompting the call for a constitutional convention for 1847, see Janet Cornelius, *A History of Constitution Making in Illinois* (Urbana: Institute of Government and Public Affairs, University of Illinois, 1969), 17–19. The 1847 Constitutional Convention produced the Illinois constitution that would prove to have the shortest life span, only 22 years, compared to 30 for 1818, 30 plus years for 1970, and 100 for 1870.
13. Frazer, "The Language of Yankee Cultural Imperialism," 62 and 64.
14. Arthur Charles Cole, *The Constitutional Debates of 1847*, Collections of the Illinois State Historical Library, vol. 14 (Springfield: Illinois State Historical Library, 1919), xv–xvi.
15. Newton Bateman and Paul Selby, gen. eds., *Historical Encyclopedia of Illinois and History of Tazewell County*, ed. Ben C. Allensworth (Chicago: Munsell Publishing Company, 1905), 2:709.
16. Newell, "Township Government in Illinois," 470.
17. Bryan E. Smith, "Today, More than Ever, Township Government Should be Celebrated," *Illinois County and Township Official*, November 1998, 4.
18. The 88 words are as follows: "The General Assembly shall provide by a general law, for a township organization, under which any county may organize, whenever a majority of the voters of such county at any general election shall so determine. And whenever any county shall adopt a township organization, so much of this constitution as provides for the management of fiscal concerns of the said county, by the county court may be dispensed with and the affairs of said county may be transacted, in such manner as the General Assembly may provide." Illinois, *Constitution of 1848*, article VII, section 6.
19. Cole, *The Constitutional Debates* of 1847, xvi–xvii.
20. Cole, *The Constitutional Debates of 1847*, xviii. Cole provided a written sketch of Wead, which appears on pp. 980–81.
21. Hezekiah Morse Wead, "Illinois & her Politicians," MS, Hezekiah Morse Wead Papers, Illinois State Historical Library, Springfield; and Thomas F. Schwartz, "'Illinois & her Politicians': The Observations of Hezekiah Morse Wead, Delegate to the 1847 Illinois Constitutional Convention," *Journal of Illinois History* 3 (Spring 2000): 31–50.

22. For examples of the lobbying and petitioning that went on as Illinois fleshed out its county boundaries, see Michael D. Sublett, *Paper Counties: The Illinois Experience, 1825–1867* (New York: Peter Lang Publishing, Inc., 1990).
23. Cole, *The Constitutional Debates of 1847,* 117.
24. Illinois, Constitutional Convention, *Journal of the Convention Assembled at Springfield June 7, 1847* (Springfield, 1847), 20.
25. *Journal of the Convention,* 21.
26. *Journal of the Convention,* 43.
27. Cole, *The Constitutional Debates of 1847,* 106–7.
28. *Journal of the Convention,* 87.
29. *Journal of the Convention,* 129.
30. *Journal of the Convention,* 140.
31. *Journal of the Convention,* 173.
32. *Journal of the Convention,* 378.
33. One outcome of the 1848 Constitution was replacing the county commissioners' court with the county court and the commissioners with judges, one elected as county judge (presiding) and the other two as associate judges. Illinois went back to commissioners in 1873, for those counties not having adopted township organization by that time. In Missouri, on the other hand, the administrative county court populated by three judges persisted until January 1985 when a new Missouri law took effect changing the county court to a county commission and renaming the three judges as commissioners, with two elected in prescribed districts and the presiding commissioner elected at large. Don Stamper, telephone interview with presiding judge of Boone County (Missouri), 2 November 2000.
34. Wead, "Illinois & her Politicians," 19 August 1847.
35. *Journal of the Convention,* 440 and 446–47. For details on Knowlton, see Cole, The Constitutional Debates of 1847, 967.
36. Cole, *The Constitutional Debates of 1847,* 981. Whitney poses a conundrum. He was a New Yorker before coming to Belvidere, the seat of Boone County, and, based solely on his nativity, should have been a supporter of townships. Cole does indicate that Whitney favored annexation of his area to Wisconsin, which was a common attitude of those in the northernmost counties where residents resented the addition of territory to Illinois in 1818 that the Northwest Ordinance of 1787 had targeted for a state appearing north of Illinois. Maybe his affinity with Wisconsin colored his vote on this matter. Or maybe he was feeling the malaise that Wead was observing.
37. Cole, *The Constitutional Debates of 1847*, 944–45; and Cornelius, A History of Constitution Making in Illinois, 30–31.
38. Illinois, General Assembly, House, *Journal of the House of Representatives,* 16th G. A., Regular Sess., 1849, 171. Hereafter I cite all of these volumes as *House Journal.*
39. Cole, *The Constitutional Debates of 1847,* 976.
40. *House Journal,* 16th G. A., Regular sess., 1849, 267–68. On motion of Rep. Crandall, the title changed to "An act to provide for township and county organization, under which any county may organize, whenever a majority of voters of such county, at any general election, shall so determine."
41. Illinois, General Assembly, *Journal of the Senate,* 16th G. A., Regular Sess., 1849, 304–5 and 339. Hereafter I cite all these volumes as *Senate Journal; House Journal,* 16th G. A., Regular sess., 1849, 510.
42. *Laws First Session Sixteenth General Assembly,* 190.

43. *Laws First Session Sixteenth General Assembly*, 191–202.
44. *Laws First Session Sixteenth General Assembly*, 202–4.
45. *Laws First Session Sixteenth General Assembly*, 204–24.
46. The Illinois version of township organization looked more to New York than to New England for guidance. Elijah M. Haines noted, in the front matter of his township laws compilation, that in the New England states a representative of the township went directly into the state legislature's lower house. The size of New York, he explained, "rendered it impossible or at least impracticable to send a representative from each town to the popular branch of the state legislature." So "an intermediate board was created in each county, called a board of supervisors, and composed of one delegate from each town." Elijah M. Haines, *A Compilation of the Laws of Illinois Relating to Township Organization and Management of County Affairs,* new rev. ed. (Chicago: E. B. Myers and Company, 1875), xvi.
47. "Township Organization," *Warsaw Signal,* 27 October 1849, 1. For many years after 1849, "Township Organization" was a common title for such newspaper items.
48. "Township Organization," *Pittsfield Union,* 10 October 1849, 2. I have used the masculine pronoun in lieu of the editor's name and will do likewise elsewhere in these early years because I never encountered a female editor in the nineteenth-century newspapers I consulted.
49. "Township Organization," *Chicago Daily Democrat,* 6 November 1849, 2; "Township Organization," *Ottawa Free Trader,* 28 September 1849, 2; "Township Organization," *Aurora Beacon,* 11 October 1849, 2.
50. "Township Organization," *Freeport Journal,* 24 October 1849, 2; "Township Organization," *Galena North-Western Gazette,* 17 October 1849, 4; "The Township Organization," *Rockford Forum,* 29 August 1849, 2.
51. Meyer, *Making the Heartland Quilt,* 175 and 192.
52. Voting patterns, even in 1849, did not always turn out as observers might have expected. In Ogle County's Grand Detour Precinct, "made up of settlers from New England, the home of the town meeting and township system," voters gave 78 votes for and 25 against township organization. Maryland Precinct in the same county, "whose people had been accustomed in their eastern homes to county commissioners and county government," went 138 to 0 for the change. Newton Bateman and Paul Selby, gen. eds., *Historical Encyclopedia of Illinois and History of Ogle County,* eds. Horace G. Kauffman and Rebecca H. Kauffman (Chicago: Munsell Publishing Company, 1909), 680.
53. Illinois, Secretary of State, Abstract of Election Returns November 6th for and Against Township Organization: 1849, Volume 1, 1818–1850, State Archives, Springfield.
54. Kelly, *County and Township Gazetteer,* 89–90 and 93–94.
55. Abstract of Election Returns, 6 November 1849.
56. Newton Bateman and Paul Selby, gen. eds., *Historical Encyclopedia of Illinois and History of Knox County,* ed. W. Selden Gale and George Candee Gale (Chicago: Munsell Publishing Company, 1899), 2:632; Schuyler County. Transcript of County Commissioners' Court Minutes. 8 December 1849; Warren County. Transcript of County Commissioners' Court Minutes. 5 December 1849. Illinois, Auditor of Public Accounts, Township Organization File, Putnam County, An Abstract of the report made by the Commissioners appointed by the County Court of Putnam County, 25 February 1850. Transcripts and report are at the Illinois State Archives, Springfield.

57. "Township Organization," *North-Western Gazette,* 14 November 1849, 2; "The County Court," *North-Western Gazette,* 12 December 1849, 2; JO DAVIESS [pseud.], "Communication," *North-Western Gazette,* 19 December 1849, 2.
58. J. H. Battle, *Counties of Cumberland, Jasper, and Richland, Illinois* (Chicago: F. A. Battey & Co., 1884), 404.
59. Meyer, *Making the Heartland Quilt,* 259–60.
60. Meyer, *Making the Heartland Quilt,* 190–91.
61. "Township Organization," *Oquawka Spectator,* 21 November 1849, 2.

3 *Confusion, Disillusion, and Diffusion: 1850–1864*

AS ILLINOIS MOVED into the second half of the nineteenth century, township organization seemed at times like township disorganization. In several counties, officials, operating under the assumption that citizens had voted to adopt, went ahead and set up their township structure only to then put the system on hold until they got direction from the courts, received new legislative guidance, and held additional referenda. Elsewhere the disgruntled and disillusioned in some 1849 adopting counties sought an immediate return to the three-man county court. Through all the confusion and disillusion, however, the township tide rushed on, so that by the onset of the Civil War, diffusion had delivered township government to 36 more counties. By the end of 1861, townships had found a home as far as three hundred miles south of Chicago (in Wayne County) and in all counties north of an east-west line through Bloomington, except for holdout Henderson County on the Mississippi River. Adoptions then came to a standstill during the war, as referenda dropped from eight in 1861 to one in 1863 before again reaching a typical prewar total in 1865. That year, marking the end of the Civil War, seems an ideal point to begin the next chapter, which covers the remainder of the century and brings aboard another 24 adopters.

Settlement at Mid-Century

Illinois stood poised for a transportation revolution in 1850. The railroads were ready to transform a state dominated largely by self-sufficient agriculturalists into a market-oriented land whose farmers would be able readily to send their products to coastal cities and beyond. Industrialization and a tremendous population boom would soon follow. So what was the settlement fabric like at the beginning of this takeoff period?

Douglas K. Meyer spent a considerable portion of his academic career looking for the answer. Over a ten-year period, he published a number of papers on the topic;[1] and in 2000 his book, *Making the Heartland Quilt,* arrived. Using the book as a guide, I will provide the reader with a snapshot of where Illinois settlement stood, on a county basis, as townships began to conquer the commission/court system.

Meyer's book culminates with a map in which he summarized numerous other illustrations and presented his interpretation of the three main culture regions that stretched across the state in 1850 (figure 8). Compared to previous depictions of these regions by other social scientists, the Meyer boundaries pinch much closer together. He had discovered in the 1850 manuscript census that immigrants from the Upland South, especially those from Kentucky, had "thrust north beyond the...National Road in overwhelming numbers" to become key players in places like Vermilion County on the Indiana border and Adams County two hundred miles to the west.[2] They and their Tennessee brethren moved up the Illinois River Valley toward Peoria and out into the heart of the Military Tract to the west. On the other hand, Yankees (New Yorkers and immigrants from the six New England states plus Michigan) pushed farther south in Illinois than Meyer had expected or other scholars had reported, especially in the Illinois Valley and to a lesser extent along the road (Hubbard's Trace) that linked Chicago with Danville in Vermilion County. Yankees seemed particularly attracted to Peoria because of its hustling economy and "New England ambiance."[3] Between these cultures in 1850, stood the Midland-Midwest Region, whose residents were from states like Pennsylvania, Ohio, and Indiana. Meyer found their territory of dominance much more restricted to the south and to the north than had Wilbur Zelinsky, for instance, in his *The Cultural Geography of the United States*.[4] Midlanders and Midwesterners, according to Meyer, tended to settle in counties where they were not the dominant cultural force, which "fostered cultural fusions and adaptations."[5] In other words, this group commonly lost its identity as it met and melded with Yankees and southerners. Meyer cited Sangamon County, location of the state capital by 1850, as the perfect example of what he called a "settlement platform," a county that personified in 1850 "midwestern cultural mixing."[6]

Using Meyer's cultural regions, we ought to be able to make better sense of the 1849 adoptions that I introduced in chapter 2. Most of the 24 early-adopting counties lay entirely within the New England Region (figure 8). On that region's southern flank were adopters like Tazewell, which had a colony of Yankees from Massachusetts settle Tremont and another colony, from Rhode Island, establish Delavan. Meyer described Tazewell County as "a cultural outpost of Yankees."[7] Fulton County, another Yankee outpost and early adopter, had attracted, as I noted, key figures from the Yankee homeland, like Hezakiah Wead. Only the three 1849 adopters on the Mississippi River, Adams, Hancock, and Pike, lay beyond Meyer's New England territory. Germans were the leading immigrant group by place of birth in Adams; Kentuckians outnumbered other groups in Hancock and Pike.[8] Of some significance may have been the colony from Pittsfield, Massachusetts, which established Pittsfield, the seat of Pike County, in 1820.[9] Perhaps the cluster

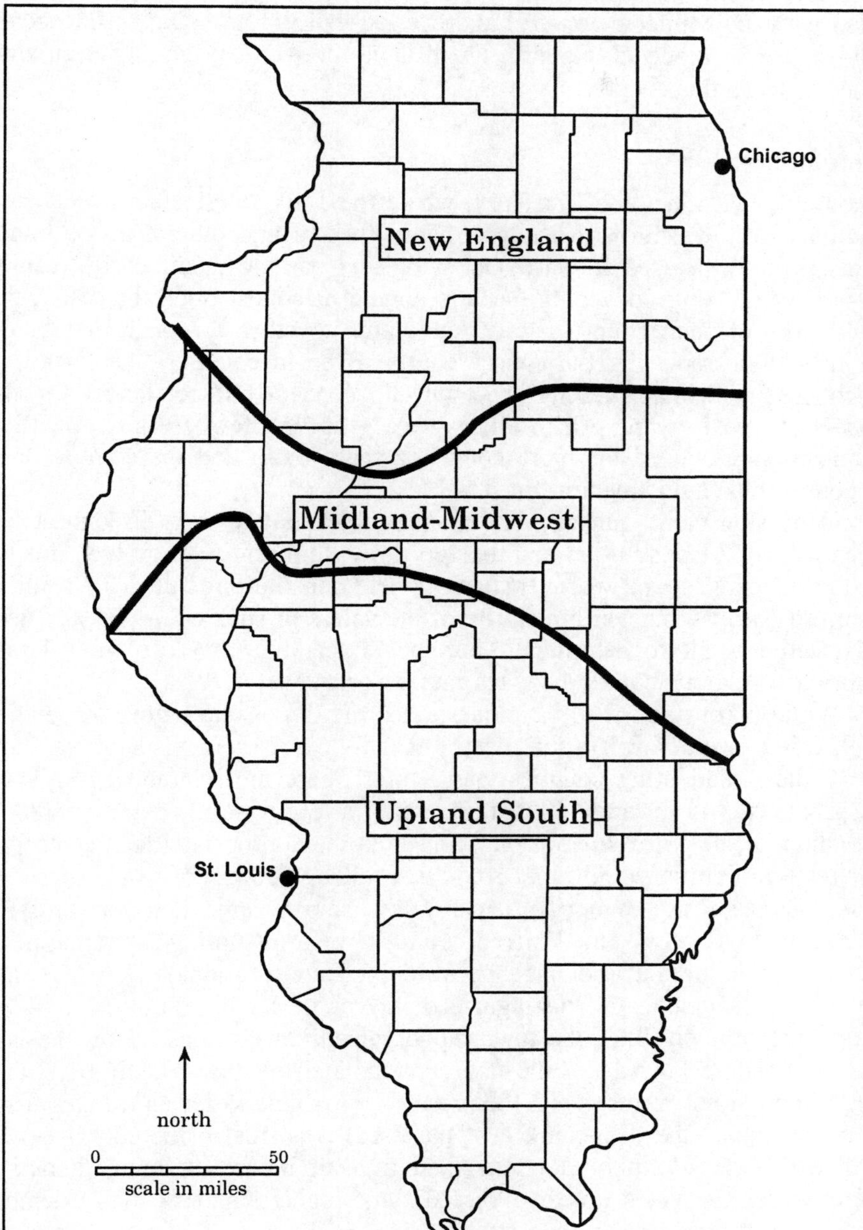

Figure 8 Culture Regions in Illinois: 1850. (Meyer)

of Yankee attorneys that had migrated to Quincy to litigate land claims associated with the Military Bounty Land exerted in 1849 a Yankee influence on Adams. As for Hancock County, I will have more to say about its circumstances later in this chapter.

Confusion

In counties like Lake and McHenry, where the 1849 referendum produced a landslide for township organization, New Yorkers and other Yankee immigrants proudly pressed forward, bounding and naming their new townships and, in April 1850, convening the first annual meetings. Soon thereafter, the supervisors of each county's townships came together for their first board meetings, thus replacing the county courts. Even in counties like Tazewell, which produced majorities for the township concept but not landslides, the process moved forward. All in all, however, the first few years of township organization produced far more confusion than Wead and his committee at the convention could have imagined.

Confusion came quickly. The main problem lay with a disconnect between the 1848 Constitution and the 1849 township organization law, involving the meaning of the word "majority." The constitution called for a referendum to pass "whenever a majority of the voters of such county at any general election shall so determine," but the Sixteenth's law required a lesser majority, "a majority of all the votes cast for or against it."[10]

While no one seems to have questioned the disconnect before November 1849, critics were vocal in the months that followed. Among the most Yankee of the nonadopting counties was Knox. There the referendum in 1849 had produced 728 for and 420 against, out of a total vote of roughly 1,480.[11] The editor of the *Knoxville Journal* consulted the appropriate documents and concluded in February 1850 that Knox had fallen "500 or 600 votes short" of having the necessary majority to carry township organization forward. He noted that Jo Daviess and Warren counties were instituting court cases to "learn whether organization had been adopted constitutionally."[12] A month later, Julius Manning, a "distinguished lawyer of Knox County,"[13] put in writing his opinion "that the township organization law passed by the last General Assembly, was not the law contemplated by the Constitution" and that "although not wholly void, the present law is null so far as it takes from the county court the management of the fiscal concerns of the county."[14] In other words, township organization had to wait before claiming dominion wherever the aye votes failed to exceed some higher standard than a simple majority of those voting on the issue.

The majority meaning was not the only aspect of the 1849 township legislation to cause confusion. In the opinion of some, the law had failed to demarcate exactly the end of the county court's duties and the beginning of those for the new board of supervisors. A circuit judge in LaSalle County set the transition date within his jurisdiction for November 1850, *after* the next general election. The *Aurora Beacon* noted, "many counties have organized and are proceeding under the new law," and wondered, "why there should be such a diversity of opinion among judges and lawyers on this subject." In Carroll County, the county court went ahead and "selected jurors, and judges of election according to the old method." If wrong, the *Mt. Carroll Tribune*'s editor felt, "the court is less to blame than the bunglers who framed the township law."[15]

Satisfactory remedies for the bungling of the Sixteenth General Assembly came first from the courts, later from the Seventeenth General Assembly, and then from the courts again. The first court case grew out of a dispute in Woodford County, a county largely within Meyer's New England Culture Region and neighbor to four counties that had adopted in 1849: LaSalle, Marshall, Peoria, and Tazewell. The legislative response was to rewrite the law rather than revise it. General assembly and supreme court actions cleared up confusion in Hancock County in 1853; and the court resolved a statewide issue affecting township referenda in 1858, as it ruled on a local dispute in Saline County.

Woodford County

Welcome B. Brown won election in November 1849 as Woodford County's first judge of the county court under the new constitution. At the same election, with over 600 voters participating, only 260 expressed a preference about township organization: 153 favored and 107 opposed.[16] Brown and his two associate judges, William E. Buckingham and William C. Poynter, met as a county court in December of 1849. The three appointed commissioners to divide the county into townships; but, when the appointees returned to the court in March, the county clerk, Edgar Babcock, refused to receive and record their report or to give subsequent notice of the annual township meetings as prescribed in the 1849 law. Some Woodford townships went ahead with annual meetings while others refused to meet, given the confusing times. Many of the officers chosen at the meetings that did occur refused to take the steps necessary to qualify as duly elected officials.

The State, in league with citizen William Davenport, sued Brown, his associates, and the county clerk to force them to certify that Woodford had adopted township organization—and to clear up this mess and similar messes elsewhere. First stop was the Tazewell Circuit Court, meeting in Metamora, on 12 April 1850. Arguing for the People were circuit attorneys Abraham Lincoln (Springfield) and Onslow Peters (Peoria). For the defendants stood Asahel Gridley (Bloomington) and Norman Purple (Peoria). On the bench was David Davis, who would one day serve as a Justice of the United States Supreme Court and, later, a United States Senator. Judge Davis ruled "the 4th section of the county and township organization law...was in conflict with the 6th section of the 7th article of the constitution" in that the former called for a majority of those voting on the question and the latter for a majority of the voters in the county. He ruled the election in Woodford "void" so Lincoln and Peters appealed to the Illinois Supreme Court.[17] Chief Justice Samuel Treat wrote that court's opinion in *People v. Brown et. al.*, saying the constitution was clear: "It does not mean a majority of those voting on the question to be submitted, but a majority of all the legal voters of the county…therefore…the township law had not been legally adopted in that county [Woodford]." Furthermore, the power of the county court over the business of the county would continue until enough voters chose to change to township organization.[18]

Observing these developments with keen interest from Knox County, the editor of the *Knoxville Journal* concluded that it was perfectly clear "the Organization law" had failed to carry in Knox, as it was "well known that there was not even a majority of the votes cast at that election in its favor." He hoped somehow Knox could get "safely out of the predicament in which she" found herself.[19] Remedy for Knox came finally in 1852, after the new township organization law went into effect.

The New Law

John Wood was a Whig candidate for the state senate in 1850, hoping to represent Adams and Pike counties. His opponents in Adams were apparently spreading a rumor that Wood had opposed township organization the year before. Wood's allies at the *Quincy Whig* ran a piece denying the allegation and informing readers that "Mr. Wood was in California when the people voted for or against organization." The *Whig* claimed Wood had never read the 1849 law, had never spoken against it, would be the "last man to do anything against it now," and would (if elected) "do all that is needful and proper" to perfect the law and make it "generally beneficial to the people."[20] Wood won.

Both houses had committees on township organization after the Seventeenth General Assembly convened on the sixth day of January in 1851.[21] Such standing committees conveyed the message to observers that Senator Wood was not the only member of the legislature to take seriously the confusion over townships. For several subsequent assemblies, by the way, the two houses kept standing committees on townships, sometimes merging them with their committees on counties.

Citizen petitioners were active in anticipation of the Seventeenth's first session. Senator William Reddick brought his colleagues a petition from Grundy County in which petitioners were "praying the passage of a law to abolish organization in said county."[22] Whiteside County, which had voted 359 to 47 for townships in 1849 and lay well within the New England Region of Illinois (but failed to adopt), sent at least five petitions, including one via Senator Hugh Wallace, in which signers asked for a law that would allow them to vote again.[23] The 1849 law had only authorized referenda in November 1849, although eight counties went ahead and voted anyway in 1850; and one of them—Vermilion—adopted and today operates under township organization despite the aforementioned technicality.

The State Archives holds numerous other petitions directed at the Seventeenth concerning township organization. In three from Fulton County, for instance, petitioners expressed dissatisfaction with township government by asking for a law repealing the 1849 law or "a law by which any county which has been organized under and by virtue of said [1849] law may absolve such organization by a vote of the legal voters of such county at any time when they may see proper."[24] So another flaw in the 1849 law was the absence of a provision to allow counties to *discontinue* township organization.

A new and improved township organization law answered most of the petitioner prayers and resolved the dispute that prompted *People v. Brown.* The 1851 law began life as a senate bill introduced on 30 January of that year by Senator Norman B. Judd of Cook County and went to the Committee on Township Organization, which consisted of Judd, and four other senators, three from early-adopting counties, Kane, Peoria, and Will, plus one from an early-majority county, McLean. Judd's original bill called for amending the 1849 law; but, when it came back from committee on 15 February, it was a new piece of legislation and carried the label "An act in relation to township organization." After passing the senate, 20-2, the committee's revised version of the bill went to the house as "An act to provide for township organization" and there, with minor changes, passed on 15 February, 60-2. The senate agreed to the changes, 18-0; and Governor French signed it into law on the seventeenth.[25]

The new law resembled the former in many ways and in exact wording in some segments.[26] Most important changes were (1) the provision in article 1 that brought the township law into compliance with the constitutional section that required a majority of legal voters to adopt and (2) a provision for discontinuance of township organization in (the new) article 25.[27] Both a referendum to adopt and one to discontinue were to come up for voter consideration if 50 legal voters of the county properly petitioned the county court/board, but discontinuance could take place, not at a general election as the 1848 Constitution specified for adoption, but at the "annual town meeting." Perhaps because the 1851 law said voters could discontinue only at the annual meeting, legislators specified a different (lower) standard for discontinuance (a majority of total voters voting that day) versus that for adoption (a majority of all the county's voters).[28] Another section of article 25 (section 3) addressed an issue that would ultimately spell the end of boards of supervisors in township counties. Section 3 allowed each Chicago ward to elect a supervisor, "in addition to the township supervisors, and the several supervisors so elected shall be members of the board of supervisors."[29] After 1972, as new legislation went into effect, county boards in township counties did not automatically seat township supervisors, because the excessive proportion of rural supervisors violated the rule of one man, one vote. If a supervisor wished to serve both as a county board member and a supervisor, she or he had to win two separate elections (see chapter 6).

Reactions to the new law were generally positive. The *Rockford Forum* liked what had come back from the capital about the law by the end of March but complained that tardiness in promulgation of this and all new laws would likely have a negative effect on the spring 1851 annual meetings in Winnebago County's townships. To the west a couple of counties, the editor of the *Galena North-Western Gazette* noted the law complied with the constitution and emphasized that approval, in the referendum upcoming that fall in Jo Daviess, "required a majority of *all the voters living in the county,* whether they do or do not vote at the election." In Canton, the *Weekly Register,* after the April 1852 referendum there to confirm Fulton County's previous adoption, seemed resigned to having to endure the township system but urged citizens "to induce the Legislature to make the law more perfect."[30]

Over the ensuing years, subsequent general assemblies regularly created new laws and modified statutes dealing directly or indirectly with townships. Having two county systems of governing meant the State of Illinois had to provide for both, sometimes producing parallel legislation. For instance, the Eighteenth General Assembly passed "An act regulating the collection of the revenue, in counties adopting the township organization law" presumably because county court counties already had such guidance.[31]

TOI considers laws affecting townships so significant that it maintains a legislative liaison to promote TOI's legislative agenda and takes other special pains to interact favorably with assembly members. To bring additional order to the applicable statutes, TOI endorsed and distributed in 2000 the "first-ever" *Township and Other Related Laws of Illinois.*[32] In the book's 815 pages, the compilers included segments, from the Illinois Compiled Statutes Annotated, pertaining to general provisions, revenue, local government, townships, roads and bridges, civil immunities, and business transactions. The township segment occupies 141 pages, and is equivalent to the Illinois Township Code, which in 1994 replaced the Township Law of 1874.[33]

Hancock County

In 1852, Hancock came close to being the first Illinois county to discontinue the township model. I gave serious consideration to reducing the list of 1849 adopters from 24 to 23 and moving Hancock's final adoption date to 7 March 1853, using a policy I applied later in the nineteenth century to four other counties—Franklin, Jackson, Richland, and Washington—that adopted, discontinued, and readopted.[34] The outcome of *People v. Couchman* is the main reason I left Hancock with the early adopters. This 1853 case was also the answer to the problem I pointed out above regarding the 1851 law's mismatch between the types of election whereat adoption and discontinuance could occur. As in *People v. Brown,* the State joined the Hancock case to assure a statewide precedent.

After adopting in 1849, Hancock County went through the process of establishing township government, a process that concluded on schedule in the spring of 1850. The *Warsaw Signal,* though claiming the names chosen for townships in the county were not to its "taste," pronounced the arrangement "judicious."[35] Waiting for a chance to proclaim otherwise, however, were hundreds of citizens who preferred a simpler system; and, with passage of the 1851 law containing the discontinuance option, they got their wish to vote again. In April of 1852, at the annual meeting of township electors, seven counties joined Hancock in voting whether or not to reverse their 1849 adoption decisions. Only Hancock, of the eight, counted fewer votes for townships (542) than it gave against keeping them (606). Warsaw (2-213) and Carthage (1-104) overwhelmingly opted for reversion.[36] A few weeks later, the defunct county court, which had been out of business for two years, convened again "for the transaction of business under the old organization." Supporters of township organization applied to the circuit court for an injunction against the erstwhile county court, and Judge Onias C. Skinner issued the injunction saying, "The County Court is not restored to the jurisdiction now exercised by the Board of Supervisors."[37] Then, working in Springfield, township sympathizers secured passage, in January 1853, of "AN ACT to authorize the voters of the county of Hancock to vote upon the question of township organization." To that date and since, this law is the only piece of Illinois legislation authorizing a specific county to vote on this issue. The law anticipated no second division of the county, authorizing the county to leave intact all previous township names and boundaries.[38] At the special March referendum, Hancock voters affirmed their allegiance to townships, but the old county court would not exit gracefully.[39]

Melgar Couchman had been judge of the county court when it had dissolved in 1850; Milton Morrill had served then as one of the two associate judges. This pair was still attempting to transact county business in the summer of 1853, alongside the board of supervisors. Wesley Manier, a local attorney, went to circuit court, with the People in his corner, to seek a requirement that Couchman and Morrill explain their presumptive behavior—in view of the March 1853 referendum confirming township government there. Judge Skinner came down on the side of Couchman and Milton, saying in effect that township organization had ceased to exist in Hancock after the 6 April 1852 vote in the town meetings.[40] *People v. Couchman* went to the Illinois Supreme Court, where in December 1853 Justice John Dean Caton authored the opinion that set Hancock straight and took care of Illinois's majority mismatch. He wrote,

> We do not think the legislature had the power [in the 1851 law] to allow a county to abolish township organization where it had been adopted [1849] in the constitutional mode, in a way less solemn or with different formalities or safe-

> guards...than the constitution required should be observed in its adoption....The constitution provides that township organization may be adopted by a majority of the voters of a county at a general election, but at none other.[41]

The justices found the Hancock vote of April 1852 had violated the constitution framers' intent that township organization decisions occur at general elections: "We are under the opinion that the township system never was legally abrogated in Hancock county, after its first adoption under the law of 1849, and that the people were entitled to judgment."[42] Several years later, Charles J. Scofield reported, "The effect of the decision was to oust the County Court from the management of the business of the County, and to recognize the acts of the Board of Supervisors as lawful and binding from the first election in 1850."[43] With few exceptions, referenda on adoption or discontinuance have occurred at November general elections since *People v. Couchman;* and Hancock townships were able to celebrate their sesquicentennials in 2000, along with townships from the 23 other early adopters.

Saline County

One other confusing aspect of referenda protocol remained unresolved until the supreme court took a case from Saline County in 1858 involving a marriage license. Unresolved was the issue of how to determine whether the vote in a county to adopt or discontinue constituted a "majority of the voters of such county," as the constitution required and *People v. Brown* confirmed. Some county courts had unilaterally and quietly decided that a majority of those voting at such an election was sufficient. They took the highest vote tallied on a particular contest at such an election. If the ayes regarding township organization exceeded 50 percent of that total, the referendum had passed. The *Bloomington Weekly Pantagraph* warned readers in 1856 that "He who votes at the election at all, and does not vote *for* township organization, *votes against it.*" Other county courts adhered more closely to the constitutional wording and would have agreed with a cautionary note published in the *Ottawa Republican* about the November 1857 vote in McLean County regarding townships. That editor felt, despite a large majority in McLean in favor, less than half of the county's voters were at the polls that day. "Staying home from elections is not conducive to advancement in public affairs."[44]

So editors, boards/courts, and voters needed guidance. Into the county clerk's office of Saline County, in Harrisburg, on 12 March 1858, strode John W. Mitchell, asking for a license to marry Mollie J. Provine. All parties later conceded that the betrothed pair fulfilled all the legal requirements for such a document, but Clerk Richard N. Warfield refused to issue one. When Mitchell sued Warfield, the People joined because much more was on the table than a license.

Defendant Warfield blamed his refusal on confusion over which community, Harrisburg or Raleigh, could legitimately claim to be the seat of Saline County.[45] At a November 1857 election to decide this important matter, Harrisburg received 725 votes and Raleigh got 689, with 26 voting on other issues but not picking one of the two seat sites. After that election, Warfield had moved his office from Raleigh (the former seat) to Harrisburg. A subsequently issued circuit court injunction had left Warfield, he claimed, uncertain as to how he should proceed. The election said go to Harrisburg; the injunction said go back to Raleigh. Several other county officers shared Warfield's confusion about where to conduct county business.[46]

Complainant Mitchell believed Raleigh to be the seat of justice. His attorneys argued the legislative act of 7 February 1857 authorizing the special county-seat election was unconstitutional because it called only for a majority of those voting on the removal, "not a majority of the voters of the county." Mitchell claimed that 40 to 50 persons, entitled to vote, did not vote. Therefore, Harrisburg lacked the necessary majority. He sought to have Warfield and all other county officers return their offices to Raleigh.[47]

Justice Pinckney H. Walker issued the supreme court's April 1858 ruling. The court admitted it was "a question of no small difficulty to determine in what mode it shall be ascertained who are the voters of the county, so as to determine whether a majority have voted in favor of a relocation." Did it require only a majority of the votes cast on the issue or the majority of all voters or something in between? The court recognized the similarity between this county-seat referendum and township organization referenda, which were occurring even more frequently. Only by assuming the legislature expected everybody to vote who was eligible to vote in such referenda could "the law be sustained authorizing township organization," wrote Walker. In other words, some of the state's "northern counties" had only adopted township organization if they applied a liberal definition to "majority of the voters of such county," as the constitution required. For Saline County, the court held "that a majority of the legal votes cast at [that November 1857] election [was] sufficient to determine the question of relocation of the county seat." On the face of that wording, a reader could infer that the seat ought go to Harrisburg; but issues of voter fraud that came to the surface during the proceedings forced the justices to order, "the county seat must remain unchanged."[48] Whether Mollie and John ever got their license is uncertain in the court findings; but, what is clear is the precedent for future township organization referenda: a majority, after *People v. Warfield,* meant *more than half of the votes cast at an election,* not more than half the county's total voters. For the next few years, therefore, the meaning of majority for adoption was to be the same as that for discontinuance.

Disillusion

While civil courts and the general assembly could help counter the confusion, township organization had to overcome considerable disillusion in order to hold onto its original conquests and spread beyond the Yankee-dominated areas. Disgruntled supporters of the commission-style government were quick to call for reversal of the early-adoption decisions. Moderate township supporters found township government took more tax money to operate than expected and more local input to get the system up and keep it running. Adamant township advocates counseled patience, as the innovation worked out its growing pains, and for better cooperation between city and countryside.

Disillusioned citizens, in early-adopting counties, vented their displeasure by condemning the innovation and praising the traditional. One venue was the local newspaper, a paper such as the *Canton Weekly Register,* out of Fulton County, a county that lay near the outer limits of New England dominance and still housed many loyal to the southern style of governing counties. The January 1851 numbers of the *Register* apparently have not survived, but (as was often the case) contemporary editors had borrowed from it, and at least one of these reprinted items is available today. So, in the *Knoxville Journal* of 4 February 1851, we can read about the *Register's* position on this issue:

> Township organization is very unpopular in this section. We do not know one voice in its favor in Canton Township. We think it has been a dear experiment, as we have in this Township about $5,000 to pay in taxes, which is not half collected, and only one week for the [township's] collector [to] go upon. Our taxes are about $2 for every man, woman, and child. Pretty hard for [a] man who has a wife and a dozen or two children.

The editor hoped the legislature would do something to get them "out of the scrape" in which they found themselves. A year later, the *Register* still carried local lamentations over their township burden: "Many of our citizens here are dissatisfied with the present order of things, and had rather go back to the old system." Out of Carroll County came the comment that "the township organization so eagerly seized upon, by our fellow citizens, has become quite a burden." That editor felt in September 1850 that "the people would gladly get rid of township organization, if a second choice were afforded."[49] In April of 1852, when Carroll voters had their second chance, they voted to retain.

Another venue for carping about fledgling township government was the petition option to the general assembly. Early in 1851, G. W. Spratt and hundreds of other Bureau County petitioners, in several nearly identical petitions, asked the house of representatives to repeal the township law because of their growing disillusion over its implementation, saying,

> Your petitioners state that they have found the laws, providing for the Government of said Townships, so imperfect and incomplete, that they have experienced great inconveniences and embarrassments in administering the government of said County...[and are] firm in the belief that the control of the affairs of said Connty

[*sic*] would be far better, to be entrusted with the County Court System...than it is under the New System.[50]

Five nearly identical petitions from Tazewell County expressed the view that "a minority of the real voters of the County" had "*inconsiderately* adopted" township organization; and they, the petitioners, found the innovation "entirely inefficient, unsatisfactory and burdensome." Furthermore, "a large majority of all the voters of the County" would "anxiously vote for the *immediate repeal* of the said township organization law in Tazewell County" and a return to the old county court system, which petitioners knew "to be much more effective, cheap and satisfactory."[51] Tazewell County was among the counties voting 6 April 1852 to retain their townships.

Supporters in early-adopting counties, as one might expect, had strong arguments for sticking with the new system until it had a chance to work out the kinks locally and find remedies in court and in the capital. Samuel E. Morris was one of nearly a hundred from Bureau County to remonstrate against the call, from G. W. Spratt and his fellow complainers, for the repeal of the township law. The Morris petition went on at length about a need for patience as the new system geared up and involved more of the citizenry:

> Time will afford a remedy and the care, study and thought that will be necessary for the people to exercise in the management of local affairs will prove one of the best primary schools for teaching the principles of self government...We believe the town system to be a reform and as is the case of the reforms in general, while the many are benefited permanently, a few who enjoyed emoluments and spoils under the old system are in favor of returning to it.

The numerous "complaints urged against it are unreasonable," said Roberts and company.[52]

A majority of editors tended, in the early adopting counties, to urge staying the course with something so new and wart-ridden. The *Freeport Journal* recalled in the spring of 1850 how Stephenson County had been "almost unanimous" in its endorsement the previous fall. The editor urged those who felt now "opposed to the township system" that, while the "precinct system" might be best for a "thinly settled county," a county like theirs required a "new mode of government" and they only needed "patience and forbearance...to bring it into successful and satisfactory operation." From Aurora (Kane County) that same month, the *Beacon* praised the possibility of "placing the management of affairs more directly in the hands of the people." The editor hoped that would "prevent the plundering that [had] hitherto prevailed among County officers." Out to the west in Adams County, the *Quincy Whig* ran minutes from a meeting of citizens at Camp Point, citizens who had adopted resolutions, one of which urged Adams residents "to fully investigate the two systems before voting against the one now adopted [township organization]. Peoria County apparently, in the opinion of the *Democratic Press,* "under the old system...had hard work to obtain credit for merchandise from [local] merchants." After a couple years under the board of supervisors, however, "the county had no difficulty in obtaining credit." Chicago's *Democrat* papers (weekly and daily) campaigned vigorously in 1852, before the April referendum in Cook County, for townships. On the day of the vote, the *Daily Democrat* urged citizens of Chicago to give the country beyond the city limits a break by voting to confirm township organization. Among other arguments, the editor reminded Chicagoans how many senators, representatives, and other elected officials then came and previously had come from the city and not the county precincts.[53]

Newspapers in counties adopting later in the 1850s, as the innovation spread from its 1849 base, also generally urged re-endorsement by voters. The *Rock Island Islander and Argus,* two years after adoption in Rock Island County, told readers they should look closely "before they do away with the township organization. As yet they know not its benefits. It has not been sufficiently tested." If a year or so later, it had failed to deliver, then they could "dispense with it."[54] One of great benefits, the *Islander and Argus* reminded in a later editorial, was the way township counties collected tax. "There is no chance for the Sheriff to speculate in county orders and grow rich off the use of the people's money."[55] Bloomington's *Daily Pantagraph* liked the way the township system brought voting places closer to the voter and thus tended to decrease the number of stay-at-home voters. The *Danville Vermilion County Press* pointed out a fact with which counties to the south would wrestle in the years to come: going back and forth between systems was a waste of resources.[56]

Illinois voters in the middle of the nineteenth century took politics seriously, even local issues. One of the big advantages that townships had over the county court approach, with its three at-large board members, was the annual town meeting, where electors could gather and vigorously debate the mundane or the monumental. Historian Charles A. Partridge, writing about Lake County many years later, helps us see why the disillusioned failed to find a bigger audience in the truly committed counties when it came to overturning the township system. Partridge reported:

> The early town meetings were interesting events. Every voter, especially if he was a tax payer, aimed to be present at the hour when the reports of township officers were read. Each expenditure was carefully considered, and every vote deliberately taken. Many of the early settlers were educated men. They had opinions and were not afraid to express them. Not a few of them were fluent and forceful debaters....Often there were spirited contests for local offices, and while the candidate was not expected to manifest undue interest in his own behalf, his friends were vigilant and argued his claims and fitness for the places aspired to with each new comer to the rolls.[57]

If townships' annual meetings were as freewheeling and popular today, township government might not face the anonymity charge that its detractors have made frequently.

Diffusion

Nearly three dozen more counties featured political townships in 1862 than had them in 1850: 59 compared to 24. Excepting new counties Kankakee (organized as a county in 1853; adopted townships in 1853) and Ford (organized 1859, adopted 1860), all of this wave of adopters had passed up the chance to make the switch in 1849. Some, like Edgar, had voted in 1849 not to adopt. Some, like McDonough, either did not vote in 1849 or failed to send referendum results to the secretary of state. Whatever the circumstance, the new kid on the prairie had spread across time, with a least one adoption occurring every year before 1862, and through space, covering the balance of Meyer's New England counties, most of his Midland-Midwest counties, and pushing well into the northern outposts of his Upland South counties.

A Winter Tale of Diffusion

Imagine for a moment a modest, midwestern, detached, single-family dwelling, consisting of seven or eight rooms on the main level and a basement. A forced-air furnace provides heat to the entire structure, except the attic and attached garage. At precisely 1847 hours (or 6:47) on a late December evening, an inexperienced cook burns popcorn on the kitchen stove. Within a few seconds, at 1849 according to the digital clock, all family and guests hanging around the kitchen have detected the distinctive odor of scorched popcorn. That detection surprises no one, however, because all in the kitchen had heard the cook's exclamation, had seen the smoke, and fully expected the smell to reach their noses. By 1850 hours, a natural draft has pulled the smoke through an open door and into an adjacent room where folks sitting by the crackling fire, dozing off after a big supper, learn of the incident. By 1852, one of the original kitchen occupants has gone to a back room and on his clothes carries the news of the burned popcorn to the little ones who were sitting around a board game. At 1854, the furnace cycles on, and within a couple of minutes has circulated smoke particles to all rooms on the main level, including the living room where a father and daughter are talking about baseball. Daughter says, "I smell burned toast." Father disagrees, saying, "I never mistake burned popcorn." She defers to his experience. One of the residents then remembers that furnace vents to the basement are not open. Once she opens the master vent control, the family patriarch, who is working in the basement on an antique mantel clock and who loves the smell of burned popcorn because it reminds him of his first job as a movie theater snack-bar attendant, learns of the scorching. A few minutes later, however, a chemically sensitive guest, feeling faint from breathing the unpleasant (to her) smell, asks for relief from the lady of the house, who complies by switching on the kitchen ventilator and evacuating the smoke through a roof vent. Standing quietly on the roof, near the vent, is a short, heavy-set, fully bearded, older gentleman in winter garb, who is neither family nor invited guest. At 1923 hours, he becomes the last to detect the odor as it wafts off with the brisk west wind. A teenager, who is in the garage tinkering with his car audio system, hears of the episode later that evening but never actually smells the scorch.

A Township Tale of Diffusion

The popcorn incident illustrates how transfer occurs in a fluid—the air—from areas of concentration, the hearth or core, to areas of lesser concentration, the periphery. Think of the scorching as the constitutional and legislative action necessary to begin the reception and adoption of township organization. Some counties were right there in the late 1840s. If just one or two in the kitchen had picked up the scent of the burned popcorn before anyone else, we could have given them special status—what social scientists call innovators, people who are out in front of the rest of us, taking chances. Others who smelled the popcorn in the kitchen next would have been our early adopters—just a little later than the innovators. But with township organization, 24 counties, everybody in the kitchen, got the whiff; and all voted to adopt immediately, in 1849.

Soon after the kitchen crew smelled the popcorn, those in a nearby room joined the in-group. Diffusion had carried molecules (the township idea) to nearby rooms, to nearby counties. Thus the natural flow of air in the house and the normal contacts between adjacent counties are carriers. A carrier speeds up the transfer from hearth to periphery.[58] Other carriers in the popcorn analogy would have included the guy who took it in his clothing to the back room and the kitchen ventilator to the roof. With cultural diffusion, as was the case with townships, carriers were newspapers or other printed matter, like pamphlets circulating beyond their counties of origin.[59] Citizens that moved from county to county or into Illinois from township states could have carried the township message, as well. Most effective of all the carriers in the popcorn home was the furnace. It was able to introduce the odor to the majority of the living space very efficiently.

Carriers frequently get help in the diffusion process from opinion leaders and change agents. The father, discussing baseball with his daughter, was an opinion leader because he was able to convince her that she smelled popcorn, not burned toast. Opinion leaders in the case of townships led the public discussion at called meetings and with messages the local newspapers carried to the voters. In the popcorn analogy, the lady of the house, by purposely opening the vents to the basement became a willing agent of change, because she sent the aroma to the only occupant who really appreciated it. With township government, a change agent might have been someone who stood to gain financially from additional adoptions.[60]

Working against innovation carriers, and their allies, are barriers to diffusion.[61] In the house, closed doors, closed vents, unheated spaces, the exterior walls and roof got in the way of the smoke molecules. Some barriers were solid enough to keep all evidence of the scorching away from occupants—like the teenager in the garage. Other barriers broke down with a flip of the switch or the normal cycling of the furnace. Permeable barriers may slow or even alter the transmission, but they do not prohibit diffusion. Open doorways and heating ducts channeled diffusion of the smoke and overcame barriers. For townships, an obvious barrier to diffusion was culture itself. New Englanders and New Yorkers could barely wait to adopt townships. Midlanders were less eager. Southerners were downright resistant. Language could have been a barrier for the foreign element in Illinois. Politics played a role in some counties because residents perceived townships as a partisan matter. Inertia played a part in the early spread, and in the strong resistance today to change. The Illinois River corridor helped township advocates channel adoption farther south and west in 1849 than we might have expected, so the river corridor served as a salient into alien territory.

At the end of the adoption process come the stragglers. The old guy on the roof learned last of the kitchen incident. Diffusionists sometimes categorize the last to adopt as a laggard, although that denotation carries a negative connotation. I prefer to refer to the three counties that adopted in the twentieth century, the last in 1923, as skeptics rather than laggards (see chapter 5). One of the three was so skeptical that it later discontinued township government.

Most students of cultural diffusion eventually get around to discussing the cumulative curve, growth curve, or S-shaped curve (figure 9). Among the first to employ such a growth curve was F. Stuart Chapin, then a sociologist at the University of Minnesota, in his 1928 book, *Cultural Change.* Chapin graphed a measure of time passage on the x-axis, increasing from left to right, against some measure of cultural change on the y-axis, with amount increasing from bottom to top. Each data point on such a graph represents the total percentage of adoptions as of a particular time. His examples included adoptions of the commission form of city government and adoptions of the city manager way of governing. Both of Chapin's graphs revealed an early phase with few adopters, a middle segment with many, and a final section with almost none.[62]

Compare the ideal S-curve with the growth curve for the adoption of township organization by Illinois counties (figure 9). The pent-up demand for township government and for its board of supervisors to run the county left no time for innovators to dally around with townships while others waited. More than one-fourth of the 99 existing counties accepted the offer at first opportunity and thereby pushed the y-intercept well up the graph—an unusual circumstance. Most adoptions begin with very few at the outset and build slowly. Even more Illinois counties joined in the 1850s and early 1860s, with the cumulative score of counties passing 50 percent in 1859 when eight adopted, making that the second largest annual total of adoptions (after the undisputed leader, 1849). So the actual growth curve for Illinois starts high on the y-axis and shoots up from there. By the time of the next

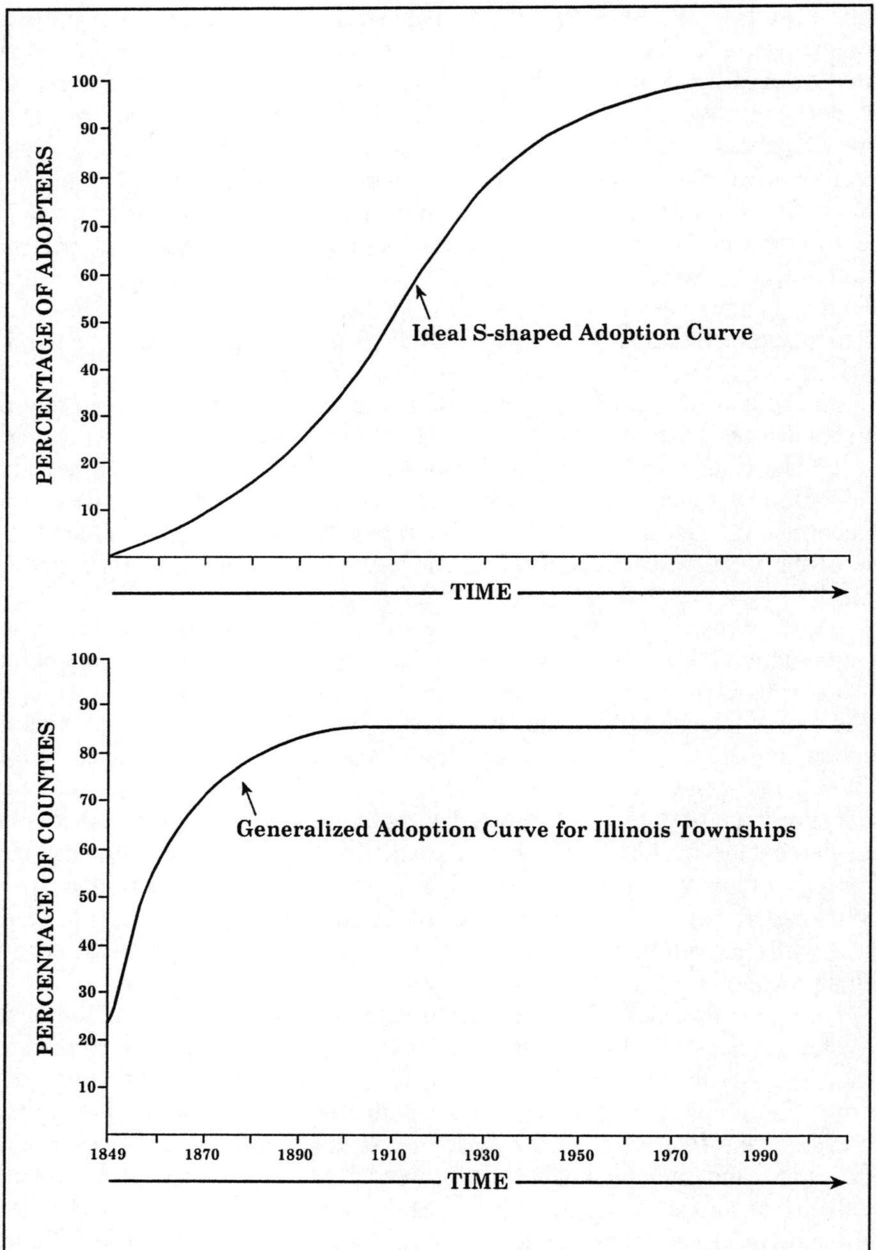

Figure 9 Adoption Curves.

constitutional revision in 1870, two of every three Illinois counties had left behind the county court; and by 1890, all but three of the counties that would ever adopt had adopted. Skeptic counties, coming aboard in the next century, were Henderson in 1906, Williamson in 1907, and Cass in 1923. Cass, ironically, lay adjacent to Sangamon County, where the constitution framers of 1847 and the legislators of 1849 and 1851 had worked making the innovation possible. The Illinois growth curve also accommodates the two discontinuances to date, one in 1876 and another in 1932. Some innovations eventually cover 100 percent of the potential adopters. Hybrid corn among commercial farmers in the Corn Belt is a good example of saturation. Many innovations, however, fall short of complete coverage; and township organization belongs here.[63]

Innovations spread outward from their hearths via a number of mechanisms. The notion of townships and a county board composed of township supervisors came into Illinois with men like Wead and Peters, so their decision to settle in Illinois produced a type of diffusion that students of the process call relocation. Once township organization had taken hold in what became the hearth (near Lake Michigan and the Wisconsin border), another type of diffusion took over: expansion. In this case, the innovation tends to remain strong at the hearth and spread mainly stepwise, county-by-county. Many counties took their cue to adopt or even to hold a referendum on the issue, after 1849, from neighboring counties. One other type of diffusion had a minor role to play in the Illinois township story: hierarchical. Larger places and more important places (perhaps politically) tend to get the word earlier about something new and to adopt cultural changes earlier. In such instances, innovations jump intervening nonadopting areas. Adams County, in 1849, is a good example of a county that adopted earlier than one would have predicted. For the most part, however, the progress of the township across Illinois was county to county.

The Early Majority

The 36 counties of the early majority occupy a zone that stretches from Indiana to Iowa, 90 miles along and either side of a line from near the point where the National Road (today, approximately, Interstate 70) crossed the Indiana-Illinois line to Rock Island (figure 10). Viewed alongside the first map in this series (figure 7) and the next (figure 13), the map for the early majority looks to be part of a great wave cruising across the land from northeast to southwest. The last map in the series (figure 15) is the backwash, as Henderson, Williamson, and Cass counties filled in gaps left as the tide traversed the state.

New adoptions occurred in every year from 1849 through 1861. Adoptions ranged from the eight of 1859 to only one in 1850, 1851, and 1858. Referenda during the early-majority period totaled 117 (of the 438 I have documented), and at least one referendum took place every year from 1850

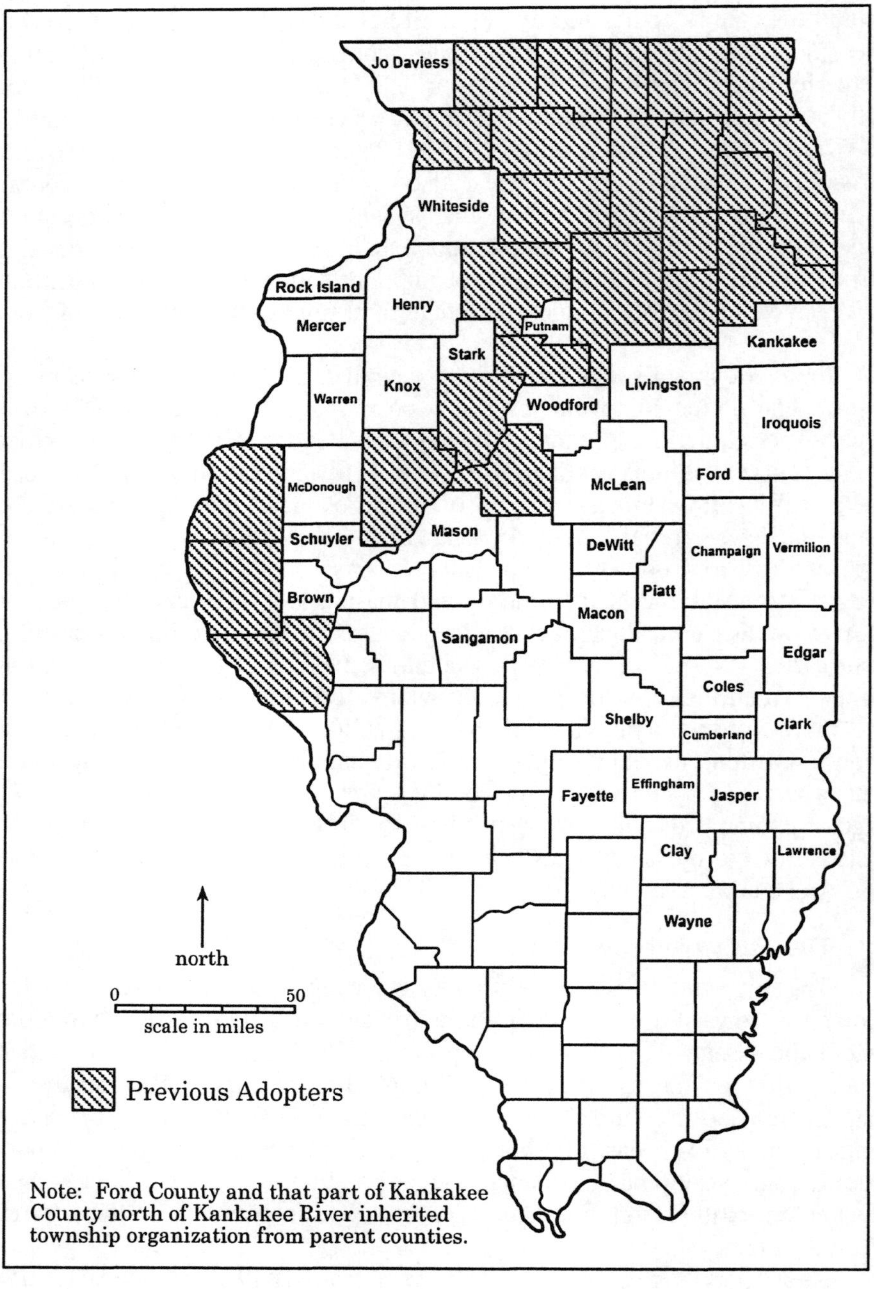

Figure 10 Counties Adopting Township Organization: 1850–1864.

through 1864 (figure 11). Of these 117, 23, or 20 percent, were the results of efforts to discontinue townships or at least to test the public's resolve to give them more of a chance.

Referenda per year fell dramatically from the 59 of 1849, a year when many counties put the issue on the ballot only because the law of 1849 said they must, to just 3 in 1851. My guess is that several more counties actually collected some votes for and against township organization in 1849 but simply failed to report because the interest level among citizens or commissioners was so low. Following new legislation in 1851, interest climbed to a second peak of 14 in 1852, partly because the 1851 law implied that electors at the April town meetings ought to vote on the matter. One Chicago paper told readers that their upcoming referendum was occurring in April 1852 because "the Legislature ordered a vote at the next election to see if township organization should be abandoned or not."[64] Referenda by year fell sharply in the mid-1850s and then reached the highest level ever (excluding 1849) at the end of the decade, with both 1859 and 1860 recording 15 each. Almost all in those two years were efforts in counties trying to join the township wave, as the word spread deeper into commission country.

Figure 12 helps reinforce the southeast-northwest configuration of counties that the adoption maps reveal. Referenda activity focused during the period on the area from the Wabash to the Iowa line, with most counties in the northeastern hearth and far southern region not voting on the issue at all. Many counties voted once, or like Crawford (on the Wabash south of the National Road) voted more than once but failed to adopt as part of the early majority. Some initially rejected then adopted before the Civil War. Others adopted and then, like the eight in 1852 (Adams, Carroll, Cook, Fulton, Hancock, Ogle, Peoria, and Tazewell) checked signals before continuing on as a township county. No counties discontinued townships in this early-majority time frame, but as we saw with Hancock earlier in this chapter, it could happen if the supporters dropped their vigilance. Hancock, by the way, failed again to discontinue in 1853. Rock Island County tried twice (1858 and 1860) to abolish what voters had adopted in 1856, after a failure to adopt in 1853. Madison and Jo Daviess led the tally by having had township organization on their voters' ballots five times between 1850 and 1864, with Madison saying no each time and Jo Daviess rejecting in 1851; adopting in 1852; and confirming in 1854, 1857, and 1860.

While the early majority was coming on board, a number of themes emerged, some of which we will return to in later chapters. The first involves the role of innovation carriers.

Carriers

Hezekiah Wead and his fellow Yankees carried the idea of townships to Illinois, through the constitutional convention of 1847, and helped make it a

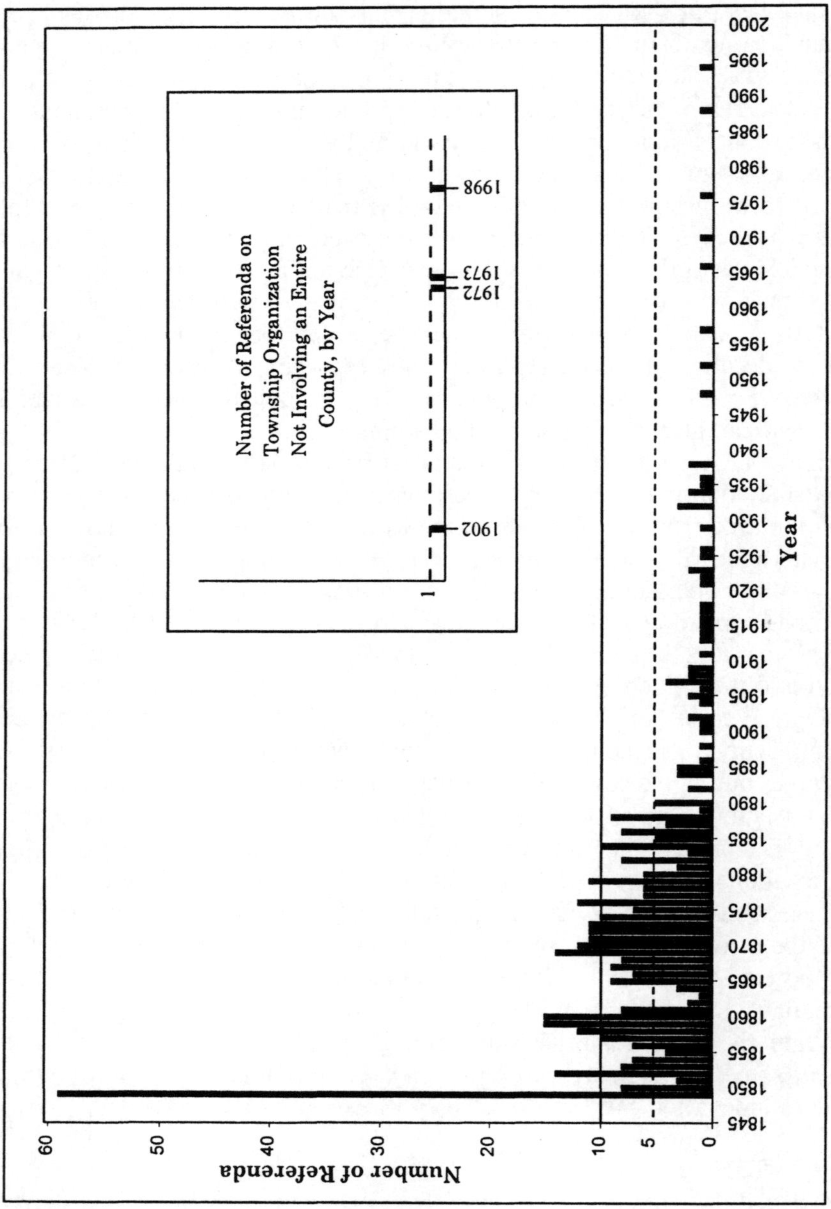

Figure 11 Number of Countywide Referenda on Township Organization, by Year.

reality in the heartland. Who or what then carried it to the far corners, to almost every county, of the state? Could we label anyone as a change agent?

Americans in the 1850s were a restless lot. The pattern Meyer derived from study of the 1850 Census of Population soon changed as foreign and domestic immigration into the state (and natural increase) continued and out migration to places like Kansas and California lured people westward or in other directions. Surely, Illinoisans moved from county to county as well; but, in the context of spreading the word about township organization, I could find no evidence that ordinary citizens personally carried the message beyond the early-adopting counties to the early majority counties, late majority, and skeptics.[65] My guess is that it did happen here and there between 1850 and the end of the Civil War.[66]

I have no doubt, however, that there was a true change agent operating for several decades on the side of township government: Elijah M. Haines. Recall that a change agent is someone who carries the message for an innovation, at least partly because the agent's livelihood depends on successfully convincing nonadopters to try and then to retain the innovation. An example today of a true change agent is the district conservationist of the Natural Resources Conservation Service, who with his or her staff, tries to induce farmers and other users of rural land to employ progressive land, water, and biomass conservation practices—like filter strips of living vegetation along waterways to trap sediment and agricultural chemicals before they reach the stream.

Haines, born 1822 in Oneida County, New York, came to Lake County, Illinois, in 1836. In late 1849 and early 1850 he was one of three on the commission to bound and name the townships for Lake County. He died 25 April 1889 after a long career as a writer, publisher, lawyer, and lawmaker.[67] Among his publishing accomplishments are compilations of township laws for Illinois, Michigan, Minnesota, Missouri, and Wisconsin. The Illinois volume went through a couple dozen editions, appearing at least as late as 1910, two full decades after his death. In the introduction to the first edition of the Illinois volume, in 1855, Haines praised the state legislature for "appropriating funds for the [purchase and] distribution of seven thousand copies of this compilation." Such a purchase and distribution, he opined, would "have the effect of doing more for the speedy correction of...errors, and the perfection of the township system in this State, than any other plan which could have been adopted."[68] Newspapers freely quoted Haines's work as editors sought to educate and convince the readers of the worth of township organization. "With this [the 1857 Haines compilation] as our guide," wrote the editor of the *Urbana Our Constitution,* "we shall proceed to give the outlines of [township organization's] features, referring the readers to the work itself for details."[69] As more counties joined the township club, more copies of his compilations came off the press. By 1863, and probably several years earlier, Haines was advertising in his monthly journal for lawyers that he had blank

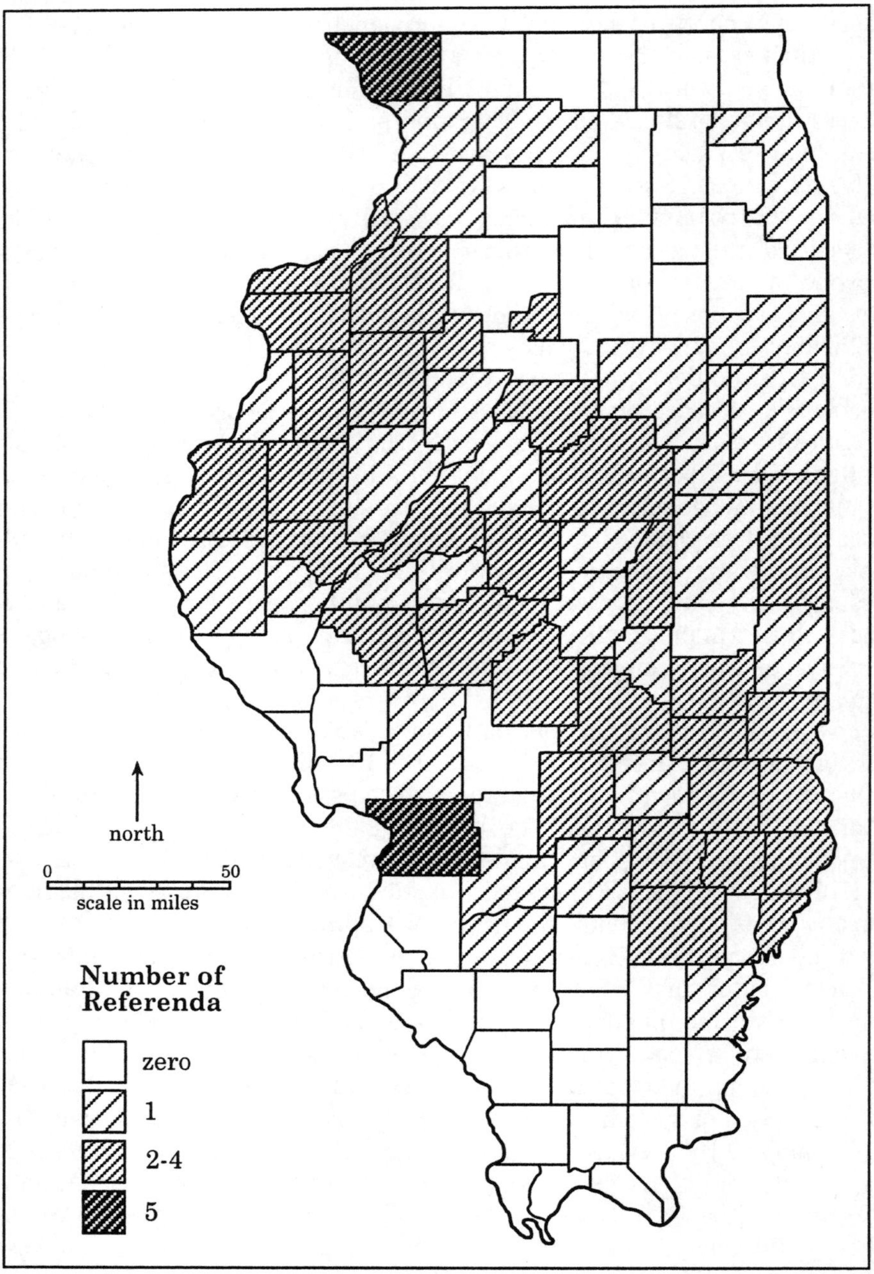

Figure 12 Countywide Referenda on Township Organization,by County: 1850–1864.

forms for sale, forms township officials could use to conduct their business.[70] Of course Haines was not the only writer and publisher to sell such blank forms, but his name became synonymous with the business of making Illinois safe for township government.[71]

Opinion Leaders

Newspaper editors were the most conspicuous of the local opinion leaders, once the township idea reached the state. Those editors who had lived in the Northeast or other states where townships had gained a foothold kept the issue in the public forum through their editorial commentary and letters they published from readers. Some editors were fervent supporters and left no township advantage uncovered. If an innovation had been working in the next county or in a county the editor held in high regard, local voters might take heed. The *Peoria Democratic Press* asked readers to note, prior to the April 1852 discontinuance referendum, how strongly the board of supervisors in Cook County had endorsed township organization in a recent Chicago meeting. In Rock Island County, a local editor began an 1858 editorial by referring to unnamed counties "that would not, for any consideration, do away with [township organization]." Two weeks later, he featured adjacent Mercer County, which had adopted in 1853 (three years before Rock Island) and where "public sentiment [was] strong against [township organization's] abolition." In the same piece, the editor cited testimony from a Mercer resident who had "opposed it when it was adopted" but was then (1858) "heartily in its favor, after experiencing its benefits."[72]

Sometimes editors, in their zeal to encourage adoption, exaggerated the facts a bit. *Our Constitution* carried information on efforts in other counties to discontinue township organization, saying, "that, while there have been perhaps a hundred instances in this state where the attempt has been made to do away with the Township organization,...in at least ninety-five of these instances the people by large majorities decided to retain the system."[73] As I noted above, far fewer than 100 discontinuance referenda occurred before the Civil War; and none resulted in reversion to the at-large election of three county court judges. Nevertheless, such validation of the township innovation must have seemed reassuring, and Champaign County adopted a few weeks later.

The *Alton Daily Courier* so favored adoption for Madison County that it listed alphabetically the counties that the editor believed had adopted by October of 1857. Of the 36 on his list, he wrote,

> They all lie in the northern and northeastern portion of the State, and are settled mainly by natives of New England, who, having there tried the system, were unwilling to be deprived of it in their adopted homes in the West. What stronger argument in favor of it need we have?

Actually, the *Courier* editor omitted seven counties that had adopted in 1855 and 1856. By 1859, after Madison voters had rejected townships in 1857 and 1858, the *Weekly Courier* listed adopters in a table and gave the date adopted, number of townships per county, and 1855 population for each county. An adjacent table carried population and area figures for survey townships in Madison County, as an indication to voters how Madison might look if it could abandon the old and join the township tide.[74]

Opinion leadership by an editor could take readers in the opposite direction; and reference to other counties worked nicely there as well. The earliest example I found of a newspaper editor using other counties as negative argumentation tool was in the *Bloomington Western Whig* in 1850. While citing some of the adverse characteristics of the township system, like the dozens (maybe hundreds) of new officials needed, the *Whig* editor called on McLean County voters to "delay its adoption at least until its operations in the counties that have adopted can be thoroughly tested." He wanted others to run the experiment, and then, if all seemed well, there would "be time enough for [McLean County] to adopt it." The editor of the *Olney Times,* out of Richland County, cited in 1857 the unhappy residents of adjacent Lawrence County, which adopted in 1856, as a good reason for their county to forgo adoption. He had begun the item by writing,

> This ridiculous humbug as the election approaches is being agitated by a few who are willing by its adoption, to advance their individual interests, at the sacrifice of the general good. We have not time, nor space at present to point out the many objections to this system, but will do so in our next issue.[75]

An opinion leader might also be a letter writer. Many writers of letters to the editor assumed in those years a penname. Joining the debate in Sangamon County toward the end of the 1850s was writer U, who focused on the roads:

> Go into those counties where the system of Township Organization has been in existence for a few years and you will find good roads. A man from Sangamon can tell, after going out of his own county, and going into one of these township organized counties, before he travels a mile into it, by his own observation of the state of the roads, that he is in a county under Township Organization. Let us try Township Organization.[76]

Logic, of course, tells us, just because there are better roads in another county, that is no reason to conclude a different system of government was the reason; but it was worth a try. Sangamon did not adopt in 1859.

Peoria attorney Onslow Peters, after his township-friendly role as a framer of the 1848 Constitution, tried to sway public opinion around Peoria toward the new system by writing pamphlets and guest articles for the local paper. His influence went beyond Peoria, however, when distant editors picked up his work. The *Freeport Journal* reprinted portions of a Peters pamphlet under the title "Town Government" in April 1850. Peters traced township prototypes to Europe and made sure readers knew the township model was then "in full and vigorous operation in New York [and] most of the Free States." A couple years later, Peters was touting townships in articles on "popular education." Freeman Miles, in a protownship letter to the *Peoria Weekly Republican,* cited Peters's third education-related article, where Peters had written, "The man who should attempt to repeal [New England's] township laws would be deemed a madman, a knave or a fool." Learning that Peters was the writer of another letter to the editor, over the penname AN OBSERVER in March of 1852, would not surprise me. That correspondent, whatever his or her name, expressed the view that "the town corporations are the last part of the laws that the New England people would be willing to give up." Peters's role as an opinion leader ended abruptly in February 1856, when, as sitting judge of the Sixteenth Illinois Judicial Circuit, he died while in Washington, D.C.[77]

Financial Comparisons

Having two different systems of county government operating simultaneously in different counties of a state or consecutively in the same county invites financial comparisons. The comparison practice persists today as modern township critics have used analysis of taxes raised and monies spent in their attempts to demonstrate the costly nature of this layer of local government. Since no Illinois county has switched systems in many decades, the modern comparisons are between counties. In 1852, when the first such comparisons appeared, almost simultaneously in three counties, they were between the new and the old. All three counties were looking forward at the time of these comparisons to referenda in April of that year to determine whether township would survive. None of the comparisons originated with the newspapers that printed them, however.

One came from a meeting of concerned citizens in Adams County. Figures presented at the meeting favored townships over the county court in terms of expenses, assessment errors, amount of delinquent property tax, and various other issues. The special committee charged with preparing the figures noted, "We could contrast the subject farther, showing corresponding items favorable to Town Organization—but will desist for the present."[78]

Other comparisons came in with letters to editors. ARGUS wrote to the *Peoria Democratic Press* at almost the same time the Adams County citizens were issuing their report. "I have examined the expenses of the County Commissioners Court [in Peoria County] in the last year of its existence," ARGUS revealed, "and find that it was in session sixty-eight days," which cost taxpayers $748.00. When the writer added sheriff, clerk, and assessor costs, the total was just over $3,500. For the most recent year of township government in Peoria County, the board of supervisors was in session 15 days, said ARGUS, costing $500.09. Travel expenses for rural board members, travel for the assessor, and clerk's salary brought the yearly total to

$1,325.50, "leaving a balance in favor of the present system of the nice little sum of $2184.04."[79]

Fulton County voters, according to the *Knoxville Journal,* confirmed township organization convincingly in 1852 (2,086-456). Both sides had taken a shot, prior to the referendum, at making the financial situation favor their position. MANY VOTERS found an advantage of "about $2,605" per annum for the old system. SAVE YOUR MONEY also came down on the side of the commission/court model, asking if anyone would "presume to say" that 182 officers could transact the public's business as well and "at as little expense as 6? No." The six officers under the county court (and commission) were the three members of the court, a clerk, an assessor, and the sheriff/collector. A VOTER challenged the figures of MANY VOTERS, point by point, found the argument favoring the former system wanting, and blamed the county clerk for helping circulate grossly deceptive "facts." This writer to the *Register* sarcastically suggested that if Fulton County voters liked having "the county officers, living at the county seat do [the voters'] business for [them]," they could dispense with all the trouble of "town meetings, levying taxes, or such matters" and go back to the old ways.[80] My guess is A VOTER was among the 2,086 voting to retain townships.

Trigger Issues

Change may require a kick-start. Someone or some group finds something about the traditional way troubling: too costly, ineffective, or scandalous. Convincing adverse criticism brings in more supporters, and a struggle ensues over whether to stick with the old and known or to try the new and unsettling. Once the Yankee counties and their nearest neighbors had gone the township route, trigger issues in reluctant counties, in the hands of an attacking force, could bring about the demise of the county court system. After all, it only took (theoretically) 50 signatures to put the township question on the ballot.[81] Here is a sample of trigger issues from the early-majority era.

Iroquois County (between Chicago and Danville on the Indiana line) adopted township organization in 1855 largely because the county court appeared to have misused funds, derived from the valuable swamplands that the county had received from the federal government, in contracting for work associated with grading and furnishing ties for the Peoria and Oquawka Railroad, which ran east and west across the county. "Charges and denunciations were freely indulged at the expense of the court. Distrust of [the court's] capacity and integrity grew apace, until it had deteriorated so much in public favor that this became the chief argument in support of township organization," wrote E. S. Picker in an 1880 county history.[82]

Edgar County adopted in 1856. One triggering element was criticism circulating then about John C. Means, the sheriff and ex officio tax collector for the county. The *Paris Prairie Beacon* took Means to task for his alleged violation of the law, saying on one occasion, "If a public officer may thus abuse his trust, to minister to his avarice, let the principle be adopted in our laws, so that the people may know what to depend on," but if it is wrong, "as it unquestionably is,...put it down and promptly rebuke the officer who is base enough to attempt its introduction into our government."[83] Under township organization, Means and other former collectors had to confine their operations to law enforcement.

DeWitt County had not voted on township organization since the more-or-less mandatory referendum of 1849, but in 1858 DeWitt adopted the innovation, and has not voted on the issue since. The county court authorized construction of a new county jail in 1857, but had so much trouble paying the bills that it had to borrow money to finish the project. "The people of the county," wrote one historian, in an 1882 review, "had meanwhile become restive and dissatisfied with the management of their affairs. A petition...was presented to the court [for a vote on the subject of townships] in September 1858." The new board of supervisors immediately reduced the assessed value of property by 20 percent in 1859—presumably bringing the needed tax relief.[84]

Champaign County, in 1858, received an infusion of funds from the return of swamplands and from unused railroad land grant lands, and soon after that windfall the county court decided to build a new courthouse. But a group of disgruntled taxpayers in West Urbana organized early in 1859 to resist the project, perhaps harboring hopes of a shift of county seat, in their direction, from Urbana. A new courthouse on the same spot as before would tend to solidify the Urbana position as the seat. At least one candidate for the county court pledged not to build a new building or make any additions to the existing structure if elected. Another denounced a new structure but was "not opposed to making such improvements upon the present building as may be deemed necessary." *Our Constitution,* an Urbana paper, urged the court to do something to protect the records stored in the courthouse, and praised the court's decision to plan carefully and spend frugally. The *Central Illinois Gazette,* out of West Urbana, believed the land funds could have gone for better roads and began agitating for a change in the way Champaign County conducted business. Its editor wrote, "We would suggest that the occurrence of the present state of affairs...exhibit[s]...the disadvantages of our present system of county government. At some future time we shall...discuss...the merits of a *township organization."* *Our Constitution* eventually joined the *Gazette* in calling for township organization, and Champaign County got both a new courthouse and a new way of governing itself in 1859.[85]

A Terrible Struggle Takes Precedence

On the eleventh day of July in 1861, the *Post,* out of Havana, Illinois, the seat of Mason County, carried a brief editorial on township organization. In it the editor said, "The article on this subject prepared for our columns by an intelligent citizen [opinion leader?], is unavoidably crowded out." He went on to warn that the *Post* might have to "advocate the adoption of township organization as a measure of relief to the people. If we decide to do it, the work will be done thoroughly." The *Post* lacked space in that issue for more coverage of townships because the editor chose instead to print President Lincoln's special address to Congress, delivered on the fourth of July, in which he explained how the government was reacting to the Fort Sumter attack and the matters relating to the rebellion.[86] War news commanded so much space in the remaining months of 1861 that the *Post* did not get back to townships. Mason and Clay counties did adopt that November, but obviously the nation had directed its attention elsewhere. Adoptions ceased until after Appomattox, and referenda fell to the lowest annual total since the onset of the issue, and until 1889. But what happened in 1865 and the next three-plus decades is the subject of chapter 4.

Notes

1. Meyer listed the relevant articles in his book' s bibliography. The first, in *Pioneer America,* appeared in 1975, the last, in the *Bulletin of the Illinois Geographical Society,* in 1984.
2. Meyer, *Making the Heartland Quilt,* 143.
3. Meyer, *Making the Heartland Quilt,* 177, and the rest of his chapter 7.
4. Wilbur Zelinsky, *The Cultural Geography of the United States* (Englewood Cliffs, New Jersey: Prentice-Hall, 1973), 118–19. Prentice-Hall published a revised edition in 1993, with the same map on the same pages.
5. Meyer, *Making the Heartland Quilt,* 198.
6. Meyer, *Making the Heartland Quilt,* 167.
7. Meyer, *Making the Heartland Quilt,* 181 and 190.
8. Meyer, *Making the Heartland Quilt,* 288.
9. Meyer, *Making the Heartland Quilt,* 181.
10. Emil Joseph Verlie, ed., *Illinois Constitutions,* Collections of the Illinois State Historical Library, vol. 13 (Springfield: Illinois State Historical Library, 1919), 80; *Laws First Session Sixteenth General Assembly,* 190.
11. Election Returns, 1849.
12. "Township Organization," *Knoxville Journal,* 20 February 1850, 2.
13. "Township Organization—An Unsettled Question," *Aurora Beacon,* 4 April 1850. 2.
14. "Township Organization," *Peoria Democratic Press,* 20 March 1850, 2.
15. "County Court," *Aurora Beacon,* 30 May 1850, 2; "Township Organization," *Mt. Carroll Tribune,* as reprinted in the *Freeport Journal,* 27 September 1850, 2.
16. Election Returns, 1849; "Township Organization," *Knoxville Journal,* 24 April 1850, 3.
17. "Township Organization," *Knoxville Journal,* 24 April 1850, 3.
18. *The People v. Brown et. al.,* 11 Ill. 479–83 (Ill. Sup. Ct. 1850); "Township Organization," *Knoxville Journal,* 9 July 1850, 2.
19. "Township Organization," *Knoxville Journal,* 9 July 1850, 2.

20. "Mr. Wood and Township Organization," *Quincy Whig,* 22 October 1850, 2.
21. *House Journal,* 17th G. A., Regular sess., 1851, 30; *Senate Journal,* 17th G. A., Regular sess., 1851, 160.
22. *Senate Journal,* 17th G. A., Regular sess., 1851, 113.
23. *Senate Journal,* 17th G. A., Regular sess., 1851, 98; Illinois, General Assembly, Petitions, 1851, Whiteside County, State Archives, Springfield.
24. Illinois, General Assembly, Petitions, 1851, Fulton County, State Archives, Springfield. The quotation comes from a petition that John Miller and 21 other men signed.
25. *Senate Journal,* 17th G. A., Regular sess., 1851, 160, 407, 415, 422, and 439; *House Journal,* 17th G. A., Regular sess., 1851, 465 and 472–73. Senator Wood voted for both the senate bill and for the compromise bill that came back from the house.
26. Illinois, General Assembly, *General Laws of the State of Illinois Passed by the Seventeenth General Assembly at Their [Regular] Session* (Springfield, 1851), 35–78.
27. Of the 371 countywide referenda on township organization occurring since passage of this 1851 law, 89 or 24 percent have been efforts to discontinue.
28. *General Laws Seventeenth General Assembly,* 78.
29. *General Laws Seventeenth General Assembly,* 77–78. Naperville residents were able to convince the Nineteenth General Assembly to pass a law authorizing that DuPage County city to elect thereafter an additional supervisor. Assistant supervisors, serving as voting members of the board of supervisors, later provided some relief for populous townships suffering from inadequate representation vis-à-vis lightly populated, usually rural, townships. Illinois, General Assembly, *[Public] Laws of the State of Illinois Passed by the Nineteenth General Assembly* (Springfield, 1855), 45.
30. "Town Meeting," *Rockford Forum,* 26 March 1851, 2; "Township Organization," *Galena North-Western Gazette,* 28 October 1851, 1; "Township Organization," *Canton Weekly Register,* 10 April 1852, 2.
31. Illinois, General Assembly, *General Laws of the State of Illinois Passed by the Eighteenth General Assembly* (Springfield, 1853), 67–89.
32. *Township and Other Related Laws of Illinois,* (Charlottesville, Virginia: Matthew Bender & Company, Inc., 2000), v.
33. *Township and Other Related Laws of Illinois,* 522.
34. An Alton editor took the position that, as of 1859, Hancock was the only Illinois county to return to the old system, but "only for a brief period." "Township Organization for Illinois," *Alton Weekly Courier,* 26 May 1859, 2.
35. "Township Division," *Warsaw Signal,* 2 March 1850, 3.
36. Illinois, Auditor of Public Accounts, Township Organization File, Hancock County, Canvass of Poll Books, 12 April 1852, State Archives, Springfield.
37. "Township Organization," *Warsaw Signal,* 26 June 1852, 2.
38. *General Laws Eighteenth General Assembly,* 486–87.
39. Newton Bateman, Paul Selby, and J. Seymour Currey, gen. eds., *Historical Encyclopedia of Illinois and History of Hancock County,* ed. Charles J. Scofield (Chicago: Munsell Publishing Company, 1921), 1048–49.
40. Bateman and Selby, *Hancock County,* 1,049.
41. *The People v. Couchman et. al.,* 15 Ill. 142–46 (Ill. Sup. Ct. 1853).
42. *People v. Couchman,* 146.
43. Bateman and Selby, *Hancock County,* 1050.

44. "Township Organization," *Bloomington Weekly Pantagraph,* 29 October 1856, 2; "Effect of Not Attending Election," *Ottawa Republican,* 14 November 1857, 2. Despite the *Republican* editor' s concern, McLean County went ahead successfully with organization following that referendum.
45. For background on events leading up to the county seat contest in Saline County, see Sublett, *Paper Counties,* 105–17.
46. *The People v. Warfield,* 20 Ill. 160–66 (Ill. Sup. Ct. 1858).
47. *People v. Warfield*, 159.
48. *People v. Warfield*, 164–66.
49. "Township Organization," *Canton Weekly Register,* as reprinted in the *Knoxville Journal,* 4 February 1851, 2; "Township Organization," *Canton Weekly Register,* 7 February 1852, 2; "Township Organization," *Mt. Carroll Tribune,* as reprinted in the *Freeport Journal,* 27 September 1850, 2.
50. Illinois, General Assembly, Petitions, 1851, Bureau County, State Archives, Springfield. The quotation comes from a petition that G. W. Spratt and 10 other men signed.
51. Illinois, General Assembly, Petitions, 1851, Tazewell County, State Archives, Springfield. The quotations come from a petition that Robert Roberts and 12 other men signed.
52. Illinois, General Assembly, Petitions, 1851, Bureau County, State Archives, Springfield. The quotations come from a petition that Samuel E. Morris and 90 other men signed.
53. "Township Organization Law," *Freeport Journal,* 15 April 1850, 2; "Township Election," *Aurora Beacon,* 4 April 1850, 2; "Township Organization at Camp Point," *Quincy Whig,* 29 March 1852, 2; "Township Organization," *Peoria Democratic Press,* 3 March 1852, 2; "City Against the County: Township Organization," *Chicago Daily Democrat,* 6 April 1852, 2.
54. "Township Organization," *Rock Island Islander and Argus,* 17 March 1858, 2.
55. "Township Organization," *Rock Island Islander and Argus,* 30 March 1858, 2. Under the court/commission system, the sheriff was the ex-officio tax collector and treasurer. "Township Organization," *Champaign Gazette and Union,* 11 March 1868, 1.
56. "Official Vote of McLean County," *Bloomington Daily Pantagraph,* 8 November 1858, 2; "The Issues of the Spring Election," *Danville Vermilion County Press,* 30 March 1859, 2.
57. Newton Bateman and Paul Selby, gen. eds., *Historical Encyclopedia of Illinois and History of Lake County,* ed. Charles A. Partridge (Chicago: Munsell Publishing Company, 1902), 642.
58. For a good discussion of carriers, see Peter R. Gould, *Spatial Diffusion,* Resource Paper No. 4 (Washington, D.C.: Association of American Geographers, 1969), 3–11.
59. One such pamphlet came out of Madison County, while that county was in the midst of a quarter-century debate on the merits of township organization. FARMER [pseud.], *Township Organization for Illinois or Local Self-Government for the People* (Alton, Illinois: Courier Steam Book and Job Printing House, 1859). I found a copy at the Illinois State Historical Library, Springfield.
60. Adoption of an innovation is not always in the best interest of the adopter. See discussions of the potential down side in these two works, the first by a geographer, the second by a rural sociologist: Lawrence A. Brown, *Innovation Diffusion: A New Perspective* (New York: Methuen & Company, 1981), 229; Frederick C. Fliegel, *Diffusion Research and Prospects in Rural Sociology: The Record and Prospects for the Future* (Westport, Connecticut: Greenwood Press, 1993), 85–87.

61. Gould, *Spatial Diffusion,* 11–23.
62. F. Stuart Chapin, *Cultural Change* (New York: The Century Co., 1928), 357–84. The key curve is on page 383.
63. Any of the 17 commission counties could adopt townships today. The last unsuccessful attempt was in 1957.
64. "To the Citizens of Cook County," *Chicago Weekly Democrat,* 13 March 1852, 3.
65. Ordinary citizens did occasionally reach out by means of correspondence to weigh in on issues confronting voters in counties other than their own. Newspapers might pick up and print a letter to the paper or to someone in town. See, for example, commentary from several outsiders as an Alton newspaper promoted the advantages of township organization: "Township Organization for Illinois," *Alton Weekly Courier,* 2 June 1859, 2. In that piece, we learned "a gentleman of Bureau County" had told a friend in Madison County that the people of Bureau "would not return to the old plan under any consideration."
66. A good example of what I was searching for in the way of ordinary citizens taking the message to other counties occurred at Belleville, the seat of St. Clair County, in 1883. There the Farmers' Club of St. Clair County invited men from adjacent Clinton County, which had already adopted township organization, to speak on behalf of the innovation at a club meeting. Some, like Andrew Eisenmeyer of Trenton, appeared in person; others sent letters of support. "Township Organization," *Belleville Weekly Advocate,* 5 October 1883, 4.
67. Elijah M. Haines, *Historical and Statistical Sketches of Lake County, State of Illinois,* 1852. Reprint. (Libertyville, Illinois: Mundelein Historical Society, 1973), i and iv; and *The Past and Present of Lake County, Illinois* (Chicago: Wm. LeBaron & Co., 1877). See, also, Bateman and Selby, *Lake County,* 215.
68. Elijah M. Haines, *A Compilation of all the General Laws of the State of Illinois Relative to Township Organization* (Chicago: Keen & Lee, 1855), 8.
69. "Township Organization," *Urbana Our Constitution,* 6 August 1859, 1.
70. "Township Organization Blanks for Illinois," *Haines' Legal Adviser and Citizens' Review,* 2 February 1863, 4. Haines also used this publication to editorialize about the worth of townships. See, for instance, "Self-Government," 1 April 1863, 4. Township opponents came to recognize Haines as a significant force behind the township tide sweeping southward across Illinois. An opponent in Morgan County called into question any facts or figures for which a proponent might cite Haines, saying "Such statements coming from men like Mr. Haines, who have fattened in the township system, should have but little weight with the voters of this county." TRUTH [pseud.], "The Other Side," *Jacksonville Journal,* 2 November 1879, 2.
71. As early as February 1850 other enterprising printers were making available for sale copies of the 1849 law as well as notes and forms for the conduct of township business, once the townships got themselves organized. See advertisements such as "Township Organization Law," *Knoxville Journal,* 6 February 1850, 3; [untitled], *Quincy Whig,* 12 February 1850, 3; and "Township Organization," *Rockford Forum,* 13 March 1850, 2. The untitled note begins with "We conclude the long township...." Map updates and revisions provided work for cartographers and printers. One example would be Rufus Blanchard, *Blanchard's Township Map of Illinois* (Chicago: Rufus Blanchard, 1863). I found an original map at the Illinois State Historical Library, Springfield.
72. "Township Organization," *Peoria Democratic Press,* 17 March 1852, 2; "Township Organization," *Rock Island Islander and Argus,* 17 March 1858, 2; "Township Organization," *Rock Island Islander and Argus,* 30 March 1858, 2.

73. "County Election," *West Urbana Our Constitution,* 24 September 1859, 2. *Our Constitution* came out of West Urbana, now Champaign, in the fall of 1859; but both before and after those few months it was in nearby Urbana, under the same name. For a closer look at the nineteenth century press in Illinois, see Franklin William Scott, *Newspapers and Periodicals of Illinois: 1814–1879,* Collections of the Illinois State Historical Library, vol. 6 (Springfield: Illinois State Historical Library, 1910).
74. "Township Organization," *Alton Daily Courier,* 28 October 1857, 2; "Township Organization," *Alton Weekly Courier,* 2 June 1859, 2.
75. "Township Organization," *Bloomington Western Whig,* 26 October 1850, 2; "Township Organization," *Olney Times,* 23 October 1857, 2.
76. U [pseud.], "Township Organization—Roads," *Springfield Daily Journal,* 29 August 1859, 2.
77. "Town Government," *Freeport Journal,* 15 April 1850, 2; Freeman Miles, "The Township Election," *Peoria Weekly Republican,* 26 March 1852, 2; AN OBSERVER [pseud.], "Township Organization," *Peoria Democratic Press,* 3 March 1852, 1; Cole, *The Constitutional Debates of 1847,* 973.
78. "Township Organization at Camp Point," *Quincy Whig,* 29 March 1852, 2.
79. ARGUS [pseud.], "Township Organization," *Peoria Democratic Press,* 31 March 1852, 2.
80. *Knoxville Journal,* 20 April 1852, 3; MANY VOTERS [pseud.], "Facts for the People of Fulton Co.," *Canton Weekly Register,* 27 March 1852, 2; SAVE YOUR MONEY [pseud.], "Township Organization," *Canton Weekly Register,* 3 April 1852, 2; A VOTER [pseud.], "Will Township Organization be Continued?" *Canton Weekly Register,* 3 April 1852, 2.
81. Not every petition drive, even if referendum advocates collected the 50 signatures, resulted in a vote on townships. The Knox County Court denied petitioners' wishes in 1851 due to confusion over what constituted a general election. "County Court," *Knoxville Journal,* 9 September 1851, 3. A dozen years later, the court in Clinton County (due east of St. Louis, beyond the river counties) ignored petitions from outlying areas requesting a referendum on township organization. The editor offered the explanation that the court would "deny all requests asked for by Union men." PETITIONERS [pseud.], "Communicated," *Carlyle Union Banner,* 29 October 1863, 2. During the 1990s and in 2000, getting county boards and election commissions to slate referenda, even after petitioners gathered thousands of signatures, was a major hurdle for abolitionists.
82. H. W. Beckwith, gen. ed., *History of Iroquois County,* ed. E. S. Ricker (Chicago: H. H. Hill and Company, 1880), 369.
83. "John C. Means," *Paris Prairie Beacon,* 16 May 1856, 2.
84. *History of DeWitt County, Illinois* (Philadelphia: W. R. Brink & Co., 1882), 70–71; DeWitt County. Transcript of County Board Minutes. 1859. Transcript is in State Archives, Springfield.
85. "A New Court House," *Urbana Our Constitution,* 25 December 1858, 2; "Tax Payers Convention," *Urbana Our Constitution,* 22 January 1859, 2; "Tax-Payers Convention—Fielding L. Scott, Esq., nominated for Associate Judge—His Pledge on his Record," *Urbana Our Constitution,* 29 January 1859, 2; "Candidate for Associate Justice," *Urbana Our Constitution,* 29 January 1859, 2; "The County Office—Duty of the County Court," *Urbana Our Constitution,* 30 April 1859, 2; "The New Court House," *Urbana Our*

Constitution, 11 June 1859, 2; "The New Court House," *West Urbana Central Illinois Gazette,* 15 June 1859, 2.

86. "Township Organization," *Havana Post,* 11 July 1861, 2.

4 Almost Irresistible: 1865–1899

WHEN THE SAGE OF WAUKEGAN, Elijah M. Haines, died in 1889, the *Chicago Tribune* devoted three full columns to his obituary, resolutions of respect from his colleagues in the general assembly, and recollections of him by his acquaintances. "Lije" had served many terms in the Illinois House of Representatives and twice was its speaker. He had been a superintendent of schools and a delegate at the 1869–1870 Constitutional Convention. In addition to his compilations of township laws, he produced *Haines' Treatise of Practice Before Justices of the Peace, Haines' City and Village Law, Haines' Probate Guide,* and other legal works. The monthly version of his *Legal Adviser* ran for over 60 years, long after his death, and the weekly for 17. Despite his best efforts, in print and on the podium, however, one-fifth of Illinois's counties had yet to adopt township organization when he died, and only five more would ever adopt—one that year, one the next, and three the next century. By 1889, Haines had seen one adopter permanently discontinue townships in favor of the three-man board of county commissioners, which in 1873 had replaced the county court. Haines had failed at the constitutional convention to have the state's new constitution mandate township organization for every county. What must have seemed irresistible to Haines was proving difficult for many to swallow as the leading edge of the township tide gurgled farther from Waukegan and township government's chief change agent.

Mapping the Late Majority

Exactly two dozen Illinois counties constitute the late majority of adopters. They joined another two dozen from 1849 and the three dozen from the early-majority period to give us 84 adopters out of a possible 102. Among the late majority, however, was Johnson County, which adopted in 1872 and discontinued four years later. Therefore, by the end of the century, there were 83 in the township column.

Figure 13 shows the distribution of the late majority, including Johnson.[1] All 24, save Logan, lay essentially south of Springfield and together form a relatively compact southeast-northwest trending area, which is especially true of a 19-county cluster from Gallatin and White on the lower Wabash to Greene and Jersey on the lower Illinois.

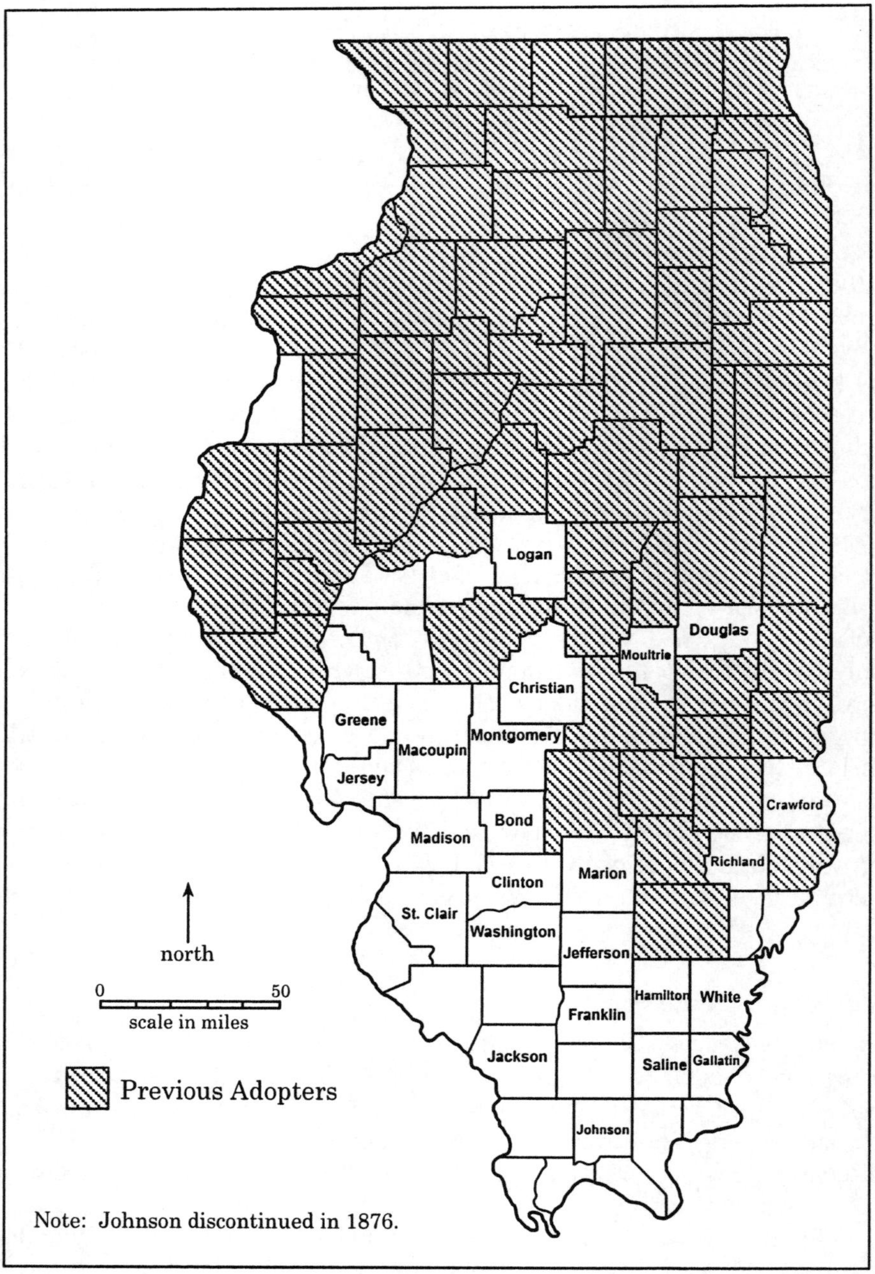

Figure 13 Counties Adopting Township Organization: 1865–1899.

Predicting when late-majority counties made their final decisions to adopt is fairly easy because they too followed the pattern described earlier, the pattern in which the innovation pushed southwestward from Lake Michigan to the Mississippi, south of St. Louis. Logan, Moultrie, Douglas, and Christian, on the north end of the late majority, were all on board the township ship by 1867. Madison, Clinton, and Marion, in the middle of the zone, had bought the idea of townships by 1875. Jackson, Saline, and Gallatin, at the outer edge of the innovation's reach, converted in the 1880s or in 1890.

A few of the 24 were a little quicker to reform county government than location would predict, and some were slower. Among the trendsetters of Little Egypt[2] were Jefferson County in 1869, perhaps, in part, because neighbor Wayne County to the east was a member of the early majority (1859), and especially Johnson in 1872, whose nearest adopting neighbor would have then been Jackson, which adopted in 1871 and discontinued in 1874. Among the most cautious southern adopters of the period was Crawford in 1867, all of whose Illinois neighbors were by then township counties, though Richland later discontinued briefly. Another wary county was Bond, east of St. Louis and northwest of the Kaskaskia River, which waited until 1888, when township counties Fayette, Montgomery, Clinton, and Madison surrounded it. Counties like St. Clair (1883), Greene (1884), and Washington (1888), laying adjacent to nonadopting neighbors, typically had a harder time deciding to adopt than did counties with only township counties on their margins.

Looking at the Laws

Legislation

Townships continued to generate considerable legal interest as the rest of the nineteenth century unfolded. At every postwar session of the general assembly, legislators tinkered with the 1861 township law, "An act to reduce the act to provide for Township Organization, and the several acts amendatory thereof, into one act, and to amend the same," and its successor, the Township Law of 1874, which incorporated mandates emanating from the 1870 Constitution.[3] Newell noted the 1874 iteration, while similar to the 1861 act, set 17 square miles as the minimum area for a township, made provision for the first election of county commissioners when a county discontinued township organization, and declared that Cook County township supervisors were not to be members of that county's board.[4] In 1877, the Thirtieth General Assembly made it possible for a city having over 3,000 in population to organize a city township; and today there are roughly twenty such entities, like the City of Bloomington Township, in which the city council serves as the township board.[5] Ten years later, the general assembly made mandatory the establishment of road districts in all counties, even those not then under township organization. Road districts of one sort or another operate in every Illinois county today.[6]

Some postwar legislation pertained only to particular counties. For instance, in 1867, the Twenty-fifth General Assembly surprised most people in Wayne County by passing a law to reduce the membership in that county's board of supervisors from 15 to 5, with the 5 to represent new districts consisting of several townships each.[7] Neighboring Clay County received similar special treatment two years later.[8] The Twenty-seventh General Assembly repealed the Wayne and Clay laws in 1871, so that each township therein would have at least one supervisor on subsequent boards.[9]

Court Decisions

The judiciary was less visible in deciding township organization cases in the late-majority years than it had been in the previous period (see chapter 3). A trio of counties, Macoupin, Logan, and Jersey, did turn, however, to the courts for resolution of referenda disputes in the 1860s and 1870s.

Macoupin County's friends of township government failed to get a ruling from the circuit court for "lack of jurisdiction" on election irregularities that may have resulted in the defeat there of township organization in 1869. Attorney George W. Hamilton then asked supporters of the innovation to ante up $150.00 so he could pursue the case at the next level.[10] Hamilton's case went no farther, but Macoupin voters approved the township model a year later.

Controversy over the Logan County election of 1865 and the Jersey County election of 1878 did reach the Illinois Supreme Court. These cases, *People v. Garner* and *People v. Ruyle,* yielded results favorable to township organization in both instances.

Logan County lagged behind most other central Illinois counties in adopting townships, as voters there declined in 1849 and 1858; and the county court refused to appoint a division committee or give way to a board of supervisors after voters in 1859 had apparently approved townships 973 to 208, with 1,426 total voters participating.[11] In 1865, Logan County voters were ready to join the postwar revival of interest in reforming local government. Voters that year gave township organization 1,612 votes versus 700 for the status quo, in an election that attracted approximately 2,900 voters. The county court acted properly in appointing a division committee and allowing the machinery of township organization to commence operations. Opponents appealed to the courts for an injunction to halt organization, in part, because, while the county court had ordered the 1865 referendum, the county clerk had failed to record the order. The supreme court, in *People v. Garner,* held that by accepting the election results and taking the necessary steps to implement township organization, the Logan County Court had acknowledged thereby its earlier approval of the referendum, whether or not the clerk wrote it in the minutes. Opponents also tried to negate the adoption by claiming lack of a legal majority because the 1,612 votes in favor may not have been more than half the voters in the county. Citing *People v. Warfield* (see chapter 3), the supreme court opted for the opinion that those voting were equivalent to the legal voters of the county at that particular moment. Opponents tried to convince the justices that the 1865 vote was illegal because the outcome of the 1859 vote was still uncertain, but the court said the 1865 referendum was entirely legal and binding. Justices also refused other objections dealing with division of the county into townships and (agreeing with the circuit court) ordered Logan County to continue down the township path.[12]

Jersey County needed a favorable ruling from the supreme court to overcome an objection to votes cast in Grafton Precinct, votes that made the difference in a close 1878 election. In *People v. Ruyle,* the court examined the election records from Grafton, decided they were legitimate, and ordered the county commissioners to offer no obstruction to township organization.[13]

A New Constitution

Illinois held its fourth constitutional convention between 13 December 1869 and 13 May 1870.[14] No other ten-year period in American history has seen as much state constitutional activity as occurred in the decade after the Civil War.[15] Revising or replacing state constitutions became common events, and Illinois joined the revision parade when its 1848 organic law just did not suit the needs of a boldly industrializing, rapidly populating commonwealth. The 85 delegates who met in Springfield over the winter and into the spring of 1870 faced critical issues like "the ever-increasing flood of private bills in the General Assembly, which had reached tremendous proportions, dwarfing the number of public bills."[16] Corruption in connection with private laws granting divorces, regulating local governmental affairs, and releasing corporations or individuals from indebtedness posed a threat to public confidence in government.[17] Township organization was not high on the list of reasons for calling the convention, but delegates did establish a Committee on Township Organization and did devote several hours of floor debate plus committee time to township issues. In the end, township government emerged relatively unaffected from its latest constitutional-convention encounter.

William P. Pierce, the delegate from Grundy County (an early adopter south and west of Chicago, where the Des Plaines and Kankakee rivers come together to form the Illinois), chaired the Committee on Township Organization. It was Pierce, on 26 January 1870, who presented his committee's majority report, which, like the township section of the 1848 Constitution, authorized adoption of township organization but did not provide a method for its discontinuance. Pierce and the six other committee members signing the majority report did suggest the innovative possibility of granting the boards of supervisors constitutional standing (mention), a standing they had not had in the previous 20 years.[18]

Elijah M. Haines, also a member of the Committee on Township Organization, rose next to offer a minority report. Excusing his report as "written somewhat hastily," Haines asked permission to read it aloud (rather than have the secretary read it). He told the convention "the old county system" of local government was "inconsistent with the wants and present intelligence of the people of the State of Illinois," a state he felt owed "much of its prosperity" to township organization. Haines wanted the new constitution to give considerable guidance to subsequent general assemblies about the operation of townships and boards of supervisors. The constant changing of the township law led to dissatisfaction, he declaimed, among the adopters and could "result in defeating the substantial benefits to be derived." Above all else, however, Haines wanted the constitution to *require* that all counties in Illinois come under township organization "whenever two-thirds of the members elected in each branch of the General Assembly shall concur."[19]

A month later, the convention assembled as a committee of the whole to debate the majority and minority reports on township organization. Several delegates participated; but Haines, though present did not further defend his call for mandatory inclusion of all counties in the township fold. Support for inclusiveness did come, however, from William R. Archer of Pike County, who said he would favor "a uniform system of county government," and from Samuel P. Cummings of Fulton County, who believed, "Township organization should be fixed, permanent. If there is anything that should be permanent and stable, it is township organization."[20] After all had a chance to contribute, the delegates present on 26 February 1870 voted to leave much of the business of township law to future legislators.[21] The framers did make sure, however, counties had a constitutional way of shedding of townships—a key element if skeptics south of Springfield were to try it. "We have provided that counties," framers said in an announcement they distributed prior to the public vote on the constitution, "having adopted township organization may by vote, dispense with the same."[22] Elsewhere, in article 10 of the constitution, they (1) changed the county court in nonadopting counties to a board of county commissioners with one of three commissioners to run for election every year and (2) left township organization in force in Cook County, but set up a 15-person board of commissioners (instead of a board of supervisors) to run that county's affairs.

Voters, perhaps taking their cue from "influential government figures and the press," voted on 2 July 1870 to approve the overall constitution and all eight propositions separately submitted.[23] One of those eight propositions was the article on counties and townships, which passed voter muster, 136,815 to 31,644, an even stronger endorsement than the constitution itself received.[24] Therefore, township organization had a niche in the 1870 Constitution, supreme court sanction, and legislative authority. The township tide seemed destined to flush out the old system, all the way to the Ohio.[25]

Referenda Redux

Nearly half of all the countywide referenda on township organization (215/438) took place between 1865 and 1899 (figure 11). Almost all the postwar nineteenth-century ballot action occurred in the state's southern half, as counties faced pressure from their own voters to modernize or economize and from elsewhere to conform to the majority mode of doing county business. In several of the postwar years, beginning in 1869, 10 or more counties voted on the issue; and it was not until the 1890s that a year went without at least one referendum on township government. The period 1870–1874 averaged 11 referenda per year, and 1866–1869 was not far behind at a rate of 9.4. Then from 1875 on, the five-year averages fell steadily to just one referendum per year in the last five years of the century. Almost certainly, the surge of referenda in the 1870s stemmed in part from the constitutional confirmation of a county's right to discontinue township organization if the format proved unsatisfactory.

Even or odd year did not make much difference to having a referendum on the ballot in the 1860s and 1870s; but, beginning in 1880, there began a definite preference for even-numbered years, for those years in which voters also could make presidential and congressional choices. In other words, from 1880 until 1899, it was more likely that a township referendum would occur in an even year (53 total referenda) than in an odd (17). Whether petition circulators planned it this way or not, I cannot document. Elections did occur annually in November of all years, beginning in 1873, in all commission counties because of the necessity of choosing one commissioner each year. Perhaps there lingered the notion that some advantage lay in getting out a large percentage of a county's electorate, and a national contest was more likely to pull voters to the polls than a strictly local contest.

How did individual counties contribute to the large number of referenda in the period, 1865–1899?[26] Many, 55 counties to be exact, did not vote at all on township referenda in the period (figure 14). In this group would fall most of the early adopters and many of the early majority. The nonvoting group also includes three counties, Massac, Pope, and Scott, that apparently *never* voted on township organization; two counties, Alexander and Hardin, which might just as well be among the never-voting because the only referendum for each was the legislatively prescribed vote of 1849; and Perry County. Perry stands out because the first of its 12 documented referenda did not occur until 1907. I am a bit suspicious of the Perry situation, vis-à-vis 1865–1899, however. First, township organization always came up short there. When this issue failed to succeed elsewhere in Illinois, there was a tendency not to report the fact to the secretary of state or auditor. I found only one of the 12 Perry referenda in the secretary of state's Election Returns and none in the APA records. Second, newspaper survival for the county seat, Pinckneyville, is quite poor for the period; and the *DuQuoin Tribune,* another county paper (with good period coverage), took little interest then in local affairs.

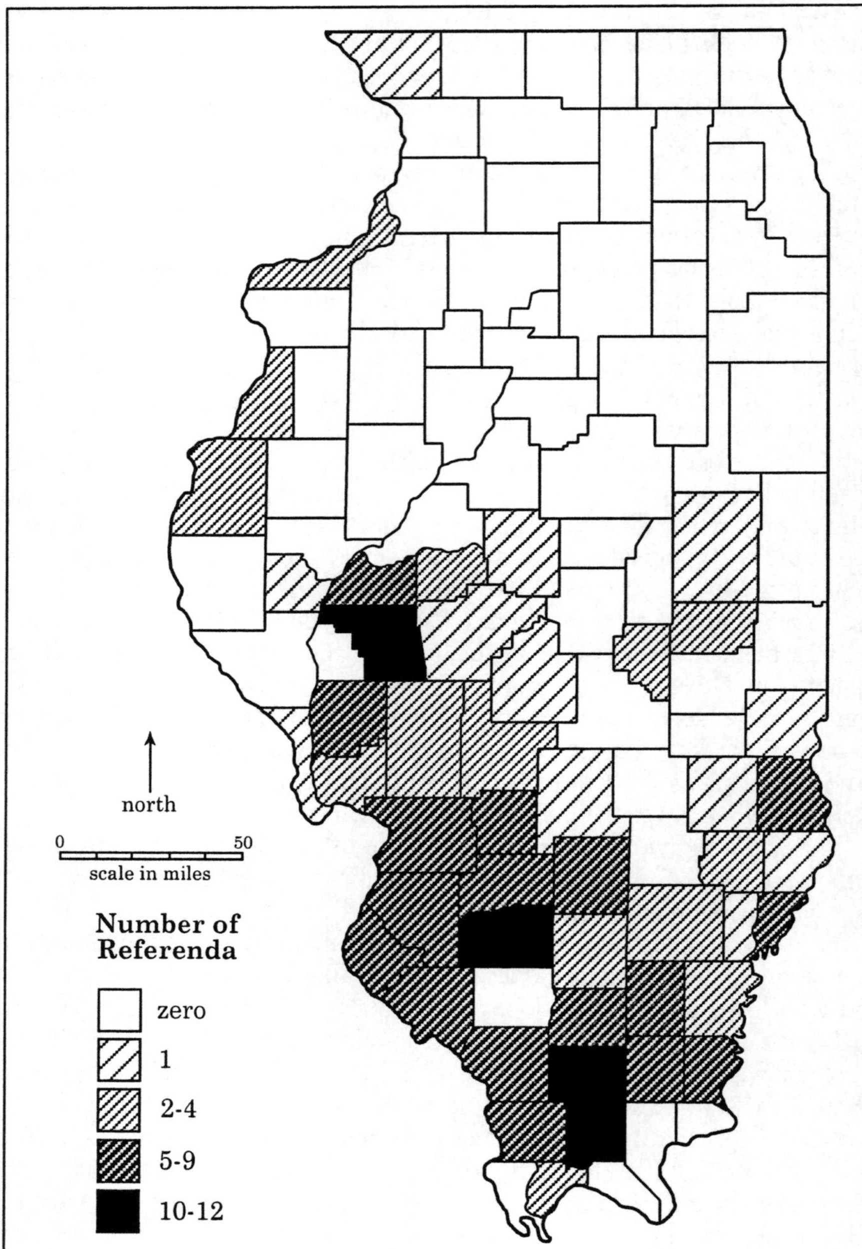

Figure 14 Countywide Referenda on Township Organization, by County: 1865–1899.

Third, even though I was able to review Perry County court/commission minutes for all 35 years of the late-majority period, there is the chance (as in Logan County in 1865) that the county clerk did not record the order to hold a referendum. Finally, Perry is an island (in figure 14) amid a sea of 43 referenda in its contiguous counties. For Perry to have resisted the urge to put township organization on the ballot sometime between 1865 and 1899 seems amazing.

Subtracting the 55 from 102 leaves 47 counties to produce the 215 referenda I was able to document between the Civil War and the beginning of the twentieth century. A few of the 47 were counties from the early majority, like Champaign, Jo Daviess, Rock Island, and Sangamon, testing the resilience of the township model. One early adopter, Hancock, site of so much legal controversy over townships in the early 1850s, put the issue on the ballot in 1865 and again in 1873. The 1865 vote was likely the outcome of a feud there over how much the board of supervisors should spend to compensate men for Civil War military service. Warsaw's *Hancock New Era* labeled as "illegal" the 1864 decision by supervisors to spend $400.00 on each man who served. Tough economic times and a desire by taxpayers to cut expenses in any way possible largely explain the 1873 Hancock vote on townships.[27] The rest of the 47 voting counties from the period were previous nonadopters or backsliders, those counties whose voters just could not find satisfaction with the township setup. Bond and Cass contributed 7 each to the total for the period; Madison, 8; Franklin, Greene, Hamilton, Jackson, and Randolph, 9; Johnson, Morgan, and Williamson, 10; and Washington, 12. All but Randolph and Morgan would eventually choose to give townships a whirl.

Losing the Numbers Game

Most of the 215 postwar, nineteenth-century, township organization referenda occurred while Haines was still living. He surely pondered the slow progress of this obvious (to him) innovation. What would Lije Haines have found if he had ventured south or immediately west of Springfield into non-township territory in the years between the Civil War and April 1889? Was this a topic of concern among the voters there? What would he have learned in places like Nashville, the seat of Washington County, or down in Benton, the seat of Franklin? How did country editors, their letter-writing subscribers, and local historians view a political apparatus that a growing proportion of the state's counties and people had embraced? Was it savior or Trojan horse? Was minority status a circumstance that Little Egypt embraced?

Getting in Step

Humans are a gregarious lot, even though we cherish our independence and individuality. As the number of township counties moved well beyond the halfway mark, one of the strongest arguments for giving up three-man rule in favor of townships and a larger county board was an appeal to voters on the basis of bringing their county into conformity. "Out of 102 counties in the State, 66 have adopted Township Organization, leaving 36 which have not done so. It will be 'old Marion's' time to wheel into line next fall," wrote the editor of the *Centralia Sentinel* in 1870. "Almost every county in the state is organized," said the *Greenville Advocate* editor in 1874. In a letter to Sparta's *Plaindealer,* on an October 1877 swing southward, Haines could have read that it was "only a matter of but few years at most till it will be adopted by every county in the State. Why fight against fate...?" "Let Greene county join the great majority and reap the benefits thereof" went an 1882 plea from a county-seat editor at Carrollton.[28]

Haines would have found some editors and letter writers taking the minority issue a step farther, by singling out their particular county as being destined for inclusion with the growing majority. Out of Carlinville, Macoupin County, a year before the constitutional convention, came an editorial lamenting the fact that there were "only two or three counties of much size in the State, still retain[ing] the old central style of government." Voters in that sizable county just south of the state capital should "adopt the better and more advanced system of government, known as Township Organization." Southeast of Carlinville, the editor of the *Litchfield Monitor* picked up a piece from the *Alton Telegraph,* as did editors in Belleville, Centralia, and East St. Louis. The *Telegraph* writer clearly thought Alton's Madison County ought to join the township lineup: "About the only large counties, remaining, that have not adopted Township Organization, are Madison, Macoupin, and St. Clair." In East St. Louis, later in 1869, Haines could have talked to an editor who felt the only significant county not in step was St. Clair: "Of all the most prosperous counties in the state, St. Clair is the only one which...still sports a county court." By 1875, the *Carrollton Gazette* was saying, "all the more progressive southern counties, in this state, have adopted township organization." Greene County and all the other nonadopters, lumped together, accounted for only one-fifth of the state's population. "Take a state map," suggested the *Gazette* editor, "and make a cross mark upon the counties named, and you will see upon what a feeble tenure the old system hangs."[29]

If change agent Lije Haines had planned his itinerary wisely, he could have attended a meeting of the Farmers' Club of St. Clair County in Belleville on Saturday, 29 September 1883. The proceedings of that meeting, as they appeared in the 5 October issue of the *Belleville Weekly Advocate,* constitute one of the best accounts of a protownship assembly I encountered for the nineteenth century. After a veterinarian displayed a stomach pump meant to relieve bloating in cattle and a local farmer showed the members "a few stocks of Bermuda grass," they took up "the subject of Township Organization," a reform that St. Clair County had already voted down several times. The meeting organizers had solicited letters of endorsement and invited men willing to stand up for the township concept based on their experiences with it elsewhere. Among the arguments they advanced in favor of the innovation were those in which opinion leaders appealed to St. Clair's pride in being a forward-thinking county. George C. Eisenmayer, of Mascoutah, on the eastern margin of St. Clair, near late-adopter Clinton County, characterized the nonadopters, "with the exception of Morgan and St. Clair counties," as being "small and sparsely settled counties, bordering on the Ohio, Mississippi, and Wabash rivers, without any railroad facilities, no manufacturing industries or other enterprises, except hoop-pole cutting—if not hoop-pole stealing—for a livelihood." Among those who sent letters was Joseph Hanke of Trenton, in Clinton County. Meeting chair August Chenot read Hanke's letter, in which Hanke praised the first 10 years of township organization in Clinton and observed that the "old system" worked well in small or thinly populated counties "especially where the people are not sufficiently educated or otherwise incapable to manage their own affairs, in regard to levying and collecting taxes by township officers."[30] One month later, St. Clair County finally wheeled into line.

Not every editor, however, interpreted the county count as a mandate for jumping into township organization. Morgan County had already voted no five times and had the issue on the ballot again in 1879 when the *Jacksonville Journal* took the position that Morgan ought to wait and see some more. "The only argument advanced in favor of Township Organization is that many states and a majority of Illinois counties have adopted the plan. The *Journal* fails to see the force of the reasoning. In many cases there is dissatisfaction, in others the trial has been too short to judge its success."[31] Morgan never did wheel into line.

Down with Old Fogy Ways

Leafing through the *Carrollton Gazette* of 23 October 1875, Haines could have read that the editor believed the 29 "border counties" (nonadopters all) represented "the fast fading sluggishness of former days." So, in addition to arguing adoption to avoid belonging to a shrinking minority, township proponents often appealed to voters' sense of modernity, of which "old fogy" was the antithesis. Township organization, said the *Centralia Sentinel* in 1869, "will certainly triumph here as it has everywhere when it has come in competition with the old fogy county system." The *Carlinville Democrat,* a frequent critic of the grandiose Macoupin County courthouse, thought 1869 was about time for the citizens of that county to vote out "the old fogy idea of county organization and adopt the better and more advanced...township organization." Bond County, said the editor of the *Greenville Advocate,* "has paid dearly for her old fogyism in many ways and in no other more than in this matter of want of organization."[32]

Country editors did not need to invoke the old-fogy bugaboo to convey their view as to which was the more modern (read: better) governmental model. The *Edwardsville Intelligencer* expressed surprise that the people of Madison County had for so long been willing "to put up with the inconvenience and disadvantages of the present system." Down on the American Bottom, a few miles southwest of Edwardsville, and a couple of weeks later, the *East St. Louis Gazette* chided the St. Clair County Court for failing to afford the voters a chance to choose "between the old and the more modern way of administering local government." Three years later, far to the southeast in Hamilton County, Haines could have read a letter from OBSERVER that prodded the editor of the *McLeansboro Golden Era* to speak out about the need for a new school and for township organization. "Let us talk with the people about advancement and progress in this age of steam—in this *Golden Era* of better times," said he or she. In Greene County, Haines could have spoken in 1875 with an editor who challenged readers to vote for township organization because it was "more in consonance with the genius and progress of the present generation than [was] the old order of things," or a year later in Bond County with the editor who wrote, "Every good man who wants reform in the county business, and prefers the nineteenth to the eighteenth century, will vote for township organization." At Nashville, Illinois, Haines might have traded change agency notes with PLYMOUTH, who submitted a spirited letter to the editor in 1882, the year Washington County adopted for the first time. PLYMOUTH claimed township organization was "a Republican measure," obviously a plus to the writer, and its adoption would show the county capable of governing itself. Here are PLYMOUTH's words:

> Will Washington county still march in the old Southern line, with Virginia at the head and Arkansas at the tail, stumbling along in the old pro-slavery ruts, or will she join the great free North, on whose banner is inscribed "Local Self-government," "No county bosses," "We know our own business, and mean to attend to it." Let us adopt Township Organization and put Washington County in the Yankee line, where she belongs.

Andrew Eisenmayer, according to the *Belleville Weekly Advocate*, said at the big 1883 Belleville meeting of the Farmers' Club simply that "All intelligent people have adopted it."[33]

Pressure from Springfield

After the Civil War, nonadopters felt increasing pressure from the general assembly to conform to the majority model. Laws dealing with county affairs had to make allowances for the two different systems of local government. Sometimes an advantage accrued to townships or to township counties. In 1869, for instance, the Twenty-sixth General Assembly passed a law allowing township counties to establish a county teacher training school immediately, while other Illinois counties had to submit the measure to a vote at the next general election.[34] Or take the case of county commissioners elected across the state in nonadopters to replace the three-man county courts, as prescribed in the 1870 Constitution. When commissioners-elect sent their requests for endorsement to Governor Richard J. Oglesby, he refused to comply because the general assembly had not passed legislation defining their duties and authorizing issuance of their commissions.[35] The *Nashville Journal* attributed the oversight to the "stupidity of the legislatures which have met since the adoption of the new constitution of Illinois." The *Chicago Tribune,* quoted in the *Jonesboro Gazette,* called it "a pretty kettle of fish."[36]

Township advocates seized upon issues like the training schools and delayed commissions to urge their counties get in step. Several 1873 editors from southern Illinois would gladly have shared their protownship feelings with Lije Haines, who was a member of the general assembly that year. In Carmi, he could have sat down with the editor of the *Times,* who was fighting the effort by some to discontinue townships in White County. Perhaps the editor could have introduced Haines to the anonymous writer of a letter that appeared on the eve of the 1873 referendum in which White voters sustained townships. Among many other observations, the writer warned, "If discontinued now, we must come [back] to it sooner or later, as the laws are generally framed to suit this [township] plan, and discourage its reverse," because "counties not under township organization are now the exception." West across Little Egypt, in Jonesboro, where Union County was trying to decide if it should adopt the innovation, the editor of the *Gazette* noted "the Revenue and other laws of the State are enacted with a view to this [township] plan of government." But it would have been in the *Greenville Advocate* where junketeer Haines might have found the most fervent complaint about the parallel but unequal legislation affecting Illinois counties. After noting a large proportion of counties (guessing 80 percent, when it was more like 70 percent) and the most prosperous of the counties had already adopted, the *Advocate* observed:

> They have a controlling power in the legislature. The statutes are made to suit the township system and they are not adapted to any other. We have found in our road laws, revenue laws, and many other statutes, such bungling inconsistencies and incoherent and impracticable provisions as have brought our legislators loud and vehement reproaches. We are inclined to the opinion that the explanation is to be found, not in the willful carelessness or incompetence of our representatives, but in the fact that they are a feeble minority.

More than a dozen years later, the *Advocate* still advocated townships for Bond County. On 24 November of 1887, the paper carried a letter from ARGUS in reference to the new road law, which targeted nontownship counties. ARGUS believed that Springfield and all the township-county legislators had a mission of cleaning up the courthouse cliques in commission counties like Bond. After roads, the general assembly was likely to attack problems with assessment and tax collection in the minority counties. "If the people of the counties not under township organization are wise[,] they will lose no time in putting themselves in harmony with the rest of the state. It may be hard on these county boards, their rulers, friends and pets, but it must come. The state no longer intends to maintain two systems of county government."[37]

Carrollton's *Gazette* editor used this legislation-oriented argument year after year to urge adoption, as year after year Greene County voters demurred. In 1874, the *Gazette* said counties like Greene were at a disadvantage in Springfield because "Under the general temper of our State laws, counties not under township organization are disparaged in their material interests in that they are put to the necessity of seeking special legislation for relief." A year later, the *Gazette* carried an editorial entitled, "Township Organization Necessary," in which a legislator like Haines might have read "the best legislation inures to the benefit of those counties under township organization," but Haines, of course, would already have known that fact. And, in 1881, in one of the best editorials on the subject for that era, or any other era, the *Gazette* editor observed, "in state legislation it is evident that special regard will be had for the prevailing [majority] system. In fact all legislations will be directed to that end. Counties belonging to the small minority...will...suffer inconveniences unknown to the great majority."[38] Greene voters still said no.

The editor(s) of the *Alton Telegraph* liked to cite numbers on the topic of legislative isolation. In response to the 1875 item, above, from Carrollton about good laws inuring to township counties, the *Telegraph* looked at the geographical composition of the most recent general assembly (the Twenty-ninth). If one set aside Cook County because of its "special law," one could find 44 senators and 132 representatives from the other 101 counties. Of those remaining legislators, 26 senators and 78 representatives came from entirely township districts, 10 senators and 30 representatives from mixed districts, and just 8 senators and 24 representatives from purely commission-county districts.

> Only a small number of the members of the Legislature have an exclusive interest in the county system and a great majority have none at all...As a consequence, legislation for counties not under Township Organization receives less attention, is not well considered, and often is entirely neglected....This evil will increase as time goes on and the counties under township organization increase[,] the counties not under township organization appear slow or eccentric.

Why, wondered the editor, had not the 1870 Constitution mandated township government for all?[39] Madison opted for the majority mode a few days later.

Subtle Hints

Commission counties felt pressure to conform from other directions in the late nineteenth century. A key aspect of their culture, their way of conducting local governmental affairs, was under attack from internal agitators, change agents, and the general assembly, as I have noted. Counties lacking townships also had to endure subtle suggestions from commercial interests with which they dealt, suggestions that they really ought to consider innovating.

If Elijah M. Haines had visited Randolph County in 1875, he could have examined a new illustrated atlas. County atlases and county histories were all the rage by that decade; and companies often issued several similar publications in a series, county by county.[40] To simplify data collection and presentation, the editorial staff adopted conventions that presumably applied to all counties—like having a county seat. In preparing the 1875 Randolph County atlas, W. R. Brink & Co. assumed that this Mississippi River border county had adopted township organization. Had Haines opened the atlas to page seven, he would have found a table titled "Statistics of Population of Randolph County, by Civil Townships,..." The data collector had used precinct totals instead.[41]

Counties purchased forms, to conduct elections, from a local job printer (usually the newspaper) or firms that served several counties. Once township organization claimed most Illinois counties, it was often cheaper for a printer to prepare a form for the bulk of counties and hope the nonadopters would buy it anyway. Evidence of such subtle pressure reposes at the Illinois Regional Archives Depository, on the University of Illinois-Springfield campus, in poll books from the 3 November 1874 election in Cass County. Both book covers feature a horizontal line, followed (to the right) by the typeset word "Township." Below and parallel to the first line is a second line of similar length, after which appears the typeset word "County." On one poll book cover, in cursive, we see "Richmond Precinct" just above the uppermost of the two parallel lines, with the word "Township" scratched. On the other book cover's top line, appears "Virginia Precinct" in cursive with "Township" untouched. Both poll books, as blank forms, originated at the National Democrat Print Shop of Peoria, an affiliate (I surmise) of the *Peoria National Democrat.*[42]

A Change Agent Speaks

Around 1880, perhaps after reflecting on all the referenda that had occurred on the township topic since the 1870 Constitution, Haines's *The Legal Adviser* informed readers that the 74 counties then in the township column would "not give up the same and return to the county system." It followed, and seemed so logical to the agent of change "that the remaining counties under the commissioner system should without further protest or objection fall into line and adopt township organization." Get the question on the November ballot said Haines or his spokesperson, call attention through local newspapers, and "urge united action in adopting township organization" because "nothing opens the avenue of intelligence of the people like township organization."[43]

A Sharp Eye on Fellow Counties

If Haines had taken the legislative junkets I have speculated about herein, news-hungry local journalists in southern Illinois would surely have pumped him for the latest in township diffusion news. What was the ratio of adopters to holdouts? Where were citizens agitating for change in county government? Had any counties reverted to the court/commission model? How would recent state laws affect township counties? Nonadopters? The list of questions might have been longer; but a common thread would have had the local editor trying to place his county in the context of other counties in the state, especially nearby counties. As diffusion expert Everett M. Rogers noted, the persuasive ability of "near peers" is more compelling than that of any other influence.[44]

Sources of Intelligence

Editors had access to the equivalent of wire service material for national and international news and might have had a correspondent looking at important state-level events like a constitutional convention; but I found no township-oriented stories, 1865–1899, resembling, for instance, the Associated Press series about townships that ran in December 2000.[45] On the other hand, some editors had lived in both types of county (township and commission), and, depending on the timing, could offer firsthand comparisons. Letters to the editor served a useful intelligence purpose by bringing an outside perspective on several occasions during the late nineteenth century. Some letters were obviously in response to an editor's request for input; others seem to have been more spontaneous. Most of the information that reached editors about the township tide in southern Illinois, as it was affecting other counties, came, however, from the pages of other newspapers. A few editors may have had the resources to subscribe to a Springfield or Chicago paper, and I did find an occasional citation in the local papers from these metropolitan sheets; but the vast majority of citations and most of the township organization news from the next county or the other side of the state derived from exchanges between editors of weekly newspapers. I mail my weekly to you; and you, in exchange, mail yours to me. I may cite yours as the source of an item, or not. I will credit your paper if I lift verbatim segments, especially if I want your paper's name to lend credibility to my stand on the issue. You, I assume, will follow the same journalistic ethic.

As the editor of the *Carbondale Jackson County Era* or *Fairfield Wayne County Press* or *Metropolis Times* or any one of several dozen other southern Illinois newspapers of that era, I would likely employ a version of the gravity model when I choose which papers to exchange and which to use as source materials about local government. If you are in the next county and your county is either now in the same category with respect to township or commission government as my county or if your county has recently made a switch to or from townships, I am likely to be quite interested in your circumstances and your editorial stance. If your county is not adjacent to mine, but still in our part of the world, and if your community's and county's names might carry a little extra status because of size or some other factor, I will more likely than not still have an interest in what I read in your exchange. If you are 10 counties away from us, unless you are Sangamon County or maybe Rock Island County, I will probably not use you as a source or compare our situation with yours on the township topic. I am apt to be particularly reluctant if you are one of the Yankee counties, when our county has strong ties to the Upland South.

Noting Who Is Voting

Editors, as human as anyone else, picked their comparisons carefully. If the editor thought his county ought to be moving in the direction of township organization, the county he chose to use as a model might be one that was voting that year or had recently adopted. If he did not want a change in county governance, he would either ignore the petitioners' plea and the order for a referendum or pick a county for comparison whose local paper condemned their experience with townships. Most editors, in that period, even in what they might have considered news items, evidenced a bias in one direction or another about township organization. A balanced piece was the exception. The *Centralia Sentinel* carried this favorable item in 1869: "Jefferson county will be under the working of township organization on the first of April next, and Marion County would have introduced such a desirable change in her local government at the same time if the Democracy had not arrayed themselves against the question of township organization as a party." Opposing adoption for Cass County a couple of years later, the *Beardstown Central Illinoian* commented on reports that citizens of several counties, now that the state's constitution made abolition easier, were "talking strongly of relieving themselves of the burden.[46]

A Lesson for the Locals

Besides keeping an eye on the choices other counties were making about the form of county government, editors watched the exchanged newspapers for reports of governmental workings in their neighbors. They compared their own county's performance with that of nearby Illinois counties as they made their pitches for change or the status quo.

Among the most commonly referenced counties was Macoupin. At issue was Macoupin's grand courthouse and mighty courthouse indebtedness. With the critical help of Macoupin resident John A. Palmer, who was the Illinois governor from January 1869 until January 1873, the Macoupin County Court in 1869 secured passage of a bill by the Twenty-sixth General Assembly permitting virtually unlimited expenditure to continue funding the construction of a replacement courthouse in Carlinville. What had begun in 1867 as a $50,000-$150,000 project quickly grew to a monumental "white elephant," as Charles A. Walker, an early twentieth century historian, entitled that chapter of his Macoupin County history. In November 1869, to plug the tax rupture, voters elected an "anti-court house" court, which "immediately repudiated as many of the acts of their predecessors as they possibly could" and ended construction. Voters finally switched, after several tries, to township organization in 1870. Forty years later, on 21 July 1910, Governor Charles S. Deneen presided at the celebration to burn the last courthouse-construction bond and close out the payback at nearly 1.4 million dollars, including interest.[47]

Here are some of the outsider observations about the Macoupin courthouse. The *Centralia Sentinel,* in the midst of Macoupin's 1869 spending spree, suggested that if Macoupin had adopted township organization "some years since, it would have saved the people the imposition of the burdensome taxation imposed for the erection of their grand and magnificent court house (significantly splendid and costly for the National Capital [*sic*]) under which they are now and will continue to groan for years to come." "Macoupin," said the *Greenville Advocate,* "stuck with the their three-man court too long; and she is crushed under a load of debt equal to that of the whole State of Illinois." Next-door neighbor to the south, Madison, watched Macoupin with a particularly keen eye. JOHN SMITH reminded readers of the *Alton Telegraph,* in a letter that appeared 7 October 1875, "We have been waiting, like Macoupin county, for a horse to be stolen before we lock the stable door. After a corrupt County Court had involved that county in a heavy debt, the voters did what would have prevented the nefarious business of the Macoupin Court House—they adopted township organization." Another Macoupin neighbor, Greene to the west, had a modest courthouse issue, which played a part in that county's swing to township organization in the 1880s. Letter writer BOAR wondered, in 1886, how the recently adopted (and that year challenged) township organization would perform in the extrication of Greene "from this pigmy of a Macoupin court house muddle." BOAR's cost comparison led him or her to a favorable prediction on the Greene outcome.[48]

Jefferson County belongs to our late majority for the state as a whole, but its 1869 decision to adopt was early for Little Egypt and led to several out-of-county commentaries about its affair with township government.[49] Looking in from the west were Washington County voters. The editor of the *Nashville Journal,* in 1871, keenly observed that men in eastern Washington, in and around Ashley and Richview, had been much more supportive of township organization at that year's unsuccessful referendum than were those closer to Nashville because, to the eastern residents, "the workings of the system" were more familiar. "Jefferson county adopted it two years ago," went the editorial, "and from what we have learned, they are well satisfied with its workings and regard it as a vast improvement upon their old democratic county court." Peering into Jefferson politics from the southeast was the

Hamilton County electorate. The *McLeansboro Golden Era* cited a *Mount Vernon Press* item of 28 March 1872 regarding Jefferson's completion of its second year under townships in which the *Press* expressed uncertainty about its overall value to Jefferson County but concluded no strong complaints probably implied the system was "popular." Four months later, the *Golden Era* editor entertained visitors from Jefferson, Judge Satterfield and Treasurer Jones, who did "not view it as so bright a picture as some." The visitors complained about the extra taxes for each township, differences in collectors' bookkeeping practices that affected the tax delinquency frequency, and excessive road and bridge expenses in lightly populated bottomland townships. Hamilton voters, according to the editor, should take heed and "make a strenuous effort to learn all about it" before the November referendum there.[50]

The township tide had exhausted most of its momentum by the time it reached Randolph County, several counties south of St. Louis. Randolph voters and editors did not find, in 1884, much success with townships in their corner of the state to recommend adoption. Jackson, to the southeast, had already adopted twice and discontinued as many times. St. Clair and Washington, to the north, had been into the innovation a year or so and might soon revert. Monroe and Perry had, like Randolph, remained true to the southern-style commissioners. Looking around, the editor of the *Sparta Plaindealer* suggested Randolph "wait a few years and watch its effect in other counties, before rashly making a change which may cost the county a great deal of money."[51] They are still waiting.

Backsliders

Township advocates kept a tally of adopters, looking for any sign of discontinuance. Until 1874, the backslider slate was clean. After November of 1874, when Jackson discontinued for the first time, and, certainly, after 1876, when Johnson did so permanently, an editor could no longer legitimately base an advocacy argument on the lack of discontinuance. So, while it had been permissible to say things like "No county of our State has ever gone back to the old system after having adopted the new"[52] in 1871 or "We do not call to mind a single county which has discarded the old system and returned to it again"[53] in 1873, it was misleading to write "No county has yet changed back to the county court system"[54] in 1875 or "Don't fail to notice that not a single county in the State has gone back to it"[55] in 1879. By 1883, some protownship editors were admitting that reversion could occur in isolated counties, but voters should not let that dissuade them from voting to adopt the new. "Counties that have once discarded the old system rarely ever go back to it again," argued the *Carlyle Constitution and Union*. "Only two or three counties in the State that have ever had Township Organization have changed from it, and that was done by court house clans in every instance," ran the advocacy line in Saline County's *Harrisburg Chronicle-Sentinel*.[56]

Discontinuance, of course, even temporarily, could provide grist for an opponent's argument; but the evidence lay clearly on the side of the advocates. Today, one of the reasons that the Township Officials of Illinois and local township officials work so hard to put down abolition efforts in any county or any single township is to avoid giving township critics the opportunity to proclaim "if one county/township can survive abolition, so can others." Once the levee starts to crumble, the floodwaters rumble through, and disaster follows.

Rogers identified five stages in the innovation-decision process and devoted considerable space in *Diffusion of Innovations* to elaboration. In order, the five stages are knowledge, persuasion, decision, implementation, and confirmation. Discontinuance can only come during the confirmation stage, and "later adopters are more likely to discontinue innovations than are earlier adopters."[57] Of the five Illinois counties that discontinued in the nineteenth century, four did so within four or fewer years of trying on their new governmental clothes. The fifth, Richland, waited two full decades before reverting to the three commissioners.

An editor in a neighboring county, himself a supporter of township government, attributed discontinuance efforts in Franklin to "politicians and court-house hangers-on" who somehow managed "to get back into the old expensive system."[58] Several decades later, a Historical Records Survey writer looked into the Franklin uncertainty over that county's governmental system and blamed changing politics (from Democrat to Republican): "The struggle between the groups favoring the county [commission] and townships forms of local government reflected, in part, and influenced, the growing political divisions of the county in the post-bellum years."[59]

Jackson, like neighbor Franklin, adopted twice and discontinued twice, over exactly the same period, 1871–1884. It must have been a busy 13 years at the courthouse in Murphysboro, the seat of Jackson County, as commissioners departed, reappeared, departed again, came back, and departed a third time. When Jackson made the first leap in 1871, the *Carbondale New Era* said the county's voters had "taken a right step, one which, we trust, never will be regretted by the people of the county." Regret came almost immediately, however; and voters reversed course in 1874. Transitioning back to commissioners over the years 1875–1877 kept residual township funds and township names in the public eye so that readoption did not produce as much shock as had the initial adoption. James A. Brownlee, writing in 1878 of the second Jackson adoption, expressed the thought, which others no doubt must have shared, that they really ought to give township organization half a chance to prosper: "It would be the part of wisdom, this time, to test for several years the new method of government. Its advantages or disadvantages cannot be fully discovered in one year, nor in two." Brownlee warned there would be "jars" in the machinery, but he believed these would smooth out after five years. "Let the people try it for not less than five years. Then will its superiority or inferiority to the other method be for ever settled, and on it the seal of approval or condemnation will set beyond the power of change."[60]

Voters in Jackson gave townships exactly five years the second time around, and then invited commissioners back to run the county in 1882. Even though township organization had lost again, voters apparently did appreciate the dispersed representation it provided the county board. Promoting its favorite candidates for the three commissioner slots, open early in 1883, *Barton's Free Press,* out of Carbondale, assured the public that Ben L. Wiley of Makanda "knew nearly ever man and every acre in the county"; E. W. Bryan of Murphysboro had "many warm friends in all the townships"; and James Alexander, also from the county seat, was "the people's candidate for commissioner all over the county." Jackson County found township names and boundaries handy when it came to dividing the county, even after the townships themselves had ceased to function as minor civil divisions. For the Republican county convention in April 1884, while townships were officially defunct, the *Jackson County Era* said, "Townships are entitled to delegates as follows..." and proceeded to name them: Elk, Grand Tower, Kinkaid, etc.[61] Later that year, Jackson made its last switch in government structure, voting 2,467 to 1,526 to bring back the supervisors and their townships.[62] After that November 1884 readoption decision, the *Era* editor had a suggestion for critics of Jackson's governmental instability: "To an outsider it would seem that our people are frequently in favor of change. But it is none of any outsider's business. The people have a right to do as they please in regard to township organization or anything of kindred nature. They pay the expenses of frequent changes and know what is best."[63]

From the spring of 1873 until the spring of 1877, Johnson was the southernmost outpost ever of township government in Illinois. What exactly happened there to prompt adoption in the first place, in the heart of the Shawnee Hills, with commission counties at all the gates, is largely a mystery today. We do know Johnson had township organization on the table at every election from 1870 through 1876, the year voters sent supervisors packing. No other Illinois county, even those that held more total referenda than Johnson, can claim six straight ballot years for township organization. Advocates succeeded on their third attempt; opponents won discontinuance on the sixth vote. Of the 1874 (first abortive attempt to abolish), the *Monitor* at Marion, in the county immediately to the north, said, "The opposition to that system [township organization] fell behind the mark last fall, but are somewhat inclined to try it over." Therefore, they did and failed once more. KENO prepared an analysis of the 1875 election for the *Metropolis Times.* His (or her) take was that the commission proponents tried a strategy that backfired:

> As you are well aware that there is a number of men that won't vote with Kirk and Looney for any thing...So Kirk and Looney, in order to beat township organization, turn for it. In order to drive men against it, but they got beat at their own

> game, and township organization has carried again...They still say that they will vote it down next time, but that has been their cry for over two years and yet she stands.

When Kirk, Looney, and comrades finally succeeded in 1876, the *Johnson County Journal* blamed discouragement (and perhaps the reversion to commissioners) on "bad financial management" and observed that, by September 1877, the county's indebtedness was rapidly disappearing.[64]

One aspect of Johnson's townships did persist, however—township names and boundaries. In 1878, the three commissioners raised tax levies in Cache, Goreville, Grantsburg, and Vienna townships while lowering levies in Burnside and Elvira townships. A year later, the *Johnson County Journal* said the voters in "the eastern part of Cache township" would appreciate another voting place before the next election. "Elvira Township," as of November of 1881, said the *Journal,* "has about $700, township funds on hand, which has been idle for four years." In 1895, county commissioners were again adjusting assessments for "Goreville Township" and "Elvira Township." And so it went, on into the twentieth century—a political relic persisted as a feature of the cultural landscape.[65] Townships might have swept all of Illinois if a three-person commission could have run countywide affairs in lightly populated counties while township officials handled business at the local level.

Washington was the fourth of the four counties that discontinued township organization after but a brief trial; and the *Nashville Journal,* a strong township advocate, was there for a play-by-play of the 1880s. The *Journal* editor lamented election defeats for a Dr. Munselle and a Dr. Henry in April of 1883, when each ran for supervisor posts at the first township elections in Washington, and warned readers that Washington County "needed just that kind of men in the first Board of Supervisors to give Township Organization a good send-off." Defeats of quality candidates would serve to deter them and others of similar aptitude from running the next time, that offices would go to weaker candidates, and that the people would tire of the new system and would "be ready to return to the old one." A few months later, the *Journal* was urging "patience with the blunders and mistakes that are inevitable from a change of system" and reiterating the need to "select only good and qualified men for town[ship] offices."[66]

Opponents in Washington did not manage to get discontinuance on the ballot in 1883, and the *Journal* took encouragement early in 1884 from the supervisors' first year in office: "On the whole, considering that legislation was a 'new and undiscovered country' to most of them, the Board did as well as could be expected." Later in that year, however, the *Journal* editor was less sanguine, noting in June "the general drift is against Township Organization." Some observers, like REPUBLICAN CO. OFFICIAL, tried to blame party politics for the defeat of the township model in November 1884, while the *Journal* took the position that Democrats as well as Republicans were on both sides of the fence. Most would have agreed that the voter-mandated switchover in 1885 cost the taxpayers for books and forms and for a second election, to choose commissioners. To switch back again to supervisors, after only a year, was in the view of the *Journal,* "not only expensive" but would disarrange "all county business." Again, the editor noted in that 30 October 1885 piece how the 1882–1884 experiment had failed in part because good supervisors had been a rarity. Finally, in November of 1888, voters did offer township organization a reprieve, allowing Ashley, Covington, Okawville, and other named townships to regain their official landscape standing, something they had unofficially retained (as in Johnson and Jackson) in the years since discontinuance.[67]

Richland was the oddity among the backslider counties of this period. Voters there had adopted in 1858, before Illinois had reached the halfway mark in signing up counties for township organization. It would be more than a decade until any of the other four backsliders adopted. Richland waited 20 years before discontinuing—much longer than did any of the other four. Richland, however, did share one important trait with Franklin, Jackson, Johnson, and Washington: throughout the initial confirmation period, all five counties had at least two commission counties as their neighbors. For Richland, those two commission-model neighbors were Edwards and Wabash. If our neighbors are prospering under a different system, then switching to (or back to) that system in our county might become a more realistic reform to sell. I have no evidence that Edwards or Wabash influenced Richland's voters, nor do I have much to suggest a reason why Richland would abruptly revert to commissioners in 1878 and then switch back in 1879. The *Olney Times,* in 1876, did not have a favorable opinion of the Richland County government: "Beyond all question the machinery of township organization is much more cumbersome and costly than the old system." The editor went on to cite recent failures to discontinue in adjacent Crawford and nearby White: "It seems impossible to do away with it." Nevertheless, by a mere 172 votes, in November 1878, Richland did do away with it; and, because it had been so long since three commissioners ruled there, the *Times* felt it necessary to explain the "old system that was in vogue before township organization was adopted."[68]

Richland's new commissioners had scant time to do much more than begin their regime, because voters immediately returned Richland to what the *Robinson Constitution* had referred to in June of that year as the county's "first love—township organization." Why did that "first love" so quickly, quicker than in any of the other readopting counties, reclaim Richland? A couple of onlooking editors offered their views on the subject, and both had used that readoption as a premise for their own arguments to voters who might have been contemplating discontinuance efforts. In White County, the *Carmi Times* said of Richland's flirtation with the old ways: "After a year's trial [Richland voters] were so disgusted at its expensive workings and the showing up that cliques can control three men easier than many, that they decided by a majority of 1,170 to return to township organization." Two days later, in Crawford County, the *Robinson Argus* said, "A year's trial was all the people of Richland county wanted of commissioners to do their business, and we have no doubt the same would prove true in this county should township organization be abolished, as some desire."[69] Richland, White, and Crawford remain in the township corral today, and none of them ever voted again on the matter.

Ballot Access/Majority Mess

Democratic decision-making ought to be simple. You vote, I vote, others vote, we tally the votes, and most votes wins. In the latter decades of the nineteenth century, Illinois voters seeking to get township organization on the ballot and then to achieve the "most votes" found the path to a clear-cut answer, as to whether the people of a county favored commissioners or supervisors, sometimes frustrating.

Ballot Access

Laws in effect during the late-majority period mandated a vote to adopt if 50 legal voters signed and presented to the county commissioners a petition (or petitions) calling for a referendum. Discontinuance took considerably more petitioners: 20 percent of the county's legal voters who voted at the preceding presidential election.[70] Either way, the pathway to the ballot appeared obvious; but that was not always true.

Commissioners, facing the termination of their privileged lot, could stand in the way of innovation; or ballots might reach the voter in a manner that favored the status quo over reform. In several instances, the court/board found a way to keep such a referendum off the ballot entirely. The *Carlinville Democrat,* really a Republican paper, accused the Democrats of Macoupin County in 1867 of ignoring the legitimate petition of 50 voters from Shipman asking for a vote on township organization, as a way to conclude the courthouse-construction conundrum. Assessor Frank "Steward is opposed to township organization," said the editor. "He wants the affairs of the county to remain in the hands of the Court House clique." St. Clair County, complained the editor of the *East St. Louis Gazette* in 1869, wrongly denied petitioners' plea for a vote that year because of the late arrival of the petitions, when, according to the editor, the supreme court had given county courts great latitude in scheduling referenda unless someone could show "that had longer notice been given a different result might have been the consequence." The St. Clair judges (commissioners) of the county court had failed to see "that the people throughout the county [were] ripe for self government." ARGUS, writing to the *Greenville Advocate* years later, accused the board of commissioners in Bond County of finding "the least excuse" for not approving a referendum, "and when they do, they oppose it at the election."[71]

What sorts of tricks could an opposing board or political party try, if the issue of adoption did reach the voters? One of the most common tactics was to avoid putting the question on the actual piece of paper that the voter used to record his preferences. Some background is in order, however. From the 1848 Constitution until 1891, in Illinois, ballots came directly from the political parties, usually from the print shop that published a party's local newspaper.[72] Only the names of a party's recommended local, state, and national candidates would appear on the ballot. For referenda, the practice was to print a choice: For X or Against X. Voters scratched the referendum choice they did not favor, leaving their choice as printed.

Omitting the question altogether was one ballot tactic. Unless a voter with such a ballot then took the initiative to write in "For Township Organization," the fact that he voted on any other candidate or issue would serve to increase the total votes cast by one, which worked against adoption. Several contemporary newspapers cited examples or warned of such ballot exclusions: *Litchfield News* in 1867, *Rock Island Argus* in 1872 and 1873, and *Carrollton Gazette* in 1875 (about 1874). The *Gazette* cautioned Greene County supporters to demand that "all the tickets be printed as the law directs" and "see that they be properly distributed."[73]

Trimming was another trick of the trade with tickets (ballots). According to the editor of one Carlinville paper, Democrats in 1868 were openly boasting that they had trimmed their party tickets so closely that no one could write on them. Similar trimming must have occurred, the piece continued, in several Macoupin precincts because those precincts produced almost no votes either way on the township referendum.[74]

Majority Mess

If the voters did manage to express an opinion on the innovation, what it took to achieve a majority remained a messy matter for years. *People v. Warfield* (1858) had produced the supreme court ruling calling for a majority of votes cast at the election instead of a majority of all the county's voters. The Township Law of 1874 had reconfirmed, saying, "a majority of the voters voting at such election shall be taken and deemed a majority of the voters of said county."[75] Newspaper editors, however, had to warn continually that a failure to vote on the question of township organization, by a man voting on anything else in that election, added to the total votes and upped the number of votes needed to adopt. The "indifferent voter who casts a ballot which he fancies to be neutral" [on an issue] was the culprit in the opinion of the *Alton Telegraph.*[76] The *Carrollton Gazette* offered hypothetical numbers and showed Greene County readers who might be wary of township organization that, while it would take more than 2,500 ayes out of 5,000 to adopt, it would require only a simple majority for discontinuance later, say 1,001 to discontinue versus 1,000 to retain. Why the difference? The framers of the new constitution set a lower standard for discontinuance (a majority of those voting on the issue) compared to that for adoption of township organization (majority of total voters voting at that election).[77]

Voters all over the southern half of the state could have read in those waning years of the nineteenth century, as they did in Cairo in 1885, that township organization had "twice been carried by a majority of the votes cast upon that question [in Gallatin County], but not having received a majority of all votes cast at that election, it thus failed."[78] In five counties, in 1869, township organization met defeat because, despite having more favorable than unfavorable votes, it fell short of exceeding the magical 50 percent plateau.[79] Here is how the Madison County clerk explained the thirteenth defeat of the township system in his county to the auditor of public accounts:

> I have the honor of enclosing to you a statement of the votes cast at the last general election in this county on the question of township organization. The votes for it are only a plurality, as the votes against it together with the ballots containing no vote on the question exceed the votes for it by one hundred and fifty-nine. I consider township organization defeated in this county.[80]

Some voters just received bad information. The *Carlyle Constitution and Union* chose to ignore *People v. Garner* and referred readers back to the 1848 Constitution, which, without the 1858 supreme court interpretation, would have called for a majority of all a county's voters and not the majority of legal voters voting at that particular election. Even constitutional officers confused the issue. In 1869, the *Nashville Journal* carried a letter to the editor from Attorney General Washington Bushnell, who, referring to *People v. Brown* (1850), confirmed the defeat of townships in Washington County that year because the votes for township organization were less than half "of *all the votes* in the county." The *Harrisburg Chronicle-Sentinel,* in lamenting the defeat of township organization in Saline County one year, blamed a rumor circulating "in some of the precincts that it required a majority of all the votes in the county." As the story went, voters in those precincts, in the very county that produced *People v. Garner,* gave up working for passage because they felt the hurdle of half the total county voters insurmountable.[81]

In an untitled editorial published between the adoption of the 1870 Constitution and appearance of the 1874 Township Law, the editor of the *Carlyle Union Banner* captured the sense of frustration other protownship editors and many southern Illinois voters must have been experiencing during those late-majority years. After citing several defeats in Clinton County, despite "favorable votes," the editor labeled the existing township law "bunglesome and contradictory." It was "so liable to a variety of interpretations," he went on, "that the proposition could be defeated by a quibble even after it had received the sanction of a sufficient number of voters, strictly, to carry it."[82]

Making the Case

Once the issue of supervisors versus commissioners reached the ballot and voters understood what not voting on that particular issue meant, opinion leaders and change agents liked to inform voters how they should vote. Those against adoption cited proliferation of public officials under township organization, the likelihood of increased governmental costs, and its inappropriateness in lightly populated counties. SCRIBBLER told readers of the *Carlyle Union Banner,* "Now we have three Judges or Commissioners, and one Assessor and Treasurer...Then we will have a horde of petty supervisors, assessors, and collectors." REASON, writing in the *Nashville Journal,* said, "It may be suitable for large cities and counties where the population is comparatively solid and nearly all over the county. But in counties not entirely 'settled up', it imposes unequal and unjust burdens upon the sparsely settled districts or townships." Township advocates liked to appeal to voter's predilection for the democratic way of governing, as did the editor of the *Alton Telegraph*, when writing these words: "All well meaning men, not only of our Union, but of the whole civilized world, agree that decentralization of power is the best principle for the safety of the commonwealth and the political education of its citizens." [83]

Populism, Salem, and Little Egypt

Midwestern farmers suffered from hardships of the economy, climate, and (in particular) increasing loss of political power to the urban East. In response to loss of power, farmers, including many from Illinois, sought to take better care of their own affairs through organizations like the Grange (National Grange of the Patrons of Husbandry) and the Populists (People's Party of the USA). Such organizations and their members looked to Washington, D.C., for stronger regulation of the railroads, which brought farmers their supplies and carried their products to market, and for greater availability of currency (the Free Silver Movement). Farmers, and those who lived in proximity to and depended on agriculture for their livelihood, took an increased interest in local governmental affairs as a way to control their taxes and lobby for improved roads and bridges for farm-to-market access. In southern Illinois, in Little Egypt, some of the disgruntled saw township organization as one way to better their lot; and many counties that might otherwise have remained in the commission column made the switch, as I have discussed here in chapter 4.

To me it seems prophetic that the national political figure we most closely associate with the populist era, the man who at age 36 became the Democrat and Populist parties' candidate for the presidency, was born just south of the Marion County courthouse, in Salem, the city that today refers to itself as the "Gateway of Little Egypt." Son of a prominent local judge, William Jennings Bryan was less than a year old when Marion County first voted on the township innovation in 1860 and just 13 when his native county finally adopted, in 1873. From Salem, he went on to study at Illinois College and practice law in Jacksonville until 1887. By then a larger stage was calling; and by the time he first ran for president in 1896, the last nineteenth century Illinois county, Gallatin, had decided to try the alternative to the three-man, at-large county board. Bryan ran again for the presidency in 1900 and in 1908; but, like the township tide, his wave of popularity had crested in the previous century. The twentieth century was not so kind to the gifted orator from Little Egypt, just as it, from time to time, proved not so kind for those who advocated the township model.[84] Chapters 5, 6, 7, and 8 will tell the twentieth century's tale.

Notes

1. I have chosen to use here the most recent adoption dates for this series of maps (see also figures 7 and 10). This decision moves Richland County from 1858 to 1879, Franklin and Jackson from 1871 to 1884, and Washington from 1882 to 1888. Further discussion of these counties and Johnson with regard to their indecision occurs elsewhere in chapter 4.
2. Little Egypt, or just Egypt, is a longstanding nickname for southern Illinois. Settlers there, in the bad winter of 1830, supposedly sent provisions northward to harder hit areas of Illinois, just as the pharaoh's court of ancient Egypt responded with grain shipments to the north in the Mediterranean Basin. Robert P. Howard, *Illinois: A History of the Prairie State* (Grand Rapids, Michigan: William B. Eerdmans Publishing Company, 1972), 163.
3. Illinois, General Assembly, *Public Laws of the State of Illinois Passed by the Twenty-second General Assembly, Convened January 7, 1861* (Springfield, 1861), 216–67. The Township Law of 1874 was not, strictly speaking, a new law but rather a revision and does not appear in the public or private laws emanating from the deliberations of the Twenty-eighth General Assembly. Instead, we find a joint resolution ordering the secretary of state to print five thousand copies of the bill and send them to the township counties and to their townships: Illinois, General Assembly, *Laws of the State of Illinois Passed by the Twenty-eighth General Assembly* (Springfield, 1874), 157. To view the printed version of the revision, see Harvey B. Hurd, ed. and comp., *The Revised Statutes of the State of Illinois. A. D. 1874* (Springfield: Illinois Journal Company, 1874), 1065–83. The State Archives holds the cursive version of H. B. 653, "An Act to revise the Law in relation to Township Organization."
4. Newell, "Township Government in Illinois," 498.
5. Illinois, General Assembly, *Laws of the State of Illinois: Passed by the Thirtieth General Assembly* (Springfield, 1877), 212–13.
6. Illinois, General Assembly, *Laws of the State of Illinois Enacted by the Thirty-fifth General Assembly* (Springfield, 1887), 265–89. In township counties, the road districts are coextensive with the townships they serve, unless voters have opted to consolidate into larger districts. Technically a separate unit of local government, road districts in township counties must have budget approval from their township boards. Some commission

counties have a unit road district covering the entire county while others have township-size road districts.

7. Illinois, General Assembly, *Public Laws of the State of Illinois Passed by the Twenty-fifth General Assembly* (Springfield, 1867), 104–5. Because the title of the legislation, "AN ACT to change the time of electing certain officers in a county therein named," gave little indication of its real intent and because the general assembly operated in what seemed a "clandestine manner," the *Fairfield Wayne County Press* took a generally negative position on the reduction in size of the county board. On the day the *Press* was finally able to reveal the law's exact contents, the editor chose to lampoon the title by suggesting some alternatives, that would have revealed no more about the contents than the one the legislature chose: "An Act to incorporate Frog Island Skating Park," "An Act to protect the Wabash and Barefoot Fishing Grounds," or "An Act to incorporate the Third National Bank of Zif." Any of "these titles would have [been] as suggestive of the nature and intent of the Bill as the one selected," he concluded. "What's in a Name," *Fairfield Wayne County Press,* 2 May 1867, 3.
8. Illinois, General Assembly, *Private Laws of the State of Illinois Passed by the Twenty-sixth General Assembly* (Springfield, 1869), 327–28.
9. Illinois, General Assembly, *Public Laws of the State of Illinois Passed by the Twenty-seventh General Assembly* (Springfield, 1871), 756–57.
10. "To the Friends of Township Organization," *Carlinville Democrat,* 18 November 1869, 2.
11. Illinois, Auditor of Public Accounts, Township Organization File, Logan County, State Archives, Springfield.
12. *The People of the State of Illinois, ex rel. Coleman Gaines v. John M. Garner et al.*, 47 Ill. 246 (Ill. Sup. Ct. 1868).
13. *The People ex rel. Oliver P. Powell et al. v. John Ruyle et al.*, 91 Ill. 525 (Ill. Sup. Ct. 1879).
14. Previous conventions had occurred in 1818, 1847, and 1862. Voters rejected the 1862 effort.
15. Ann O'M. Bowman and Richard C. Kearney, *State and Local Government,* 3rd ed. (Boston: Houghton Mifflin Company, 1996), 59.
16. Cornelius, *A History of Constitution Making in Illinois,* 43.
17. Cornelius, *A History of Constitution Making in Illinois,* 51.
18. *Debates and Proceedings of the Constitutional Convention 1869,* 293.
19. *Debates and Proceedings of the Constitutional Convention 1869,* 293–94. Statewide uniformity, as Haines advocated, was not in 1870 a new element of Illinois constitution crafting. At the 1862 Constitutional Convention, Delegate John Burns (representing Marshall, Woodford, and Putnam counties) offered a resolution for the Committee on Township Organization, in which Burns asked members to consider "inserting a provision in the constitution of this state making the township organization law uniform throughout all the counties, or abolishing the system entirely." The convention journal does not reflect further debate of his resolution, and the article that would have governed county and township matters gave counties the option of having or not having township organization. Illinois, Constitutional Convention, *Journal of the Constitutional Convention of the State of Illinois, Convened at Springfield, January 7, 1862* (Springfield, 1862), 75 and 1094.
20. *Debates and Proceedings of the Constitutional Convention 1869,* 876 and 881.
21. *Debates and Proceedings of the Constitutional Convention 1869,* 873–82.

22. "The New Constitution," *Carbondale New Era,* 24 May 1870, 1.
23. Cornelius, *A History of Constitution Making in Illinois,* 64.
24. *Debates and Proceedings of the Constitutional Convention 1869,* 1894–95.
25. A useful offshoot of the convention's interest in townships was a report from the auditor, issued as the result of a resolution by Delegate Pierce, giving the names of counties having adopted township organization and the names of their townships. The adopter count then stood at 66. Illinois, Auditor of Public Accounts, *List of Counties that have Adopted Township Organization, and List of Organized Townships, etc.,* C. E. Lippincott, Auditor (Springfield, 8 March 1870). Among the newspapers to publish parts of the report was the *Springfield State Journal,* which added a comment as follows: "The only wonder is that every county in the State has not long since adopted the system of Township Organization, as being in every respect superior to and more satisfactory than the effete system of County Commissioners' Court." "Township Organization in Illinois," *Springfield State Journal,* as reprinted in the *Centralia Sentinel,* 31 March 1870, 3.
26. The reader may recall how I reported in chapter 1 that I am confident I am underreporting the number of referenda. Of all the periods I cover in this monograph, 1865–1899 is the one most likely to suffer from an undercount, because, for some counties, newspaper and other key sources are either missing or unreliable.
27. "Supervisors' Court," *Warsaw Hancock New Era,* 24 November 1864, 2; A [pseud.], "Reason for Voting Down Township Organization," *Carthage Gazette,* 29 October 1873, 2.
28. "Township Organization," *Centralia Sentinel,* 31 March 1870, 4; "Township Organization," *Greenville Advocate,* 21 October 1874, 2; Wellington Wilcox, "Township Organization," *Sparta Plaindealer,* 30 October 1877, 1. This letter to the editor was the third of that month to the *Plaindealer* in which Wilcox strongly urged Randolph County voters to see the township light. Haines would have viewed Wilcox as an Egyptian ally, and we can count him as an opinion leader. Subsequent and equally fervent Wilcox letters appeared in 1878 and 1879, but the county always chose to remain with the commissioners. "Shall We Have It?" *Carrollton Gazette,* 9 September 1882, 1.
29. "Township Organization is Death to Court House Cliques," *Carlinville Democrat,* 9 October 1869, 2; "Township Organization," *Alton Telegraph,* as reprinted in *Litchfield Monitor,* 21 October 1869, 1; "The County Court," *East St. Louis Gazette,* 23 October 1869, 2; "Township Organization," *Carrollton Gazette,* 23 October 1875, 2.
30. "Township Organization," *Belleville Weekly Advocate,* 5 October 1883, 4.
31. *Jacksonville Journal,* 28 October 1879, 2.
32. "Township Organization," *Carrollton Gazette,* 23 October 1875, 2; "The Elections," *Centralia Sentinel,* 4 November 1869, 6; "Township Organization is Death to Court House Cliques," *Carlinville Democrat,* 9 October 1869, 2; *Greenville Advocate,* 23 November 1876, 4.
33. "Township Organization," *Edwardsville Intelligencer,* 21 October 1869, 1; "Good County Roads," *East St. Louis Gazette,* 6 November 1869, 2; OBSERVER [pseud.], *McLeansboro Golden Era,* 3 February 1872, 2; "Our Following," *Carrollton Gazette,* 16 October 1875, 2; *Greenville Advocate,* 5 October 1876, 4; PLYMOUTH [pseud.], "Township Organization," *Nashville Journal,* 20 October 1882, 8; "Township Organization," *Belleville Weekly Advocate,* 5 October 1883, 4.
34. *Bunker Hill Union Gazette,* 15 April 1869, 2.
35. "County Commissioners," *Carrollton Gazette,* 22 November 1873, 2.

36. "County Commissioners," *Nashville Journal,* 26 November 1873, 2; "County Commissioners," *Jonesboro Gazette,* 22 November 1873, 2.
37. "Township Organization," *Carmi Times,* 31 October 1873, 2; *Jonesboro Gazette,* 4 October 1873, 2; "County or Township? Which?" *Greenville Advocate,* 30 April 1873, 2; ARGUS [pseud.], "What We are Coming To," *Greenville Advocate,* 24 November 1887, 9.
38. "Township Organization," *Carrollton Gazette,* 31 October 1874, 4; "Township Organization Necessary," *Carrollton Gazette,* 25 September 1875, 2; "Township Organization," *Carrollton Gazette,* 6 August 1881, 1.
39. "Township Organization," *Alton Telegraph,* 21 October 1875, 1.
40. For a scholarly look at atlases such as this, see Michael P. Conzen, "North American County Maps and Atlases," in *From Sea Charts to Satellite Images: Interpreting North American History through Maps,* ed. by David Buisseret (Chicago: University of Chicago Press, 1990), 186–211.
41. *An Illustrated Atlas Map of Randolph County, Illinois* (Edwardsville: W. R. Brink & Co., 1875).
42. Cass County. Election Papers. 3 November 1874. Originals are at the Illinois Regional Archives Depository, University of Illinois, Springfield.
43. "Township Organization Illinois," *The Legal Adviser,* as reprinted in *Greenville Advocate,* 14 September 1882, 1.
44. Everett M. Rogers, *Diffusion of Innovations,* 3rd ed. (New York: The Free Press, 1983), 170.
45. On consecutive days, 20–22 December 2000, several Illinois newspapers carried an Associated Press investigative series of articles about Illinois township government. Most papers did companion pieces, which ran those days or later. Editorials about the need for townships, guest columns, and letters to the editor followed. See chapter 9 for a thorough review of this press event.
46. *Centralia Sentinel,* 25 November 1869, 6; "Township Organization" *Beardstown Central Illinoian,* 25 January 1872, 3.
47. Charles A. Walker, supervising ed., *History of Macoupin County[,] Illinois[:] Biographical and Pictorial* (Chicago: The S. J. Clarke Publishing Company, 1911), 157–70.
48. "Township Organization," *Centralia Sentinel,* 28 October 1869, 6; "County or Township? Which?" *Greenville Advocate,* 30 April 1873, 2; JOHN SMITH [pseud.], "Our County Debt and Township Organization," *Alton Telegraph,* 7 October 1875, 1; BOAR [pseud.], "Some Facts Favoring Township Organization," *Greenfield Weekly Argus,* 16 October 1886, 4.
49. Dissatisfaction over indebtedness and excessive taxes was apparently a key to the township takeover in Jefferson County, rather than affection for township organization. In fact, a local paper began receiving so many post-adoption inquiries from citizens about the workings of the new system that the editor, in a preface to a summary of the highlights of the township law, commented, "Many people voted both for and against it not knowing very much about it—whether it was for the best, or *vice versa.*" "Township Organization," *Mount Vernon Free Press,* 11 November 1869, 2.
50. "Township Organization," *Nashville Journal,* 2 December 1871, 6; "Township Organization," *McLeansboro Golden Era,* 18 April 1872, 3; "Township Organization," *McLeansboro Golden Era,* 1 August 1872, 2.
51. *Sparta Plaindealer,* 17 October 1884, 2.
52. D [pseud.], "Township Organization," *Nashville Journal,* 30 September 1871, 1.

53. "Township Organization," *Centralia Sentinel,* 23 October 1873, 4.
54. TOWNSHIP ORGANIZATION [pseud.], "Township Organization," *Waterloo Dollar Advocate,* 21 October 1875, 3.
55. "Township Organization," *Greenfield Weekly Argus,* 13 September 1879, 4.
56. "Township Organization—Two Opinions," *Carlyle Constitution and Union,* as reprinted in the *Belleville Weekly Advocate,* 14 September 1883, 4; *Harrisburg Chronicle-Sentinel,* 2 November 1883, 1.
57. Rogers, *Diffusion of Innovations,* 188.
58. *Harrisburg Chronicle-Sentinel,* 5 October 1883, 1.
59. U. S., Works Progress Administration, Historical Records Survey, Illinois, "Franklin County," January 1941. At the State Archives, Springfield.
60. "The Election," *Carbondale New Era,* 11 November 1871, 2; James A. Brownlee, *History of Jackson County, Illinois* (Philadelphia: Brink, McDonough & Co., 1878), 21.
61. "Political Conventions," *Carbondale Jackson County Era,* 15 March 1884, 2.
62. Jackson County. Proceedings of the Board of Commissioners. 1880–1888. 4 December 1884. Transcript at Illinois Regional Archives Depository, Carbondale.
63. *Carbondale Jackson County Era,* 15 November 1884, 3.
64. *Marion Monitor,* 30 September 1875, 2; KENO [pseud.], "Johnson County Election," *Metropolis Times,* 13 November 1875, 2; "The Outlook for Johnson County," *Vienna Johnson County Journal,* 14 September 1877, 1.
65. Johnson County. County Commissioners Record B. 1872–1883. 15 July 1878. Transcript at Illinois State Archives, Springfield; *Vienna Johnson County Journal,* 7 November 1879, 4; *Vienna Johnson County Journal,* 11 November 1881, 1; Johnson County. County Commissioners Record. 1892–1900. 12 July 1895. Transcript at Illinois State Archives, Springfield. In a 1977 county history of Johnson, the author compiled her list of cemetery inscriptions by township: Elvira, Goreville, Tunnel Hill, etc. Rose Parker Bucciferro, *Parker's History of Johnson County, Illinois* (Hartford, Kentucky: McDowell Publications, 1977).
66. *Nashville Journal,* 6 April 1883, 6; *Nashville Journal,* 26 October 1883, 6.
67. *Nashville Journal,* 7 March 1884, 4; *Nashville Journal,* 20 June 1884, 5; REPUBLICAN CO. OFFICIAL [pseud.], "Township Organization," 26 December 1884, 4; *Nashville Journal,* 30 October 1885. At its special October 1885 session, the Washington County Board of Commissioners ordered tax rates for 10 named townships, despite the fact that the townships had technically disappeared. "County Board," *Nashville Journal,* 23 October 1885, 4.
68. "Round-About," *Olney Times,* 15 November 1876, 1; "Township Organization," *Olney Times,* 13 November 1878, 1. I find it interesting that this explanation of a governmental system is exactly the sort that numerous editors provided citizens when a county was considering or had just adopted township organization.
69. "Neighborhood News," *Robinson Constitution,* 26 June 1879, 4; *Carmi Times,* 11 November 1879, 2; *Robinson Argus,* 13 November 1879, 1. Both the *Argus* and the *Times* editors clearly had a bias in favor of continuing townships in their respective counties.
70. Hurd, *Revised Statutes 1874,* 1066 and 1068. In chapter 9 I discuss the evolving requirements for petition signatures and referendum majorities.
71. "Township Organization," *Carlinville Democrat,* 31 October 1867, 4; "The County Court," *East St. Louis Gazette,* 23 October 1869, 2; ARGUS [pseud.], "What We are Coming To, "*Greenville Advocate,* 24 November 1887, 9.
72. Cornelius, *A History of Constitution Making in Illinois,* 67.

73. "Election," *Litchfield News*, 25 October 1867, 2; "Township Organization," *Rock Island Argus,* 1 November 1872, 2; "For Township Organization," *Rock Island Argus,* 31 October 1873, 2; "Be Mindful," *Carrollton Gazette,* 16 October 1875, 2.
74. "Township Organization," *Carlinville Democrat,* 12 November 1868, 4.
75. Hurd, *Revised Statutes 1874,* 1067.
76. *Alton Telegraph,* 28 October 1875, 2.
77. "Township Organization," *Carrollton Gazette,* 11 November 1884, 1. When the framers of the 1870 Constitution sent the township article to the Committee on Revision and Adjustment, the required majorities for adoption and for discontinuance were the same. Why they came out of that committee as two different majorities is a mystery. George D. Braden and Rubin G. Cohn, *The Illinois Constitution: An Annotated and Comparative Analysis* (Urbana: University of Illinois, Institute of Government and Public Affairs, 1969), 497.
78. *Cairo Citizen,* 19 November 1885, 8.
79. *Nashville Journal,* 18 November 1869, 2.
80. B. E. Hoffman to C. E. Lippincott, 9 November 1874, Illinois, Auditor of Public Accounts, Township Organization File, Madison County, Illinois State Archives, Springfield.
81. *Carlyle Constitution and Union,* 15 December 1868, 2; Washington Bushnell, "Township Organization," *Nashville Journal,* 25 November 1869, 3; *Harrisburg Chronicle-Sentinel,* 9 November 1883, 1.
82. *Carlyle Union Banner,* 30 October 1873, 4.
83. SCRIBBLER [pseud.], "Letter from Trenton," *Carlyle Union Banner,* 12 October 1871, 4; REASON [pseud.], "Township Organization," *Nashville Journal,* 4 November 1881, 4; "Township Organization," *Alton Telegraph,* 3 November 1871, 2.
84. "Birthplace of William Jennings Bryan," *Visitor's Guide to Salem, Illinois: Gateway of Little Egypt,* n.d., <www.ci.salem.il.us/visit.htm> (29 January 2001); "William Jennings Bryan" and "Populist Movement," *Encyclopaedia Britannica,* n.d., <www.britannica.com> (29 January 2001).

5 Skeptics, Critics, and H. S. Hicks: 1900–1939

On the first Tuesday of November in 1923, in a central Illinois county between Springfield and the confluence of the Sangamon River with the Illinois, the township tide died. That day saw Cass County, after 12 unsuccessful attempts, finally give township organization sufficient votes to consign commissioners to posterity. Every single subsequent Illinois effort to adopt or readopt the township model has failed. Diffusion gave way to survival, to holding the line, to plugging the dike, to persistence. Cass was one of three skeptics to adopt in the first decades of the new century. Other skeptics, lagging years behind even the late majority, were Henderson, south of the Quad Cities, and Williamson, south of nearly everybody in Illinois. Critics in those same decades had perhaps the best chance ever to dispense with township government, and they took their best shot. By then the serious criticism of townships and supervisors came not from country editors or anonymous letter writers but rather from academics, efficiency experts, and state officials—elected or appointed. Criticism, to be sure, targeted not only Illinois townships but also townships elsewhere and even counties as being wasteful and redundant. Rockford resident H. S. Hicks was a special township critic. In the 1930s, he became a change agent whose sole purpose in life seemed to be tormenting townships and other little governmental units.

Setting the Scene

Countywide referenda, from 1900 through 1939, totaled 39—nearly one per year and far fewer than the 215 I documented for the previous period, 1865–1899. Continuing a phenomenon that first appeared in 1891, entire years frequently passed without a vote on township organization: 1903, 1910, 1912, 1919, 1920, 1924, 1927, 1928, 1930, 1931, 1933, 1937, and 1939 (figure 11). Nearly three-quarters (28) of the 39 referenda were the result of efforts to adopt or readopt. The other 11, mostly in the 1930s, were discontinuance projects. No year saw more than four referenda, and just nine years saw more than one county vote on this issue. The three adoptions, during the skeptic period, were in November 1906, November 1907, and November 1923.[1]

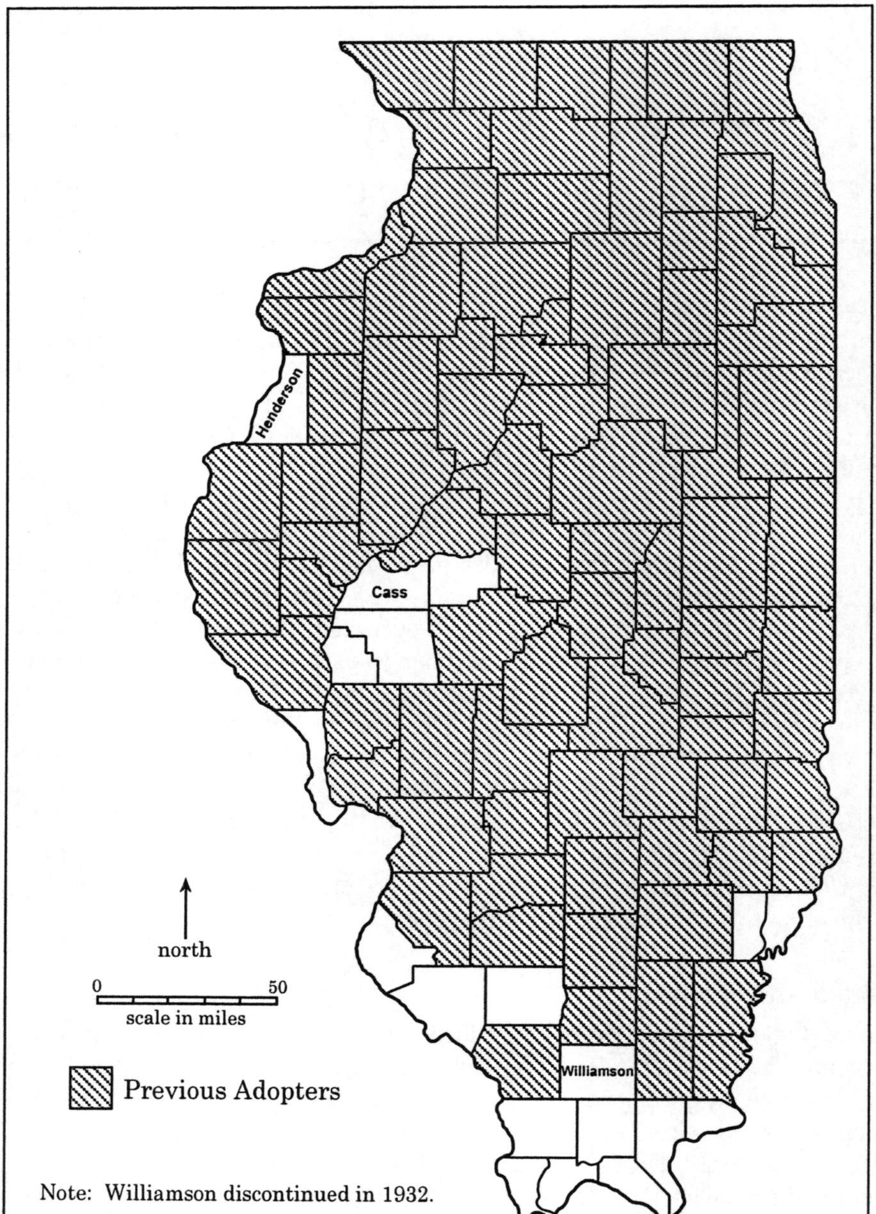

Figure 15 Counties Adopting Township Organization: 1900–1939.

Figure 15 shows the location of the last three adopters, along with the 83 other counties then in the township column. The three skeptics, although not contiguous, provide us one more vestige of the township movement's push toward the southwestern edge of the state. All three lie next to features that, as I have noted before, seemed a focus for resistance for the Yankee invader: big rivers and the Shawnee Hills. Henderson had held out, in isolation from commission counties, for a half-century since its nearest neighbors, Hancock (in 1849), Mercer and Warren (both 1853), and McDonough (1856), had traded commissioners for supervisors. Williamson and Cass, on the other hand, left behind in 1907 and 1923 adjacent neighbors that to this day eschew townships. Henderson was surprising because its voters resisted for long; Williamson and Cass were surprising because their voters adopted at all, so late in the diffusion process.

Exactly one dozen different Illinois counties placed the issue of township government on their countywide ballots over these four decades (figure 16).[2] The 12 voting counties arrayed themselves in two distinct clusters. One cluster consisted of 10 counties, from Henderson on the north-northwest to Williamson on the south-southeast. A majority of that 10 had more than one ballot contest; and Perry, with 12, joined, in just a few years, the top 10 counties overall in terms of total number of referenda. A second cluster, consisting of Ogle and Winnebago, on the Rock River, far to the north, evidenced (in part) the impact of H. S. Hicks and other local reformers. The two Ogle County referenda and the one in Winnebago came during the turbulent and trying 1930s, when township critics were most determined and conditions seemed ripe for an ebb tide.

Besides the three skeptical (but eventual) adopters, four other commission counties voted whether to adopt townships during that period. Monroe, whose citizens had voted six times in the nineteenth century, added two more referenda in the twentieth: 1907 and 1909. Morgan, second only to Williamson in terms of total countywide referenda, came into the new century with 14 referenda behind it, added one in 1902, another in 1907, and the seventeenth many years later. Perry held all 12 of its referenda between 1907 and 1929. Randolph held its first of 11 in 1869, tenth in 1904, and last in 1905.

Women Get to Vote

The *Marion Leader* had predicted in 1905 that township office aspirants would favor township organization "to a man." Maybe they would, maybe not; but unless a citizen was male, in 1905, that citizen could not legally vote either for or against the measure. In fact, over four hundred countywide referenda on township organization occurred before the first female vote counted.

Illinois women gained the right to vote in stages, beginning in 1891 with the option to vote on school board members and any other school official except the state superintendent of public instruction and county superintendents

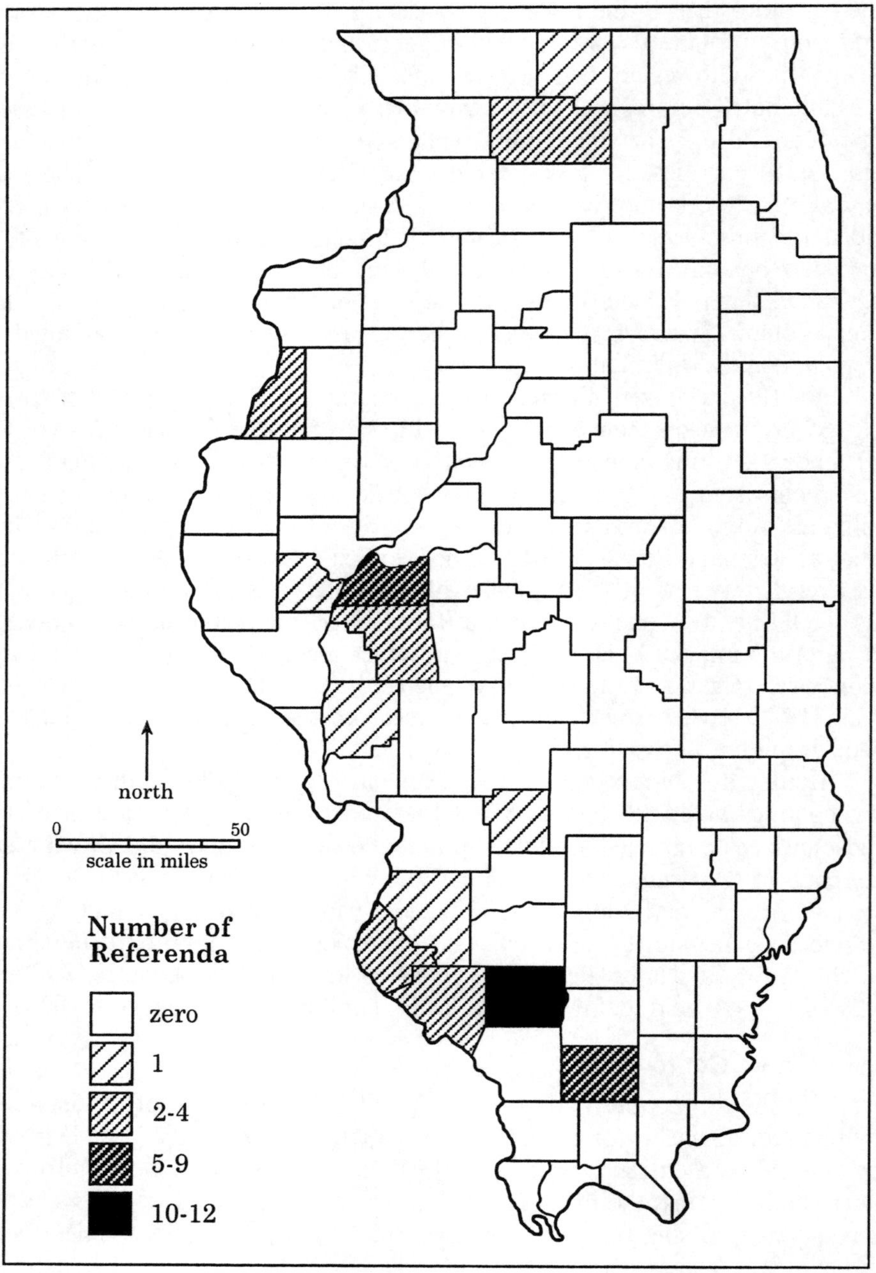

Figure 16 Countywide Referenda on Township Organization, by County: 1900–1939.

of schools.[3] Next came a major suffrage law, which Governor Edward Dunne signed in June of 1913, giving women the right to vote (if otherwise qualified) for presidential electors, various county officers, all municipal officers, and "upon all questions or propositions submitted to a vote of the electors of such municipalities or other political divisions of this State." The law further extended the vote specifically to township "supervisor, town clerk, assessor, collector and highway commissioner." Women gained also the right to vote at annual and special township meetings. The new voters were to register for voting "in the same manner as male voters," but they were to receive distinct ballots and place them in ballot boxes separate from the men.[4] Because of the separation of ballots, observers could learn how women, as a group, voted. The first chance for women to vote under the new law came on 4 November 1913, when Illinois's 17 commission counties held the election that occurred every fall to elect one member of the county threesome. Ironically, while women in 1913 might vote on dumping the commissioner form of county government, they could not vote on the men running for the commissioner slot because the county commissioner was a constitutionally mandated officer in Illinois and thus off limits to female voters.[5]

The overriding issue in the November 1913 election was not the county commissioner slot or township organization; it was liquor. To be more exact, the issue was local option, by community, as to the continued and future existence of saloons: the wet/dry issue. Long-suffering, franchise-denied women came out in goodly numbers to hammer (at the ballot box) saloons in places like Thebes (in Alexander County), which went from wet to dry "by a majority of 40, with 200 women voting." Observers, on the scene, credited women as being "responsible for the result." All told, the drys that day won or retained 18 communities, while the wets held five (but gained none); and there was one tie. Women voted, said a *Chicago Tribune* article in closing, and incorrectly, "only upon the local [liquor] option issue."[6]

Perry County happened to be holding one of its semiregular referenda that November on townships; and Perry women got to express their preference on their "Official Woman's Ballot," which asked if they were "For Township Organization" or "Against Township Organization." For the *Pinckneyville Advocate,* the big stories after the election were about the facts that women showed up and that Pinckneyville voted to become an "arid district," putting, in the process, 10 local saloons out of business. Women of Perry County cast 1,733 votes on the township question in 1913, and they proportionally gave the reform less support than did the men. Only one in five women asked for a change in county government, versus one in three men.[7]

Compared to the Perry County press coverage of the 1913 township referendum, newspapers in nearby Williamson County, the next year, made sure readers were well aware something new was happening at the upcoming November election. "As this is a public question," said the *Marion Egyptian Press* in September 1914, "the women will be allowed to vote the same as men." A month later, the *Herrin News* let readers in that coal town know "Women as well as men may vote on this."[8]

Williamson women began to read advertisements that appealed directly to them. The *Egyptian Press* carried a public letter in which local Progressive George W. Young sought their support of his candidacy. Young commented on how confusing all the political parties must seem to the new voters, and he noted that his Progressive Party had been a major force in passage of the 1913 suffrage law. A few days later, the same paper carried another promotional item that asked potential women voters: "Do You Want to Be Disfranchised?" The ad writer left little doubt how women should view the existing township system: "The women will probably lose the ballot in county affairs unless they go to the polls November 3, and vote for the continuance of township organization." Once the county commissioners retake control, "for whom women get no vote," women would have to let men do the choosing of county board members. "Therefore, women to save themselves from losing the ballot in county government," should stick with supervisors and townships.[9] Township advocates also printed and distributed a couple thousand broadsides carrying a similar message to Williamson County women. Promotional activities may have helped account for the stronger support of women, 758-165, for township organization, than was evident among the men, 3,592-3,451, that November.[10]

Even though women had already voted twice in a commission county on the topic of townships (Perry in 1913 and 1915), women in Cass County who, in 1917, wanted to exercise their franchise found the women's ballot differed from the men's regarding propositions up for consideration. Men got to vote on two issues: a county tuberculosis sanitarium and township organization; women could only vote on the sanitarium.[11] Why did the distinction exist? Cass was still a commission county. Since women could not vote for commissioner, perhaps county officials decided women had no business voting on such a significant question as organization. Perhaps women's traditional role as caregiver qualified them to help decide a question like the sanitarium but did not carry over into courthouse politics.

Universal women's suffrage came in August of 1920 with the Nineteenth Amendment to the United States Constitution. Illinois tied with two other township states, Michigan and Wisconsin, for the honor of being the first to ratify the amendment.[12] When Cass County again considered township organization in 1923, women clearly had the vote. No longer did papers print separate specimen ballots for men and women or provide results by gender. A letter writer, supporting the township concept for Cass in 1923, properly couched his appeal by saying he hoped "each and every voter [would] make it his or *her* business to vote on the question" (italics mine).[13]

A Peaceful Demise for Chicago's Townships

Women had no say in the termination of Chicago's townships in 1902, a termination that had been coming for a long while. The 1870 Constitution replaced the Cook County Board of Supervisors with a Board of Commissioners whose 15 members the voters were to elect at large, 10 from the city and the rest from the outlying areas of Cook. That 2:1 ratio corresponded to the approximate population distribution in 1870 between city and rural Cook County; but by the early twentieth century, due to the rapid growth of Chicago, it meant Chicagoans had too small a piece of the board.[14] Legislation from Springfield chipped away at Chicago's political townships, especially near the turn of the century. An 1899 law, and subsequent legislation, made possible township consolidation and transfer, given voter approval, of all township-official functions to Cook County.[15]

The *Chicago Tribune* strongly supported the end of township government in the city, not just on its editorial page but also in what passed as news articles. When a 1901 township-consolidation bill cleared the Illinois House of Representatives, the *Tribune* highlighted the efforts of Chicago's Citizens' Association, then campaigning to educate voters on a question that might be on their ballot the next April. The Association had printed and was distributing a bulletin urging the Cook County Board to refuse to levy taxes for township expenses in the fall of 1901. In its article about the bulletin, the *Tribune* printed, without township commentary, charges from the Association such as this: "The North [Chicago] Town[ship] failed to publish its Supervisors' account until April 15, although the law required it to be filed on March 20 with the clerk. The salaries paid the officers are illegal and are in excess of those shown to be paid in all except the West [Chicago] Town[ship]."[16] When the 1901 annual township meetings took place in the city, the *Tribune* referred to them as "funereal" because of the gloomy atmosphere that prevailed. Of the South Town meeting, which lasted all of 10 minutes, the reporter wrote, "The meeting might prove a historic one, as it is considered unlikely that another will be held....The necessity for calling the citizens together for approval of the doings of the officials has been reduced to a minimum."[17]

A year later, voters did call for an end of townships as functioning governmental units in the city. The *Tribune* lauded the adoption of "the proposition to abolish these now useless offices," declaring that this might be a first step in taxation and other governmental consolidation within the city's limits. "The Supervisors never had anything to do that any one in recent years could discover." Among expected improvements were a big saving in taxes and a reduction in scandalous township affairs. Every city ward approved the termination; and, in a section of the article labeled "Empty Honors in Towns," the *Tribune* listed all the newly elected township officers. None would be able to fulfill their duties. On the editorial page, the paper emphasized its antitownship stance by praising the voters for ridding the city of "these expensive and useless petty offices [which]...ought to have been abolished twenty years ago."[18]

Officially, the vote was 64,721 for abolition of township offices and 10,516 against the measure. Mayor Carter Harrison, Jr., ended the half-century of Chicago township government by issuing a proclamation on 5 April 1902 to disable the seven townships within his jurisdiction.[19] Some fear apparently persisted that the elected officials would contest the referendum and proclamation, but lacking a statewide support/lobby group or a powerful political friend, like Elijah M. Haines, they chose to let pass the chance for resuscitation.

Another Constitutional Convention

One final scene-setting item remains: the 1920–1922 effort to update the state's constitution. Dissatisfaction with the 1870 Constitution had been building, even before the turn of the century. Industrialization, the booming growth of Chicago, and the seeming impossibility of amending the existing constitution helped fuel a widely held belief among the politically savvy that a constitutional convention was necessary. Governor John Peter Altgeld, in 1892, issued such a plea. The state senate, in 1893, passed a resolution calling for a convention. Subsequent resolutions by senators and representatives continued to appear, then to die for lack of support, as the nineteenth century came to an end. Senator Edward C. Curtis, a Republican from Kankakee, pushed the topic of a convention via resolutions he introduced in 1909, 1911, 1913, and 1915. Organizations formed to push for a convention during the 1910s. The Progressive movement urged a variety of reforms, which only a new constitution would permit in Illinois. Democrat and Republican parties began, during World War I, to include a convention call in their state platforms; and, in 1917, the newly elected governor, Frank O. Lowden, in his inaugural address, said the time had come to put the issue of a convention before the people. Once more Senator Curtis rose to introduce a resolution on a convention-calling referendum, and this time it passed both houses. On 5 November 1918, the people said, by a wide margin, "call the convention." A year later, on 6 January 1920, 102 elected delegates met in Springfield to form the fifth Illinois Constitutional Convention.[20]

County government, whether township-driven or commission-focused, played a minimal role in the call for action; but the delegates, needing, of course, to deal with local matters, created a Committee on County and Township Government.[21] That committee made its initial report to the convention on 18 May 1920, as committee chairman Arthur M. Smith of Jo Daviess County assumed the chair of the convention floor for the committee's presentation. Smith then called on committee member Lewis A. Jarman of Schuyler County to share the committee report with the convention delegates assembled.

Smith's committee had, according to Jarman, concentrated on article X of the 1870 Constitution, except for the portion thereof dealing with Cook County.[22] The committee was recommending that a new constitution leave section 5 of article X, the one dealing with township organization, little changed, asking only for those alterations necessary in making "it applicable to the present conditions." One such modernization was to have been omission of the 1870 prescription that no two townships have the same name. Here is how the proposed section 4 of the Local Governments Article came from the Smith committee to the floor of the convention:

> Any county now or hereafter organized under a Board of Commissioners may organize under township organization whenever a majority of votes cast upon such question in such county shall so determine, and any county now or hereafter organized under township organization may organize under a board of commissioners whenever a majority of all the votes cast upon such a question in such counties shall so determine and the election upon such questions shall be submitted to a vote of the electors of any county at a general election in the manner provided by law; and the affairs of the county shall be transacted under the respective organizations as the General Assembly may provide.[23]

Note that adoption or discontinuance could only take place if voters, by a majority of those voting at the election, so decided and that the legislature was to provide most of the details as to how the two systems would function. Delegates approved this segment without debate.[24]

The Smith committee had reservations about both the supervisor and commissioner systems of doing county business. In a new section (section 8) of the Local Governments Article, the committee suggested authorizing the general assembly to devise new ways of managing county affairs, if a county's voters approved. Jarman went on to explain that the committee felt "it might be best to provide a uniform organization of county government which would be much better than the present forms of government, the township organizations [*sic*] and the commissioners' form of government." Committee members supporting such a reform believed "that as the efficiency and desirability of the new form of government was demonstrated by one or more counties, other counties would accept the same."[25]

Debate did occur at the convention level, the following day, on section 8. Delegate (and former governor) Joseph W. Fifer of McLean County deemed it "a dangerous experiment," saying, "It will be a wide open door for all the cranks and experimenters in the State of Illinois." Fifer had heard no complaints about county government in general: "We can go back to the old commissioner form, but I doubt if any county in Illinois that now has the board of supervisors [as had McLean since 1858] would want to go back."[26]

Albert E. Taff, of Fulton County, rose to defend the Committee on County and Township Government, of which he was a member, by noting the committee had spent four times as long on section 8 as any of the other

14 sections. Both the commissioner form and township form "have very antiquated propensities," he continued. Under neither form was there one official "upon whom you [could] place responsibility for any act of county government." The legislature had only those two choices and could not abolish either form. Taff lamented the lack of scholarly attention to county government "throughout the United States," compared to city governments. Taff said this section 8 was the only avenue "by which you can in the future obtain any improvement in your county government." The committee had also, apparently, considered proposing abolition of both commissioner and township forms but ultimately decided "it would possibly get a great deal of criticism from counties over the State, word going out we were abolishing all their forms of government." He closed by urging the delegates not "tie the hands of the counties throughout the State to the two antiquated forms of government."[27]

Herbert F. Lill, also of the committee, reinforced Taff's comments by citing Lill's experiences as a member of the St. Clair County Board of Supervisors. He said the condition of county government across the river from St. Louis had "astounded" him over the previous two years: "There is nothing in our whole government which is similar to county government. There is no head nor tail to it." His board had 51 members and might, after the 1920 Census, jump to 65 or 70, as population increase drove the formula that added or subtracted assistant supervisors. Lill elaborated on how cumbersome such a large board had become. Furthermore, while the rural St. Clair townships were careful to pick good custodians of their funds and, therefore, to send careful men to the county board, the typical assistant supervisor, from an urban township, was likely to be incompetent when it came to county business. Three commissioners, he felt, made for a board too small to be truly representative of the people's interests. Maybe five or seven commissioners, with a president of the county and an advisory officer (like a county manager), would be a good compromise.[28]

With the approval of a wording amendment, which clarified the issue of uniformity (across all classes of counties) as to whatever new form the general assembly might ultimately authorize, the delegates accepted section 8 on 19 May 1920 and moved on to subsequent sections.[29] What we have seen in the paragraphs just preceding, however, is only an inkling of the critical commentary that counties and townships were receiving during these early decades of the century. I will have much more to offer on the subject later in chapter 5.

Local governments came up for final consideration on 28 April 1922. Again, delegates went through what had become article VIII one section at a time. Township organization and county commission organization remained the preferred methods of conducting county business; but the Smith committee's suggestion, that the constitution permit future general assemblies to create different (and presumably better) governmental models for the 101 counties (omitting Cook), remained firmly intact. Delegate Rodney H. Brandon of Kane County mildly protested the convention's lack of willingness to prescribe what that improvement should be and offered an amendment that would have required legislators to come up with a single system to fit all counties (save Cook). When fellow Kane County Delegate Lee Mighell asked Brandon to specify "which system he offers upon the people of this State," Brandon said the committee should have done its work properly and "this article should embody the system." Brandon later withdrew his amendment and voted along with 58 others to approve. None disapproved.[30]

By a huge margin, on 12 December 1922, voters said no to the final product of the constitutional convention. Opposition was widespread and effective. Cornelius argued one of the best outcomes of this three-year event was the advice it provided on how *not* to hold such a convention.[31] Township organization, as prescribed in 1870, modified by the general assembly, and interpreted by the courts and local officials, would continue to be the preferred method of conducting the people's business across much of the state. A new and fairer method for choosing county boards in township counties (save Cook County), which a 1922 Constitution might have anticipated, would have to wait.

The Skeptics

Perhaps I should refer to the last three adopting counties not as skeptics but as skeptical adopters. After all, those counties that never adopted were skeptics, too. What I call Cass, Henderson, and Williamson, however, matters less than the circumstances that compelled such widely distributed counties to revoke the commissioners' mandate so long after so many other counties had taken that route.

Henderson County

If Illinois had to remake its county boundaries, there probably would not be a Henderson County. Together with neighbor Warren County to the east, the combined area of the two, 923 square miles, would form the largest county west of the Illinois River and place it among the top 10 counties in the state. In fact, for 16 years, from 1825 until 1841, they were one—Warren County. Then, yielding to agitation for separation from the Henderson half, the Twelfth General Assembly created Henderson by taking the western 381 square miles from Warren. In the enabling act, the legislators designated the tiny Mississippi River port of Oquawka as Henderson's seat of justice.[32]

Robert P. Sutton, in an unusual effort by a professional historian to do justice to the local-history genre, attempted to explain why one Military Tract county became two in 1841. He cited an 1886 history of Henderson that said residents of what became Henderson were unhappy having their county seat at Monmouth and, failing to obtain a more "central location" (read: location farther west) for the seat, got themselves busy and got themselves a new county. Sutton did not tackle the even more perplexing issue, however, of why Warren would adopt township organization in 1853, putting Warren among the earliest 40 adopters, and Henderson would still be a commission county in 1900.[33]

Meyer's 1850 settlement maps provide a modest clue as to the reason for Warren's early adoption and Henderson's lengthy skepticism. Warren fell, in 1850, partly in the New England Culture Region and partly in the Midland-Midwest. Henderson was entirely Midland-Midwest.[34] My guess is that cultural baggage, river focus, rolling terrain, small population, and irregular boundary (on the west) all played parts in keeping Henderson a commission county for so many years beyond what its latitude and neighborhood would have forecast.

Henderson voters turned down townships four times in the nineteenth century: 1849, 1858, 1868, and 1890. In 1906, they got another chance. Announcing the 1906 petition drive, the *Oquawka Spectator* said, there was "strong sentiment in some sections of the county in favor of township organization." The *Spectator* editor seemed unable to understand why otherwise sensible taxpayers would accept the extra expense but agreed that "if the people want to foot the bill and have it, that is their privilege." He suspected "some prosperous townships" wanted a bigger share of the taxes that they contributed to county coffers.[35]

Why circulate petitions in 1906, after the earlier defeats and after so long with the commissioner system? Perhaps there was a trigger for the renewed interest in townships. The *Warsaw Bulletin,* from neighbor Hancock County to the south, suggested one such trigger in an article the *Spectator* reprinted in early October: "Some things have developed in the county," said the *Bulletin,* "which have caused a prejudice against the present mode of government, and among these was a discovered shortage in the accounts of certain county officials." The *Bulletin* editor cautioned his Henderson neighbors, however, that, as in Hancock "a few years back," township organization was not necessarily the ideal cure for a county's financial ailments. Was the "cumbersome and expensive" township system a better bargain just because it was more "representative"? The *Bulletin* did not offer an answer.[36]

Another 1906 Henderson community with a newspaper was Stronghurst. It lay on the Chicago–Kansas City main line of the Santa Fe Railroad, between Galesburg, Illinois, and Fort Madison, Iowa. The *Stronghurst Graphic* carefully explained the township system to readers in its 18 October issue, offering only a closing comment that "it is more expensive than the simpler form but many claim that the advantages to be derived more than make up for the additional expense." A couple weeks later, the *Graphic* took a straw poll of farmers attending a local institute event. W. A. Spears was opposed. He tossed out the analogy of "Media township," where the amount of work the roads received then was "in an inverse ratio to the number of road officials employed." Several other farmers disagreed with Spears, citing the opportunity for better understanding of sectional issues (within the county) because of the expected proximity of township officials to their constituents. That poll outcome was 14 favoring and 3 opposing townships.[37]

Yes votes prevailed on election day. Stronghurst voters were even stronger township supporters than had been the straw poll participants, going 180 to 21, 89 percent versus 11 percent, for the change. Elsewhere in the county, except for Raritan, which gave township organization 92 percent of its votes, Henderson voters were much less supportive. Six precincts, forming a crescent from Oquawka, at the northernmost point, south along the river to Gladstone, Carman, and Lomax, then east to Terre Haute and Decora, produced more negative than positive votes.[38]

Challenges to the referendum's outcome were in the wind over Henderson as the process of division into townships began. The *Spectator* reported on 21 November that "a scouting party" of township organization's enemies from the southern part of the county was in Oquawka to see what they could do legally to overturn the outcome. They were hoping apparently to use the familiar tactic that had caused many such referenda on the issue to fail: mandating a majority of all voters in the county. The *Spectator* of 19 December 1906 questioned whether the 1870 Constitution's "majority of the legal voters of such county voting at any general election" took precedence over the 1848 "majority of the voters of such county." Judge John A. Gray, at the March 1907 term of the circuit court heard the complaints and responses. Attorneys from Carthage and Monmouth presented the challenge, while the state's attorney of Henderson plus lawyers from Stronghurst and Monmouth defended the outcome. The judge sided with the defense; division proceeded; and, in the spring of 1907, Henderson welcomed 11 civil-political townships to the map.[39]

A year after division, the thwarted challengers were back with petitions in hand. R. A. Lomax, a part-time resident of nearby Burlington, Iowa, and of Texas, whose family name graced a Henderson village and township and whose family had major financial interests in the Illinois county, was one of the disgruntled fans of commission government who argued for a reversal of the 1906 decision. Lomax came to Stronghurst in August of 1908 to circulate petitions and make his case, claiming that the riparian townships were carrying a bridge burden "out of proportion to the amount of taxable property in the townships." In a long letter to the *Graphic* a couple months later, Lomax called the upcoming discontinuance referendum "an opportunity to correct a mistake." He based much of his argument on cost comparisons between the old Henderson (pre-1907) and the new Henderson, and he tried to explain the financial situation that had led to the 1906 call for reform there. Finally,

Lomax discounted township-supporter claims about that system providing "better roads even if it does cost more."[40]

The *Graphic* provided space the week after the Lomax letter to opponents of the Lomax position. One writer, W. A. M. Crough, traced townships back to New England, citing his principal source, Elijah M. Haines, as "one of the great lawyers of Illinois." Crough stretched the truth a bit when claiming that 92 of the 102 counties had already gone the township route and when stating, " A vast majority of those counties have used this form of government more than 125 years." Lomax, said Crough, really lived in LaPort, Texas, where Lomax had "been enjoying the enervating climate of Texas so long that he [had] gotten a little behind the northern procession." Crough blamed the county commissioners for allowing the tax collector to make such a mess of Henderson finances in 1905 (the trigger issue). Another writer, COMMITTEE, had a letter in the same *Graphic* as Crough attacking Lomax and other "non-residents of Illinois" as men who had been fortunate before 1907 to avoid paying their share of local taxes. Land in their southwestern corner of Henderson, said COMMITTEE, had seen assessed value rise from $.25 per acre to $3.00 or $4.00. Several times COMMITTEE chided Lomax for trying to govern Henderson County from Texas. The letter closed with the writer(s) urging voters to " make no retrograde movement" and continue having a say so about how officials spend tax revenues.[41]

Taxes prevailed over Texas, 1,388-964, in November 1908. Oquawka, Gladstone, Lomax, and Carman townships continued to poll more negative votes than affirmative; but voters in the seven remaining townships had liked what they had seen so far. Raritan Township led the way for reapproval, with 91 percent voting to make no retrograde movement. The *Graphic* editor commented about how the overwhelming confirmation would serve to inhibit further effort "to change back to the form that has been abandoned by all but ten counties in the state." While in error on the number, the editor was correct about the fact that Henderson, even in the 1930s, would not turn back the clock to the decades when it stood alone in its corner of the state and resisted the township tide.[42]

Williamson County

No Illinois county, and probably no other county in the nation, has voted as many times as has Williamson regarding township government. Of its 18 referenda on the issue, eight took place in the four decades between 1900 and 1939. None have occurred since. On its southern margin lie the Shawnee Hills and farther south are the bottomlands of the Ohio and related rivers. Its political neighbors to the south, all lightly populated and rural, were then and are today staunch in their commitment to their county commissioners. On Williamson's other margins are township counties, who themselves have questioned on occasion their predecessors' decisions to abandon commissioners for townships—in the face of the evidence that the southernmost seven counties of Illinois were then (and are today) succeeding sans townships. Williamson felt the pull of both systems and tried both, making changes twice during the skeptic period.

Township-minded reformers had brought the issue unsuccessfully to the ballot in Williamson thrice in the 1890s: 1894, 1895, and 1896. As the 1900 referendum approached, the *Herrin News* lamented how little attention the matter was receiving, calling it "one of the most important issues to be voted for." The *News* urged readers to move Williamson from has-been to progressive, to match the county's natural resource abundance (bituminous coal). Investors, believed the editor, would relish such a positive step. *News* readers and other voters did give the reform nearly two hundred votes plurality, but the 2,024 positive voters fell a thousand voters short of being a majority of the more than six thousand that went to the polls.[43]

Five years later, proponents again achieved a small plurality but fell short of a majority. The Herrin editor informed readers as to how much the 1905 outcome displeased him. While Herrin and Marion had gone for the innovation, said an editorial on 10 November, "the natives on the bordering precincts killed the proposition as if had been a poisonous reptile." Voters in the northern half of Williamson seemed to be on board the township train while "natives down in the hills seem[ed] to be the sort who think well enough is well enough." Of course, the hill folk in the south could look across the county line at Union and Johnson counties where townships were nonexistent. Not much honor accrued, said the editor in later opinion piece, to Williamson for being among the largest of such an "unprogressive" bunch as the commission counties. Eventually, "the people are bound to come around to choosing it and throwing overboard an inconvenient, unpopular and less efficient system."[44]

Williamson voters threw the commission system overboard on 5 November 1907. Beforehand, the *News* was trumpeting the "higher calling of townships over precincts and commissioners." Afterward, when voters, by a mere 200 votes, gave township organization the nod, the *News* editor could not resist saying Williamson was finally joining the other "civilized" communities of the state. Later still, the editor listed all the new jobs that needed filling for each township and asked "Why not run for office...when the new township organization displaces the antedated precinct organization?" Blairsville Township, west of Herrin, got the message apparently; 29 aspirants vied there in 1908 for the 10 jobs.[45]

The successful Blairsville candidates hardly had time to settle into their new township jobs when commission proponents tried to fire them and all other township officers in the county. The *News* continued to fly the township standard, claiming that the reform displeased "only a few warty old politicians" who could no longer manhandle the citizenry. Herrin voters were, according to the editor, to remember that township organization would save them an annual trip "over to Marion" to pay their property taxes. With their new township collector doing the collecting, taxpayers could pay their share right in Herrin Township. One more reminder seemed appropriate: Williamson was now safely on the side of progress with its "modern system" of doing its government business. Voters gave it, in November 1908, a cushion of almost six hundred.[46]

I devoted space earlier in chapter 5 to the role and effects of women voters on the 1914 Williamson referendum. The *Egyptian Press,* in alerting Marion readers to that upcoming effort to abandon the township experiment and to the fact that women could vote on the issue, reminded all that the law only called for a "majority of the votes cast upon the question" to carry the day. Therefore, it mattered not how many voters ignored the question; and, theoretically, a return of two votes to discontinue out of three votes cast would force a revision to commissioners. The *Herrin News,* by 1914, had changed editorial sides and was calling it an "expensive form" of local governance. Countering rumors circulating around Herrin, the editor carried word-for-word excerpts of laws pertaining to taxes for roads and bridges, which made it clear that eliminating townships would have no effect on where road-and-bridge taxes from Herrin Road District would go. The law, whether Herrin Township persisted or not, specified that local taxes stay in the road district where the taxpayers paid them, just as was the case with school districts. Once again, as in 1908, Herrin Township and Williamson's other 11 civil townships survived, this time with a mighty assist from the ladies.[47]

Not even the ladies could save townships in 1932, as Williamson County, on 8 November that year, became the most recent Illinois county to give up the township ghost. Trouble had been stewing for years, even before the Great Depression took its terrible toll. In fact, almost exactly three years before the 1932 referendum, on 5 November 1929, Williamson's supervisors had to leave the courthouse without their usual pay for the November session because, according to the *Marion Republican,* "the money was all gone."[48]

As the effects of the stock market crash, bank closings, and a drastic slowdown of the economy reached southern Illinois in the early thirties, Williamson's leaders continued their struggle to keep the county solvent. The Williamson County Tax Review and Equalization Board urged township assessors to avoid drastic reductions in assessments, while the Williamson County Board of Supervisors listened to suggestions for reforms that included holding meetings quarterly instead of monthly and issuing warrants (to pay bills) only when "the money [was] on hand for payment." One study, by the county's Finance Committee, found that Williamson had been running an annual deficit for a decade because of failure to collect delinquent taxes and unfunded mandates from the State. I. A. Palmer, of Herrin Township, introduced a resolution at the supervisors' October 1931 meeting calling for confiscation of property in lieu of delinquent taxes owed the county. Supervisors had to deal with Sheriff G. J. Frick, who, after holding eight thousand dollars in worthless county promises to pay, issued a statement saying his office would serve no further legal papers on the citizenry of Williamson unless the county made good on its obligation to him. "I mean cash," Frick declared. Not even tax anticipation warrants, promises by the county to pay principal and interest when property tax dollars came in, were finding investors in the fall of 1931.[49]

The new year, a presidential election year, did not bring an improvement. An audit of county offices revealed expenditures had exceeded revenues by more than $23,000 in 1931 ($122,676 versus $99,257). The decline in the coal industry made matters especially bad in townships west and north of Marion. Some relief came from emergency state grants, the Quakers, and the Red Cross; but critics of township organization saw the ongoing Williamson woes as a chance for reversion to a simpler governmental model, which would mean replacing a board of 26 with a board of three. By June, nearly six months in advance of the general election, local agitators had their completed petitions on file in the courthouse. One of the reforms they would thereby accomplish would be the automatic repeal of the township pauper tax, implemented in December 1931. Many of the 26 supervisors were among the five thousand petition signers. An exception was W. J. James of Southern Township. James, at a board meeting where the supervisors also discussed appointing a road superintendent, said he feared that the centralization of power in just three commissioners would preclude lightly populated areas like his township from having a say in county affairs; and he demanded a "microscopic inspection" of the petitions. State's Attorney Arthur Melvin, however, told the supervisors they had "no power to pass on the legality of the petition[s]," so the referendum should be on the ballot. At their September meeting, the board made it official.[50]

Local editors were quick to comment, once the board acted on the petitioners' request. Carterville's *Herald* was not sure the switch would truly save the county any money. After all, commission counties were doing "little better than they are here." If the new commissioners "had the gizzard," wrote the editor, a strong central government might somehow lead the county out of its "muddle." A month later, the *Herald* was still leaning toward keeping things as they had been since 1908. Cost comparisons, seeming to favor the commissioner way, had not swayed the editor much: "Our opinion is that these statistics are not to be relied upon, for counties face different tasks." Over to the east a couple of townships, the editor of the *Johnston City Progress* was in the mood for retrogression, however: "The supervisors have done their best but the handicap of the antiquated form of township organization has been too great." Ironically, what had been new and refreshing in 1907 and 1908 had become, by 1932, old and unneeded.[51]

A. M. Townsend had little use for township government and said as much in similar letters that the *Carterville Herald, Johnston City Progress,* and *Marion Republican* carried early in November of 1932. Townsend began by making clear that readers/voters knew the board would automatically shrink from 26 to three if the reform were successful. Then Townsend rolled out cost comparisons that Townsend said showed supervisors and tax reviewers were costing the county "eight to nine times" what the commissioners would require in salaries. "Township organization is a refuge for old man extravagance. It is the bitter enemy of the taxpayer." Otherwise, why were "ten counties" voting on whether to abolish it "at the coming election?"[52] The number voting was actually three: Greene, Williamson, and Winnebago.

Of the three, only Williamson agreed with Townsend, so, on the same day that Franklin Delano Roosevelt defeated Herbert Clark Hoover, in a national reform movement, Williamson voted to rejoin the commission counties, bringing Illinois thereby to the current tally of 17 counties without and 85 with civil-political townships. The vote was 9,117 to 7,168. The *Republican* and *Herald* prepared lengthy post-election informational articles about how the county would transition to the commission setup. Henceforth, for instance, Sheriff Jim Frick would collect the taxes and no longer need to beg the board for salary relief. Treasurer J. A. Burkitt would trade tax collecting for assessment supervision. Three commissioners, one to serve one year, one for two, and the third for a full three-year term, took up the challenge of remedying the Williamson situation.[53]

N. F. Atwood was the Williamson commissioner who drew the long straw after the 2 January 1933 election in which he, E. A. Fry, and H. A. Carey received the most votes for the new board of commissioners. By lot, Atwood won the right to serve three years; and he almost made it through his full term before the township tide gurgled up again. Fry and Carey had already seen their abbreviated terms as commissioners abruptly end, after a November 1933 defeat for Fry and a similar fate for Carey in November 1934.[54] Perhaps dissatisfaction with the lack of apparent progress, after the demise of the supervisors, was growing, even before 1935, among the voters.

Agitators for the 1935 township referendum needed only 50 signatures to get voters to consider a township revival for the county, compared to the much larger number their opponents had needed to get discontinuance on the ballot in 1932. Commissioner Hosea Borum defended himself and fellow board members by noting they had cut annual spending by $25,000. On the other hand, "insufficiency of county funds" had prevented them from recovering, during the commissioners' brief tenure, "the expense of making the change in county government three years ago." Later, at a Rotary Club forum to discuss the two forms of government, Borum estimated the earlier (and then unpaid) transition expense at $15,000. He said another transition would cost at least that much again, "with no funds in sight to pay the expense." County Democrats came out in favor of keeping the commissioners, while Republicans took no position, as a group, although the petition circulators had been Republicans. The Trades Council endorsed the status quo. Picking up the Borum line about transition expenses, the *Republican* editor urged staying the commissioner course. Opponents of changing back to supervisors and townships derived confidence from the fact that township advocates would need a majority of all votes cast on November fifth, not just a majority of those voting on the question. Commissioners prevailed, and Atwood won a second term.[55]

In 1938, Williamson voters went to the polls for the eighteenth and (so far) final time on the matter of township government. Republicans and Democrats shared petition-circulation duties that year. A likely trigger for this late 1930s reversion effort was the political controversy then swirling around the board of commissioners. The *Johnston City Progress* said people might not remember the board's accomplishments but would not soon forget "the 'fight'nist' county board we ever had." Board members had ousted Treasurer A. O. Ledbetter and had moved the fight across the hall of the courthouse to tell County Clerk S. E. Storme how he ought to do his job. The *Progress* editor thought the board's fighting days were about over, given the November referendum that loomed. One Johnston City citizen, said the editor, thought he would vote for township organization because it would put a couple dozen to work on the board, "instead of just three men."[56]

Both the *Progress* and *Republican* ultimately supported readoption of townships. "County affairs," ran a *Progress* editorial, "can be administered better by supervisors from each township rather than by three commissioners from the most thickly populated areas." The *Republican* said, "A vote in favor of change is one for giving each community representation on the county board where now but three can have membership." According to the *Republican,* there had been little organized effort to get out the protownship vote and "the proposal face[d] the possibility of being defeated by persons overlooking the separate ballot."[57]

That is exactly what happened. All three little-ballot issues, including one on statewide banking and another on public policy) had more positive than negative votes, but lost in Williamson because so many voters did not bother with them. Township organization needed 12,481 to surpass 50 percent of votes cast. It got 9,882.

Actually, Williamson County never really obliterated its civil-political townships from the landscape. People have continued to use the names and to understand the former townships' boundaries. See, for instance, the reference in the *Marion Republican* to "East Marion" Township in a 1935 article; the vote outcome for "Carterville Twp." in the November 1938 election in the *Carterville Herald*; or a 1939 book in which the author frequently referred to nonexistent townships, as in "Thomas Roberts, of Corinth Township, was the first Williamson county man to take active part in state affairs." Look at a 1975 plat book and its countywide maps. All the township names and their boundaries are as they were 1908–1933. Check the voting precincts as of 1993 for the county: Blairsville, Carterville, Grassy, Herrin, West Marion, Southern, Lake Creek, East Marion, Creal Springs, Corinth, Crab Orchard,

Stonefort. The townships live, but Williamson folk apparently prefer the specificity of named townships without the complicated machinery of governing them.[58]

Cass County

As late as 1878, there were seven contiguous counties to the west of Springfield still holding out against the township tide. Adoptions by Jersey (1878) and Greene (1884) reduced that number in the nineteenth century to five: Calhoun, Cass, Menard, Morgan, and Scott. Much of that cluster remains intact today, as the northernmost outpost of a southern tradition.

Woven through the Cass County uncertainty about its mode of local government was a lengthy contest for dominance between the twin towers of Cass politics. The western tower was the bustling Illinois River community of Beardstown, typically Democrat in politics and a bit on the wild side culturally. To the east, a little off center but much closer to the heart of Cass was Virginia, typically Republican and a bit on the conservative side. Here was a classic case for confrontation over the county courthouse location. Beardstown was quite a bit larger, but Virginia was more central to the county's area and to the politics and outlook of the farmers and other small towns. Beardstown won the right to serve as the seat of justice between 1837 and 1839 and then between 1843 and 1867. Virginia prevailed from 1839 to 1843, and it has been the seat since 1867.[59] Even after the 1867 shift to Virginia, Beardstown remained a threat, because of its greater voter base, to wrest prestige and property value from Virginia if Virginians let up their guard. One way to challenge the Virginia dominance was to agitate for township organization, which just might serve to diminish the luster of the county seat and the three-man commission that convened there.

Agitation had occurred many times in the nineteenth century, but the most recent township organization referendum had been back in 1880. The 1901 effort actually began because of a petition originating in the Ashland neighborhood, which lay between Virginia and Springfield—at the other end of the county from Beardstown. Predictably, however, the *Beardstown Illinoisan-Star* favored the change, while the two Virginia papers, *Enquirer* and *Gazette,* opposed it.

Our only window to that year's *Illinoisan-Star* is through the jaundiced eye of the Virginia papers because none of the Beardstown paper of that era survived. According to the *Enquirer,* the *Illinoisan-Star*'s editor was trying to convince his Democrat faithful to vote for change because a supposed "ring" had control of county politics "to the great detriment of the taxpayers and of 'this part of the county in particular.'" In other words, the western tower was not getting its just due. The *Enquirer* lit into the *Illinoisan-Star* by asking the Beardstown editor if he hoped "to inaugurate a sectional war" with such rhetoric. Did he not know how "thoroughly filled" Cass County was then with Democrats?" Any ring would be foolish to favor one locality over another, given (in the view of the *Enquirer*) such widespread distribution of that party's faithful. The *Enquirer* editor admitted township organization might be "well enough for a large county containing two or more considerably large cities," but it would not work in Cass. "Sensible republicans and sensible democrats" in the editor's opinion, would "unite to oppose it." In that same issue, the *Enquirer* did give considerable space to correspondent J. J. Paschal for a careful review of the township system, which Paschal began by noting how few counties still used county commissioners to run their affairs. Paschal made clear how the laws provided for redress if townships, supervisors, and all the other attributes of the new system failed to fulfill expectations. "The issue is worthy of the voters' intelligent consideration," Paschal ended.[60]

Meanwhile, over at the *Virginia Gazette*, the *Illinoisan-Star* found few fans that fall, and neither did township organization. The *Gazette* editor wondered if the fact that, under township organization, Beardstown would send an extra one or two members to the county board had anything at all to do with support there of the innovation. Even Beardstown's Republicans seemed supportive of change. As for Virginia and other areas the *Gazette* served, the concern was increased taxation if the reform should prevail. That issue was so strong, said the *Gazette,* that "several counties in the state...[had] been seriously considering the propriety of getting rid of this cumbersome system of county legislation."[61]

Virginia views prevailed in 1901, as 1,240 voters said no to townships and only 436 said otherwise. Ashland, where the petitions had originated, went for reform, as did one Beardstown precinct. All the other 14 Cass precincts said 1901 was not the time for their county to reduce the west-central Illinois cluster of township-averse counties from five to four.[62]

Nor were the war years propitious for township promotion in Cass. By 1917, when the issue next appeared on the ballot in Virginia, Beardstown, and other Cass districts, Cass was among just 17 Illinois counties still resisting. Township proponents got themselves organized with a meeting in mid-October at the courthouse and decided to focus their campaign on educating the public with "facts and figures." The *Chandlerville Times* had advertised the October meeting and suggested organizers consider using as their slogan something that a "progressive farmer" had shared with the editor: "Be not the first but whom the new tried, nor yet the last to lay the old aside, so vote for Township Organization." Although the wording is rough, the message is clear to me: at least someone in Cass was feeling that his county was lagging behind the majority.[63]

Opposing forces organized that year as well. Someone placed a display ad in the *Enquirer* that probably appealed to the businessmen among the readership. In the ad, opponents of township government cited the "crowning success" of Governor Frank O. Lowden's new administration in the reduction of state bureaus and bureaucrats. "Efficiency in county government will not be reached," said the ad, "by voting for more offices." Township organization meant "at least fourteen more offices." County Commissioner George F. Kuhlman came forward as a township opponent in a long letter to the *Enquirer.* He took a dim view of the charges that township proponents were leveling against him and the other two commissioners. Agitators for change liked to refer Cass voters to Schuyler County, a township county across the Illinois, as a model of local governance, said Kuhlman. Just look at the roads in Schuyler: "Not a foot of oiled roads in the County, and we have one hundred miles in Cass County." Moreover, Kuhlman continued, there are "no better dirt roads in the State than we have in Cass County." Critics were unfair, said the commissioner, to charge certain unfortunate expenses against the cost of maintaining the county's poor. Was it the fault of the commissioners that the poor farm had lost hogs, cattle, and a horse to disease and other maladies? Was the board to blame for a fire that necessitated a big repair bill at the poor farm? Could township proponents fairly lay on the board the cost of battling the recent small pox epidemic?[64]

The Cass County Township Organization Committee kept up the pressure for reform with paid advertisements and at least one letter to the editor. "Why Should Cass County Come Under Township Organization" was the title of display ads that ran in the *Enquirer* on 24 October and the *Chandlerville Times* two days later. The ads said township organization would save money by reducing the number of elections; handling the poor more economically; cutting fees that the treasurer then received for collecting various local taxes; cutting board member salary totals; and making less likely scandalous arrangements for doing the county's business, such as the one involving a printing matter that had recently gone to trial. Voters ought to know, said the ad, that only 17 Illinois counties "still have county government like Cass." That same committee sent a letter to the *Enquirer* and paid for an ad in the *Times,* in response to the Kuhlman letter that I discussed above, in which voters could read point-by-point refutations of his claims.[65]

By a vote of 1,373 to 1,449, Cass voters said no to townships in 1917, and said likewise a year later, despite another protownship campaign. Proponents, in 1918, ran advertisements in county newspapers, partly to counter claims that township opponents were making. No, Cass would not have a poll tax if supervisors formed the county board. Yes, even if Cass adopted, Cass could revert to commissioners. No, Chandlerville would not lose the hard road it was expecting to gain. Yes, road districts around the county would still receive State aid for roads and bridges. The 1918 vote in Cass gave townships a plurality, 1,819-1,407; but advocates of change fell 18 votes shy of a majority. At least the innovation was making progress toward acceptance.[66]

Five years later, every precinct, all 20 of them, from the levee at Beardstown to the high prairie at Ashland, gave a majority for a system that had claimed its first Illinois counties 74 years earlier. Totals showed 3,521 voters were "For Township Organization" in 1923, while only 761 opposed the reform.[67] What could have turned, so quickly, a county of the unconvinced or unconcerned to a county of believers? Taxes truly were the trigger for this last of the skeptics.

Pressure for reform came first in March of 1923 and by September had yielded 10 times the necessary petition signatures. In a letter to the county's assessor and to deputy assessors, the Cass County Farm Bureau, a farmer service/lobby organization, released information that its officials had collected from the State Tax Commission (STC) about assessed valuation on railroad property and personal property like livestock and equipment. STC figures showed Illinois generally assessed railroad property too lightly, compared to property of other owner classes; and Cass County was above the state average with respect to personal property assessment. The Farm Bureau pleaded with the local assessors to "equalize [reduce] this assessment yourselves so that Cass County pays no less or no more than her Just Proportion of taxes" particularly, of course, on farms. When petitions began circulating in August to call for a referendum on township organization, the Farm Bureau pledged, as was its policy, to stay out of the fray. Meanwhile, the STC had held hearings on the Farm Bureau's call for equalization of Cass's tax assessments. In mid-September, the Commission announced it would not "order reassessment of farm lands in Cass" but urged instead "every possible effort" at readjustment locally, especially property in Beardstown and the riverine "duck hunting lands." Wetlands along the Illinois and Sangamon rivers attracted big-city and other out-of-county hunters who paid well for the privilege of downing ducks and other flyway fowl. The Illinois Agricultural Association (Farm Bureau's statewide organization) charged that these profitable acres of Cass County ought to have a higher assessment and therefore pay a higher share of property taxes. Whether it was duck lands, railroads, or something else that fueled the fire, 479 voters signed the petitions that went to the county board in late September 1923.[68] Only one time since then has an Illinois county, under a board of commissioners (but never under a board of supervisors), held a referendum on adoption of township organization; and Cass was the last to say yes.

The campaign for votes generated considerable press interest, especially in the *Republican-Gazette,* whose editor strenuously advocated the township format as a source of tax relief. Owner J. E. Hansell did all in his power as a journalist and publisher to see that the proposition passed. He employed a banner heading, "Commissioner Rule Means High Taxes," over an article claiming that Cass County would have to declare bankruptcy "if township organization loses." Hansell invited an unnamed "Business Man" to author an article that offered salary comparisons between 13 supervisors operating in Schuyler County and one of the Cass commissioners then in office. In the last regular weekly edition of the paper before the referendum, Hansell authored (with a byline) a long piece in which he warned of "eleventh-hour moves" by "those whose position will not permit an argument that will stand up under the critical analysis of the voter." In other words, if a flyer or a newspaper item appeared after Hansell's last pre-election issue, he feared he would not have a fair chance to refute. To counter that problem, Hansell published a special election-oriented paper on 3 November, the day after his regular weekly appeared. Among the items in the special was a mockup of the corrected ballot, telling readers how to vote.[69]

Hansell and the other proponents of change had an ally in the *Chandlerville Times* but an opponent in Beardstown. The editor of the *Times* looked north across the Sangamon at Mason County and found a township county that supposedly had gotten by in 1922 with only $1,250 in supervisor salaries compared to more than six thousand dollars for the three Cass commissioners. "Let's go along with the other 86 [really 85] counties in the state. The change is sure to be beneficial." The *Times* also ran a sample ballot showing voters how to mark it "For Township Organization." Over in Beardstown, however, the *Illinoisan-Star* editor used his paper's daily coverage to hammer home advice of a different sort, with headlines like these: "Vote 'No' Next Tuesday," "Vote for Beardstown, Vote No," "Tomorrow, 'Vote' Against." The *Illinoisan-Star* maintained that, if criticism of commissioners were true, then voters should replace the officeholder and not the office. Beardstown voters ought to remember also how much influence the river city had because of the 7,200 residents therein. Expect, said the editor, for "smaller towns, and the country precincts of the county…to cast a heavy vote in favor of abandoning our present form of government." In the end, the Beardstown editor admitted "the interests of Beardstown in this election may not be identical with those of the remainder of the county." Beardstown voters failed to heed that advice, however, giving township organization 776 votes versus 380 for the old way, and helped provide what afterward Hansell called "a majority greater than ever given a proposition voted on in Cass county."[70]

Township opponents tested the voting public's will in Cass one more time, at the depth of the Great Depression. Leading the Cass County revisionists was the *Beardstown Illinoisan-Star,* with the fighting words "Abolish Township Government" heading its 1934–1935 platform.[71] As for other county papers, their editors were relatively neutral.

The *Illinoisan*-Star kept the issue before it readers throughout the campaign of 1934. In one September piece, the editor referred to the state's Retrenchment Commission and to a new state law providing for less stringent abolition requirements, important developments to which I will return in the section of this chapter to follow. The *Illinoisan-Star* dealt thoughtfully in another pre-election piece with frequently asked questions regarding the transition back to the commission format. What offices would cease to exist if voters opted for the switch? Who would assess property, if not the township assessor? How would the highway commissioners' duties change, if any?

Could the townships remain intact as voting precincts? The *Illinoisan-Star* publicized costs for Beardstown Township, ran comparisons of Cass County with commission counties Morgan to the south and Menard to the east, cited the antitownship efforts of H. S. Hicks, discussed the referendum on township organization upcoming that fall (also) in St. Clair County, and ended with a specimen ballot for Cass showing those (for whom words might not suffice) exactly how to vote.[72]

Newspapers in Virginia and other agriculturally oriented Cass communities had little, editorially, to say that fall on the township topic. Perhaps their editors sensed the growing mood in Illinois and the nation that was critical of this extra layer of local government. The *Ashland Sentinel* provided ample space for supervisors and the county's auditor to defend the township system but issued no opinion as to the preferred outcome. [73] *Chandlerville Times* readers read what the supervisor of Sangamon Valley Township had to say, but not an editorial word came from the editor.[74] J. E. Hansell no longer owned or edited the *Virginia Gazette*. His successor made it a point in 1934 to inform readers that although the paper had, under Hansell, been township government's "chief sponsor...and responsible for the change," the paper would take no sides "in the coming contest."[75] And, besides running several of the same supervisor-authored letters as fellow journalists in Ashland and Chandlerville, the *Gazette* held firm to its neutrality position.

Township organization did not enjoy as much voter confidence as the idea had in 1923, but the ratio of those voting for continuance to those voting against was almost 2:1. Out of 16 precincts, the township side carried all but one, a Beardstown precinct. Editorializing afterward, the *Illinoisan-Star* admitted defeat and thanked the participants in the debate for bringing forth new facts. "The vote," said the editor, "definitely settles the question."[76]

The Critics

Township government weathered a severe storm during this four-decade period and emerged from it with the framework of a statewide defense mechanism in place that would continue to serve the cause of township persistence through the rest of the twentieth century and into the next millennium. Critics hammered at the shortcomings of the township system, decade after decade, with their blows reaching a peak in the 1930s. In fact, if critics were ever to have picked off township counties, one by one, and returned them to the commission column, the decade of the Great Depression was their best chance during the entire century and a half under scrutiny here. I have discussed above the successful Depression-era discontinuance (and the two subsequent affirmations) in Williamson and the 1934 failure to discontinue in Cass. Several other counties voted on the matter of discontinuance in the

1930s, but not nearly as many counties voted as the critics must have expected or desired. Despite the critics' best efforts, including the considerable labors of change agent H. S. Hicks, Illinois's township counties, except for Williamson, stayed the course and moved toward World War II even stronger than they had been in the year 1900.

In the 1900s

If I had been a township clerk in 1910, looking back at my first decade in office for Mythical Township, I would probably have felt confident about the future of my office and that of the thousand-plus clerks who similarly served. Although we had failed to convince voters in Cass (1901), Morgan (1902 and 1907), Randolph (1904 and 1905), Monroe (1907 and 1909), and Perry (1907 and 1909), we had managed to pick up Henderson and Williamson and to defeat efforts in Bond and Henderson to put township officials in those counties out of a job. We had not lost a single county to the commissioners since Johnson reneged more than a quarter century ago, and who had ever heard of Johnson County anyway? By 1910, we had in place a tiny organization to help look after our mutual interests, the Illinois State Association of Highway Commissioners and Town Clerks (ISAHCTC).[77] By 1910, things could have been running so smoothly for my township and for me that I might have overlooked the work of Mason H. Newell.

Newell seems to have been the first of what became a long list of academically connected township critics to emerge in the early twentieth century. In a 1904 article, a version of his master's thesis in public administration at the University of Illinois, he wrote that his research had led him to prefer the commission system, with its three-man board, to the multitude of supervisors and their townships. Commissioners cost less and were more likely, according to Newell, to investigate and deny the "thousand and one petty claims of every conceivable character" that might come before the board. Newell felt that a larger board of supervisors provided too much anonymity for the individual supervisor and therefore meant its members had less accountability to the voters. The town meeting, while originally perfect for New England towns, had outlived its usefulness there and in Illinois, as township populations had often risen to the point of making a full-fledged town meeting impractical. Fewer citizens were choosing to attend the annual meeting, and those who did attend were usually interested more in their own welfare (or that of their relatives) than they were in the good of the townships. Reduce the number of supervisors, suggested Newell. Pay better those that remained. Turn welfare over to the county, as well as the task of choosing township highway commissioners. Newell apparently believed the need for self-government at the local level left room for townships themselves to remain but that much streamlining needed to occur.[78] Maybe W. Rufus Kendall, the Belmont Township (Iroquois County) Clerk, who organized the first meeting of highway commissioners and town clerks, had read or gotten wind, however, of Newell's critical comments.

In the 1910s

John A. Fairlie and H. S. Hicks were players in the political arena of the 1910s. Had it not been for the recommendations of Fairlie's Joint Legislative Committee, township government might have escaped the century's second decade relatively unnoticed. True, there were several local campaigns, in counties like Perry and Cass, where editors and letter writers took the township notion to task. Yes, Illinois established its State Tax Commission in 1919 to look into tax questions affecting, among other entities, local governments. Yes, H. S. Hicks would likely never have emerged as the greatest township critic of all if not for his legislative service of the 1910s and 1920s. As a matter of fact, Rep. Hicks would not have attracted the attention of even the most observant and wary township clerk between 1910 and the end of 1919. But Professor Fairlie would.

Born in 1872, John A. Fairlie was a political science professor at the University of Illinois from 1909 until 1941 and an expert on local government at a time when these governments were beginning to receive careful professional scrutiny. He served as chief clerk of the Illinois Joint Legislative Committee of the Forty-seventh General Assembly (1912–1913) and directed the Illinois Efficiency and Economy Committee (1913–1915). Later Fairlie edited the *American Political Science Review* (1917–1925) and served for one year as president of the American Political Science Association (1929–1930).[79] While the Efficiency and Economy Committee focused on reorganization options for the state's executive branch, and not at all on local government, the main task of the Joint Legislative Committee (JLC) was "to take up the matter of making a general revision of the laws pertaining to County [Commission] and Township Organization and those relating to Roads, Highways and Bridges."[80]

The JLC issued a two-volume report in 1913, and Fairlie was its author. He also published that year, in the *Annals of the American Academy of Political and Social Science,* a synopsis of the JLC report and the list of his recommendations about county and township government. The JLC had concluded that Illinois's "decentralized and unorganized" methods of governing its counties was "entirely inadequate to the complex social and industrial problems" of the 1910s. Fairlie and committee admitted that fundamental constitutional changes were necessary for real reform. For one thing, "annual township meetings" should not be a constitutionally mandated requirement. Both the JLC and the Efficiency and Economy Committee went on record as advocates of a constitutional convention to address that and other issues. Meanwhile, Fairlie said modest improvements were possible by way of statutory reform, such as these. Eliminate township-level assessment and tax collection by letting the county perform those tasks. Use districts for choosing county board members, not the at-large system characteristic of commission counties. Authorize county boards to call unilaterally for referenda on "the question of adopting or discontinuing township organization" so that a reform-minded board would not need citizens or a newspaper to take up the cause of change. Make uniform the number of signatures that petitioners needed collect, so that calling for a referendum to discontinue would no longer need more signatures than one to adopt. Give municipalities exceeding ten thousand residents precedence over any previously existing townships.[81]

Few of the JLC recommendations took root; but, had I been Mythical Township's clerk, and received one of the thousands of questionnaires the JLC sent to Illinois township and county officials, I might have paused to wonder what was in the wind. My questionnaire would have asked, "Has any proposal been made to return to the county [commissioner] system?" "Would a board of county commissioners be more, or less, satisfactory than the board of supervisors?" "What is your opinion of the town assessor, and do you favor a county assessor, as recommended by the special tax commission?" and "What is your opinion of the effectiveness and value of the town meeting?"[82] After perusing a list of questions like these, I might well have suggested that the ISAHCTC put the JLC project on the agenda for our next annual conference.[83]

Like Fairlie, Herbert Stephen Hicks was born in 1872. Hicks lived in Rockford as a child, and then attended the preparatory high school, in Normal, associated with Illinois State Normal University (ISNU). The prep school prided itself, around 1890, as the logical step before enrollment in certain distinguished colleges, like Dartmouth. Hicks thrived in Normal and in his last year served as co-editor of ISNU's yearbook. Instead of leaving Illinois for the East, however, Hicks headed in the opposite direction, to enroll in 1892 at Stanford University. While there, according to the *Rockford Register-Republic,* Hicks was "a prominent figure in activities of school bodies and athletic organizations" as well as a friend of fellow matriculant Herbert Hoover. Hicks, Hoover, and another student whose surname began with the same letter, became the "Three H's" at Stanford. He returned to Rockford after graduating in 1896, passed the Illinois bar exam that year, and began the practice of law.[84]

In 1914, H. S. Hicks (as he referred to himself professionally) embarked on the pathway that would lead him to become township government's most strident critic. He ran, as a Progressive, for a seat in the Illinois House of Representatives, and won the right (along with two other men) to represent Rockford and the rest of the Tenth District in the Forty-ninth General Assembly. Voters sent Hicks back to Springfield to serve a second house term (1917–1918) as an Independent, and a third term (1919–1920) as a Republican.[85] His house service must have been satisfactory to his constituents and to Hicks; and he gave no indication, during those six years, of his distaste for township organization. Among the several bills that Rep. Hicks sponsored and shepherded through the legislative process, only one dealt directly with townships. In 1917, Hicks's H. B. 340 served to consolidate municipal elections in certain counties under township organization so that voters would henceforth choose city officers and township officers on the first Tuesday in April.[86] Looking back, nothing about H. S. Hicks should have raised concern for a town clerk watching out for his township interests during the first three terms of Hicks's general assembly service.

In the 1920s

Except for discussion during the constitutional convention, the 1920s was a decade of relative calm before the townships had to face the dirty thirties. Cass County finally came on board the township ship in 1923. Although never succeeding, Perry County kept trying as proponents kept hoping voters would see it the modern way. In addition, for the first decade since the 1850s, no Illinois county voted on the issue of discontinuance. As a veteran township clerk and keen observer, I might have reported all was well, even though Mr. Hicks had raised my eyebrow a notch and the League of Women voters had made a wave or two, too.[87]

When Hicks returned to Springfield, early in 1921 for commencement of the Fifty-second General Assembly, he did so as the newly elected senator from the Tenth District, set for a four-year term. He ran in November 1924, won again, and served out that second senatorial term. Senator Hicks chose not to run in 1928, however, the year one of the other Stanford "H" men of the 1890s won the presidency.

Hicks was a modestly successful legislator, if we can draw such a conclusion from the percentage of his introduced bills that became law. In the Fifty-third General Assembly, for instance, he batted .235 with four of the 17 he introduced making it through the senate, the house, and Governor Len Small to the statute books. Quite a few of the 51 Illinois senators (19) had a higher success rate; but 12 had no successes at all, and the overall batting average of the senate for that two years was just .210 (113 laws out of 536 the senators introduced). His four successful bills in the Fifty-third dealt with cities and villages (two bills), state reformatories, and schools. As a township clerk, with my ear to the political ground for any legislation adverse to the township cause, I would likely have cheered the failure of his bill (S. B. 125) dealing with township organization, however.

Senator Hicks introduced S. B. 125 on 21 February 1923, to make it easier for a county to discontinue township organization, and it followed one of the suggestions Fairlie and the JLC had made 10 years earlier. Here, in part, is how the Hicks bill read:

> Upon petition of at least one-fifth of the legal voters of any county having adopted township organization, to be ascertained by the vote cast at the last preceding

> Presidential election *or upon a resolution duly passed by the county board,* the county shall cause to be submitted to the voters of such county, at the next general election the question of continuance of township organization (italics mine).[88]

The italicized segment of the Hicks bill was the change he was proposing and very similar to one of the Fairlie/JLC recommendations in the 1913 report. Instead of gathering hundreds or thousands of petition signatures, township abolitionists would, if the bill had passed, needed only to convince a majority of the board of supervisors that a referendum was necessary.

When, on 22 March, Hicks brought his bill back from the Committee on County and Township Organization, it contained an amendment that would have worried the clerk of Mythical Township even more. In place of "one-fifth" of the voters, the committee had inserted "one-tenth," thereby halving the number of signatures that discontinuance advocates would need to collect. As amended, S. B. 125 passed the senate on 12 June 1923, 39-0. Clearly, the senate, predominantly Republican, saw the need for easing the restrictions on elimination of governmental townships.[89]

The bill arrived at the house on the same day as it passed the senate, and went to the House Judiciary Committee. By a vote of 14-0, the committee reported the bill on 18 June but with a major alteration. The committee struck the phrase "or upon a resolution duly passed by the county board." Therefore, the Fairlie/Hicks idea of making the county board of supervisors an alternative to the petition route to referendum died in committee. After a third reading before the house on 18 June, the entire Hicks bill disappeared; but its author, the senator-lawyer from Rockford, was a batsman that township clerks and other officials would have to face in the coming seasons.[90]

Some time after he gave up his Tenth District seat, Hicks became the chief clerk for the State Tax Commission (STC). The State Archives holds a spotty collection of materials from the STC, and no records from 1929 or 1930. I cannot confirm that Hicks was an employee of the STC until his name appears on an STC document dated 22 January 1931. In any case, the STC had, since its creation in 1919, engaged in the business of supervising real estate and property tax assessment, assisting local assessors, prescribing rules and regulations associated with assessment of railroad property, and equalizing property assessment across the state. Townships, road districts, and small school districts became prime STC targets; and chief clerk H. S. Hicks was up to the task.[91]

Women began voting on all candidates and issues in the 1920s, because of the Nineteenth Amendment. To enlighten women, who might never have voted or attended to political matters, the League of Women Voters came into being in the early years of that decade. Then, as now, the League developed agendas for study by its members, conducted those studies, and sometimes took positions on the issues after study and debate. The *Bulletin* (of the) *Illinois League of Women Voters* carried an article in its June 1922 number entitled "County and Township Government." In it, author Victor S. Yarros told League members that the county was the most poorly run unit of government in the country and noted that "Not a little of the foregoing will be found to apply to the township system as it is known in the Middle West." A Chicago resident, Yarros used a number of Illinois examples to make his case. Townships, he said, had too few functions and too many elected officials. Since the average citizen cared little about town meetings, he suggested ending them, as well as reducing the number of officials and forcing townships to operate effectively. That same number of the *Bulletin* carried a series of questions for women to ask as they fanned out across Illinois to pursue their government research agenda. Who are township officials? What are their duties? When will the annual town meeting occur? Who is likely to attend the meeting and to vote? What sort of business will transpire at the meeting? Are townships and their future a local topic of discussion?[92] In the coming decade, the League would continue to question whether townships belonged on the Illinois landscape.

In the 1930s

I am not sorry that I missed the Great Depression, but I have always greatly admired the endurance that my parents and other Americans demonstrated between 1930 and 1939. Government, at all levels, was part of the solution to the hardships they endured but also part of the problem. Local government received a tremendous amount of criticism, mainly because of the role these units played in taxing their populations. All sorts of individuals and organizations eagerly provided advice on how to clean up the situation and, in the process, reduce taxes and maybe improve governmental services. H. S. Hicks, eventually an agent of the township-critical Illinois Chamber of Commerce, was the most obvious individual critic; but many other names deserve mention, including a former governor. Illinois's League of Women Voters kept governmental efficiency on its study agenda throughout decade. State commissions conducted and sponsored research on governmental topics and recommended changes. State legislators reacted to these recommendations and other criticism with legislation and more investigations. Townships and township officials continued to persist.

H. S. Hicks devoted much of his life, after leaving the state senate, to counteracting what he often called "waste in local government." Although he still referred to himself as an attorney, he may never have practiced law in the 1930s. When he died in 1936, he had to his credit nearly a dozen reform-oriented publications; and his name should then have been as much on the minds of township officials as it was with those who shared his passion for governmental reform. If anyone, of all the critics that township organization has inspired, from the 1840s to the present, deserves recognition as a change agent bent on reversing the township tide, it must be H. S. Hicks. He carried a message and made his living being the messenger. That message was to eliminate townships as well as the various other little taxing units that had littered the landscape for so long. Many township critics stood with him in the 1930s, but he stood out above them all.

Omer N. Custer, chairman of the State Tax Commission and former state treasurer, signed the STC's minutes of 16 July 1931. Hicks, the chief clerk, was present and attested to the authenticity of the chairman's signature. The Commission had met that day to discuss "the subject of making a survey of the counties of the state to ascertain the results of local assessors." Custer and the other two commissioners decided that the chief clerk should seek data pertinent to the comparison of "Real Estate assessments in each township of the state." If Hicks had not already begun his crusade against townships, this charge from the STC made it official.[93]

In the next few months, the fruits of Hicks's research began to reach the public eye. Under the auspices of the STC, probably early in 1932, he distributed a mimeographed report entitled "The One Room Schools in Illinois." In it he argued that, as population declined in rural areas, a higher percentage of farm property taxes would have to go to support the one-room schools. He suggested reducing the number of school districts to only one per county, so that schooling costs would decrease, as duplication ceased, and so the much-coveted railroad property tax revenues could help fund all the county's education efforts. *Why?* was the title of his other STC brief from that period. Hicks had been comparing property taxes in those Illinois counties touching Wisconsin and Indiana with property taxes in nearby out-of-state counties. He found Wisconsin counties were getting by with less tax burden on the ordinary property owner than was the case south of the Wisconsin-Illinois line. When he looked at the Indiana-Illinois border counties, he found taxes were higher per capita in both states than in the Wisconsin counties. Hicks could not understand why his state and its neighbor to the east depended so much on property tax to support local governments while Wisconsin was tapping other sources of tax revenue, like excise and special taxes. "An intensive study of why these wide differences exist in this group of three of the best States in the Union might be helpful in pointing the way to tax relief for the property owners in Illinois and Indiana."[94]

In these two STC reports, Hicks avoided any reference to the civil-political township. That circumstance was soon to change, however. The *Illinois Journal of Commerce* carried a short Hicks article in October 1932, the title of which was a question that should have gotten the attention of township officials: "Is the Township an Obsolete Form of Local Government?" While admitting that cutting expenditures would help reduce property taxes in Illinois, Hicks said he believed those who had "given the subject of taxation serious study" knew a better way to begin was to reduce the "increasing number of minor political divisions which now exist." Responses (to a Hicks inquiry) from county clerks had yielded information that allowed Hicks to count 17,157 tax-levying entities in the state, of which over fourteen hundred were townships. Local activists were already at work, as I have discussed earlier, that fall of 1932, trying to reduce the number of taxing districts via the referendum route. "The first point of attack," wrote Hicks, "has been the Township." He defended his position by elaborating on how 16 commission counties were getting by nicely without the township. In certain township counties, by comparison, each township could levy whatever tax rate the officials desired, because there existed "no limit fixed by law," unlike "all other taxing units of the State," for which there were statutory limits. Hicks provided data that showed in the 16 commission counties, the per capita tax levy for 1931 (excluding that for schools, municipalities, and other minor subdivisions) had been \$2.21 and in 10 township counties was \$4.96. Anyone wishing to argue that townships were outdated in an "age of quick communication and rapid transportation" could point to the fact that 16 Illinois counties were then "functioning satisfactorily without the additional cost of township taxes." It is only fair to note, however, that, while the 16 may have been getting along without townships and township taxes, Hicks did not support his statement about them performing any more "satisfactorily" than the other 86 counties.[95]

The Hicks project that led to his *Illinois Journal of Commerce* article found a different and probably wider audience via a newspaper story that originated with his hometown paper, the *Register-Republic,* in September 1932. A reporter used Hicks's study and new quotations from Hicks to build an argument in which the writer claimed seven counties were voting that November to discontinue. For the *Register-Republic*, Hicks provided several paired comparisons of counties where one half of the pair was a commission county and the other a township county. "The counties of Menard and Kendall are nearly similar," Hicks told the reporter, "as any two in the state...[yet] the per capita cost of operating Menard this year is \$4.08 and Kendall \$6.54." Wabash bested Boone; and Perry was cheaper than Clinton, in similar head-to-head cost comparisons. Hicks said in the interview that he had found no "movement or desire" in the 16 commission counties "to make any change," while there was "a vast amount of dissatisfaction in counties operating under the township system." In the mail that very day, Hicks had gotten letters, he said, "from all parts of the state asking for information regarding the steps necessary for a change." This Rockford news article went out on the press wire; and, over the next few weeks, appeared in papers as far away as Williamson County."[96]

Hicks probably left the STC in late 1932 or early 1933.[97] By 1934, he was working for the Illinois Chamber of Commerce. He apparently published nothing during 1933; but, beginning in 1934 and extending beyond his death in 1936, he authored a string of township-condemning articles and booklets. His two articles both appeared in the state Chamber's *Illinois Journal of Commerce.* The booklets were either a joint effort between the state Chamber and a local chamber or strictly a local effort.

First of that final flurry of Hicks publications to appear was a study of his home county, in collaboration with the Rockford Chamber of Commerce, a study that provided the model for similar countywide studies that followed. The Winnebago item began with a discussion of the "multitude of governmental agencies," to which the "average citizen" supposedly responded with astonishment "when he learn[ed] there [were] over 17,000 separate and independent minor political subdivisions existing in Illinois, each with the power to levy taxes." Winnebago had then 153 taxing units compared to only 50 in a similar (in area) Blair County, Pennsylvania. "Winnebago County," he said, "does not need fourteen townships nor does it need fourteen road districts—neither do we require one hundred and two school districts to meet our educational needs." Using maps, graphs, photographs, and line drawings, Hicks set out to show how much taxpayers could save in Winnebago County if they would only take the proper action. Hicks felt ignorance of local government lay at the heart of this 1930s dilemma, and he went so far as to write a quiz that he administered to "businessmen, club women and high school seniors" in his and nearby counties. The results were dismal. No respondent, for instance, could tell Hicks how many townships were in the respondent's county. Hicks suggested new laws to aid in the elimination of tiny road districts and similar school districts; but, because protocol already existed to permit elimination of townships, he urged "a campaign of education" to show the people "the advantages which they may secure through the abolition of the township as a unit of local government."[98]

A pair of short articles in the *Illinois Journal of Commerce* constituted his next educational effort. The September issue carried a synopsis of his Winnebago booklet and came out at the time petitioners were trying to get the abolition question on the November ballot across the state. "Everything but the actual [Winnebago] figures," said the *Journal* editor, "can easily be applied to your county—and to your township taxes." Hicks called large boards of supervisors, such as still existed in New York, Michigan, Wisconsin, and Illinois, relics "of the Colonial period." Elsewhere, he claimed, efficient boards of commissioners were devoting more or less "full time to the business of the county." Hicks cited former governor Frank O. Lowden as a reform-minded man who had been advocating elimination of townships. Lowden, Hicks, and "the thinking citizens of Illinois" were trying to remodel local government to right the wrongs Hicks had discovered in Winnebago. In a sidebar, "Some Questions Answered," Hicks explained, with useful detail, how voters must petition for an abolition referendum and what would transpire if they voted out townships and supervisors. The following month, a Hicks article about railroad taxes appeared. In it, he lamented "the gross injustice and inequity" of the taxing system. At the root of the problem, of course, were the townships. Some townships collected half their tax revenue from the railroads while townships elsewhere in the same county, without a railroad, "got nothing from this important source of revenue." Hicks said abolition of townships, as taxing units, would help remove the inequity because then all parts of the county would benefit from railroad taxes that the county levied and collected. Eliminating one-room school districts, to provide for a single countywide educational district would solve the same problem that he had found with separating rail-poor and rail-rich townships. Little hope existed, however, unless taxpayers understood that such change would not impair services and would lower taxes. "To secure this result," said the change agent, "a statewide campaign of education is necessary."[99]

The Illinois Chamber of Commerce published, co-published, or assisted Hicks with dissemination of his five studies of governmental waste dated 1935 or 1936. All but one of them focused on a single county, as had the 1934 Winnebago booklet. The exception was his 1935 look at road district waste in a dozen counties.[100] The single-county efforts each carried a foreword from J. Paul Clayton, who chaired the Taxation Committee of the Illinois Chamber of Commerce. Clayton explained how the chamber had a longstanding interest in tax problems and how the chamber would loan "the services of Mr. Hicks" to local chambers so that he could conduct his fieldwork and do his writing. Hicks used his time well. The booklets flow smoothly and hammer home his points about unnecessary townships, township road districts, other small road districts, one-room school districts, and small high school districts. Even though Hicks had a transferable framework for the county studies, much of each booklet was new material. Except for a repeated map and related text allowing comparison of the number of Illinois governments with numbers in other states, graphics, tables, and text are particular to each county. In the Macon, St. Clair, and Sangamon booklets, cartoons by E. Ward Blaisdell liven the pages. Hicks used Blaisdell's drawings of farm animals and other items of personal property to illustrate how variable the tax assessments were in the same county.[101]

Herbert Stephen Hicks died on Monday, 3 February 1936, in Decatur, as the result of injuries he sustained the previous day in the line of duty. He had been northwest of Decatur on Sunday morning taking pictures of Warrensburg High School and other structures for the Macon County study. Returning to Decatur, Hicks crashed head-on into a vehicle turning left across his path and crushed his chest against the steering wheel. Because he had almost finished the work on Macon, the state and local chambers published it for him, as was also the case with the Kane study. The Macon booklet's foreword said that work and other publications of his would "remain as monuments to his tireless efforts on behalf of Illinois taxpayers." Clayton, in the foreword to the Kane study, called Hicks one of Illinois's "ablest and most conscientious citizens, and probably its ablest student of taxation." In an editorial tribute to a native son, the *Register-Republic* praised him for his public service and willingness to take a stand on something in which he believed: "An acknowledged expert in his province, he provoked thought because he was a thinker, and disagreeing critics admired the fullness of his acquirements even if they disputed his dicta."[102]

Did Hicks, the change agent, have any effect on the political landscape of Illinois? The only defection to the commission column during his campaign for reform occurred in 1932, just as the crusade was beginning. I find nothing in Williamson's 1932 decision to revert to commissioners that I could attribute to his effort. He may have played a background role in getting abolition on the Winnebago ballot that same year; and surely his work played a part later in the decade as several counties checked signals at the ballot box in 1934, 1935, 1936, and 1938. Township-sized road districts remain in most Illinois counties; but a few counties, like Williamson, have since gone to unit (countywide) districts. On the other hand, Hicks may have been at the vanguard of the movement to consolidate school districts, a process that mainly occurred right after World War II. From over seventeen thousand, the number of Illinois tax levying districts has dropped to around seven thousand, still too many, say some, but a lot fewer than when H. S. Hicks began his long struggle to save Illinois from too much self-government.

Hicks was not alone in his 1930s crusade against little governments, as academics of various stripes and other well-read and well-respected individuals took a look at township government in Illinois and elsewhere—and generally disliked what they saw. Agricultural economist T. B. Manny believed civil townships, as units of local governance were "universal failures." He felt the problem lay in the fact that township boundaries did not match the social fabric of rural areas and recommended abolition of townships and establishment of large rural units, exceeding even existing counties in area.[103] Frank O. Lowden, former Illinois governor, proudly noted that during his gubernatorial tenure in the 1910s, Illinois had abolished most township tax collectors and saved taxpayers, since doing so, more that two million dollars. Township road commissioners and assessors ought, Lowden thought, to follow the collectors out the door.[104] Paul W. Wager, a University of North Carolina professor, called for the immediate elimination of the township, outside of New England: "The fact that twenty-five states have never found any need for townships and eleven others have never given them any important powers would seem to suggest that townships are not indispensable in the states in which they exist."[105] Arthur W. Bromage, a University of Michigan political scientist, found the township unsuitable to the 1930s in many ways. He felt it should immediately lose its right to send its supervisor to sit on the county board. Where the township still was a unit of local justice, citizens could serve their needs better by opting for county trial justices and sheriff patrols. Township roads ought to be under county control, and Bromage called for "scrapping of the township upon the political junkheap."[106] M. H. Hunter and H. K. Allen, professor and doctoral student in the Department of Economics at the University of Illinois, looked at various dollar indicators in Illinois's township and commission counties. Although

Hunter would not state flatly that all counties should dispense with townships, the professor said that findings pointed toward the conclusion "that the township is neither a necessary nor an economical unit" and "that township organization add[ed] to the cost of government without the addition of a commensurate service." Allen, in a brief summary for the *Illinois Journal of Commerce* of his and Hunter's research, said, "It is pretty generally agreed that [the township] does not adequately meet the requirements of existing economic and social conditions."[107] L. D. Melton, Director of the Research Division for the Oklahoma Tax Commission, reported on the indirect method that Oklahoma had used in the 1930s to abolish townships and noted how much money taxpayers had saved—without apparent decline in services. Melton felt Oklahoma's method might transfer successfully to other states and "force townships from the scene."[108]

While academics and others were criticizing local government during the 1930s, the League of Women Voters was busy doing its homework. The Illinois Chamber of Commerce had concluded, as had many academics, that townships ought to disappear, in order to improve efficiency and save taxes. From its programs of study, however, the League drew less draconian conclusions and never, during this decade, really came down on the side of abolition.

Marietta Busey Tawney was a key figure in the Illinois League's governmental studies. Tawney chaired the League's Department of Government and Its Operation and, within it, the Structure and Functions of Government Committee, one of three committees in that department. Her papers provide good insights into the ways the League handled the township issue. First, members excerpted research that others were conducting, such as the 1932 Griffenhagen study for the Illinois Commission on Taxation and Expenditure (see below), the Hicks press release that led to the *Rockford Register-Republic* story, and other articles around the state about the seven counties supposedly voting on abolition in 1932 (see above). Second, chapters of the Illinois League conducted studies of local governments in their areas and funneled to Tawney their findings. She, in turn, summarized and made recommendations for the state convention of the League. Third, the League adopted positions regarding topics that others and their own members had studied. In the Program of Work for 1934–1935, the League supported, under Tawney's department, "The elimination of unnecessary governmental units." Fourth, the Illinois League assisted the National League of Women Voters with content and distribution of informative pamphlets, such as *Know Your County,* from which League members could learn more about townships and about the sorts of questions they should be asking in their local investigations. Finally, the Illinois League, as has become even more common recently, sponsored a public debate (about the 1932 abolition referendum in Winnebago County). As the decade came near its end, Tawney's Department of Government and Its Operation was still advocating elimination of local government units, but never specifying which ought to disappear.[109]

Had Frank O. Lowden been Illinois's governor during the 1930s, the State might have given township government even more official grief than it actually received from Springfield. But Lowden was only an outsider by that time, another critic making speeches and writing antitownship essays. His successors in the governor's mansion had other battles to fight and left it to the legislators, their commissions, and their consultants to decide whether townships ought to remain strong. As it turned out, townshippers had little to fear from the general assembly, leading me to believe that the affinity, which today characterizes the connection between township government and state legislators, was already in place by the 1930s.

At least three commissions scrutinized townships during the decade: the State Tax Commission, the Illinois Commission on Taxation and Expenditures, and the Retrenchment Commission. I discussed the STC earlier in this chapter, in connection with the 1930s work of H. S. Hicks. Other than its routine monitoring of township budgets, receipts, and expenditures, the STC apparently took no major part in the criticism of township government after the departure of Hicks for his Illinois Chamber of Commerce position. Both the ICTE and RC were ad hoc creations of the general assembly, both gave township government a close look, and neither survived the thirties. Townships, of course, did.

The ICTE came into existence on 29 June 1931, when Governor Louis F. Emmerson signed into law S. B. 624. That commission was to "study the organization for State, county and all local government functions" and suggest ways to simplify and eliminate overlap "so as to reduce the cost of government." ICTE findings were due to the governor and general assembly in 1933; and, on 24 January of that year, its *Report and Recommendations* appeared.[110]

The ICTE had created a Subcommittee on Local Government and asked its members to consider "the matter of merger and simplification of local governments in Illinois." Given a budget of between ten and twenty thousand dollars, the subcommittee hired Griffenhagen & Associates, "Specialists in Public Administration and Finance," to conduct a detailed study and prepare a research report. That report came back to the subcommittee in two parts, late in November of 1932. One part covered Chicago and Cook County; the other looked at the rest of the local governments in the state.

Griffenhagen, a national consulting firm with much experience in such matters, said in the foreword to its report that simplification of governmental structure was that firm's guiding principle and that more than "90 percent of the proposals for simplification and correction of the structure of local government" (in their report) was feasible without constitutional reform. The foreword went on to comment on how the Illinois taxpayer of that difficult period cared little about how government operated but a lot about taxes.

Change, the report said, would mean objections from those who feared a negative impact and from those who looked upon change as having "a sinister motive." In other words, Griffenhagen was warning the subcommittee that resistance was inevitable. Abolitionists, who in recent years have taken on township officials and TOI in attempts to dismantle the township structure in various Illinois counties, would probably find the final sentence of the foreword to part 1 of the report as pertinent now as Griffenhagen did in the Great Depression.

> It is true that those who are the beneficiaries of the existing system and those interests that will be adversely affected by the simplification of the structure of government are likely to be more vociferous in their attempts to perpetuate the existing scheme than the beneficiaries of constructive improvement plans are in support of such plans.[111]

Griffenhagen found little about township organization to like. The town meeting was "unworkable and unwanted." Management of county affairs by a horde of supervisors, though "stubbornly retained," was "poorly adapted" to the modern era. "Plainly stated," the report said, "township government…is irresponsible government, administered by untrained personnel, and not well conducted." Function by function, Griffenhagen recommended turning over township duties to the county, to the state, to anybody, leaving nothing for any township official to do except the supervisor who would ex-officio serve on the county board. Given guidance from the United States Supreme Court and subsequent general assembly action in the 1960s, that ex-officio duty would disappear as well. Meanwhile, before a constitutional overhaul, Griffenhagen wanted the general assembly to make it easier for a township county to rid itself of township organization, by reducing the number of petition signatures that citizens needed to collect in order to call forth a referendum. Griffenhagen suggested 100 signatures, instead of the prevailing 20 percent of the number who had voted in the county at the last presidential election.[112]

The Subcommittee on Local Government concurred with much of what the consultants proposed, as did the ICTE. Both the subcommittee report and final ICTE list of recommendations urged the general assembly to provide for the 100-signature stipulation. Here is how the ICTE report read: "It is consequently recommended in the report that it be made easy for those counties now operating under the township form of government to adopt the non-township form by a vote on petition of 100 legal voters."[113]

Rep. Elmer J. Schnackenberg, a Republican from Cook County, the man who had chaired the Subcommittee on Local Government, introduced H. B. 296, on 9 February 1933, to the Fifty-eighth General Assembly. His bill called for requiring just 100 signatures on a petition to hold a referendum on discontinuance. Co-sponsoring the bill were two other ICTE members, Rep. M. F. Browner of Pulaski County and Rep. Michael F. Hennebry of Will

County. After H. B. 296 had gone to the Committee on Efficiency and Economy, it came back for subsequent readings before the house as a whole. For some reason, perhaps due to pressure from township officials or the ISAHCTC, on 14 March, Schnackenberg offered an amendment, upping the number from 100 signatures to a figure equal to at least 5 percent of the number voting at the previous presidential election. With that amendment in place, the house voted 114-0 to approve. In the senate, the township proponents had an influence, too, because senators further amended H. B. 296 so that the 20 percent rule for petitions to call a referendum on discontinuance came back to life for counties wherein 75,000 or more residents voted in the previous presidential election. Counties falling below that figure would need only 5 percent to call a vote. Senate approval, house reapproval, and the governor's signature came in June of 1933.[114]

Did the change make township abolition easier after June of 1933? For the one Illinois county (Cook) where 75,000 or more residents voted in the 1932 election, there was no change in the law that had prevailed since the nineteenth century. For the rest of the 102 counties, the 5 percent rule now prevailed; and 5 percent of anything is a 75 percent reduction from 20 percent of the same thing. Therefore, in Cass County, for instance, where 8,367 voted in the 1932 presidential election, the proponents of the commission system in 1934 would have needed only 419 signatures compared to 1,674 before H. B. 296.[115] Of course, if the Griffenhagen and ICTE recommendations had survived the amendatory process in the Fifty-eighth General Assembly, only 100 would have sufficed in Cass or any of the other 84 township counties. Overall, H. B. 296 represented a modest victory for the forces aligned against township organization.

The Fifty-eighth was not through with studying local and other governmental units. On 10 May 1933, as H. B. 296 wended its way to passage, Rep. James T. Burns, a Democrat from Kankakee County, was introducing legislation to authorize yet another commission. Burns's H. B. 906 (as signed into law on 8 July) created the Retrenchment Commission, calling for the commissioners to recommend "consolidation or abolition" of governmental offices. It did not mention any units of government as targets, however. A report from the commission was due during the next general assembly's biennium. Perhaps out of deference to its retrenchment mission, the budget for the RC was only $1,000, compared to the $25,000 with which the ICTE worked a couple years before.[116]

Burns was a Kankakee attorney, who pursued his legislative duties with vigor for four terms in the house. During 1933 alone, he introduced 60 pieces of legislation, including H. B. 906. Of the 60, 21 became law. Burns had been a governmental reform activist before suggesting the RC. In a 1932 letter to Marietta Busey Tawney, responding to her letter regarding his position on local government matters, Burns said he was familiar with Lowden's "stand with respect to efficiency in government." Burns admitted to Tawney that his intention was to effect efficiency by eliminating several township positions, including assistant supervisors, trustees, clerks, and treasurers. He confided to her that he expected "a stiff fight against the abolition of any office and the reduction of any salaries by so much as a dollar." If the Illinois League could help his cause, he said it would "be well for you to get on the scene [get into action] right away."[117]

When the RC held the last of its 14 sessions and concluded its work on 15 September 1934, Chairman Burns issued a press release. "There was a good deal of talk," he said, "on both sides, as to the possibility of abolishing all township government." His commission had also discussed consolidation of counties and elimination of numerous municipal officials, executive departments at the state level, and the lieutenant governor." The editor of the *Beardstown Illinoisan-Star* took the Burns release as a sign that the legislature might save counties even more trouble by just voting to eliminate all the state's townships in the same manner that the Fifty-eighth, with H. B. 296, had made "it easier for any county to vote on the question of abolishing township organization." The board of supervisors, said the *Illinoisan-Star,* had already done all the retrenching it could in Cass County under the "present form of government" and yet had "not materially lowered the tax bill."[118]

A year after his press release, Burns issued the final report of the RC. In his foreword, he chastised government officials for extending their power at the expense of the taxpayer and suggested Illinois's governmental costs could decrease by 50 percent "without impairment of efficiency or curtailment of worthwhile services." Among those who had appeared before the RC were township critics, like John A. Fairlie and M. H. Hunter from the University of Illinois, and advocates, like "Messrs. Ayers and Gillespie from the Association of Town Clerks and Highway Commissioners." The RC report called for cuts from Springfield on down. Even the secretary of state was excess baggage. In its section entitled "Consolidation," the report urged county consolidation, saying simply, "There are too many counties." Township abolition and adoption of the commission form should pose much less of a problem, said the RC, because Illinois citizens already had the right to petition and vote for discontinuance. Above all else, the RC urged the calling of another constitutional convention.[119] As far as I can discern, the RC had even less impact on township business than had the ICTE.

Survival

To read in *Township Perspective,* the current name for the official TOI magazine, about town clerks and other officials serving their townships for decades is commonplace.[120] Mythical Township's clerk might well have been at that post for the first four decades of the twentieth century. If so, maybe in the late 1930s, the time would finally have come to step down.

What sort of parting thoughts might I, as that town clerk, have had, as the world stumbled toward World War II?

I would likely have left my post heaving a great sigh of survival relief. Rep. Burns had departed the general assembly without affecting us in any way that I was able to notice. Hicks was no longer harassing us. The ICTE and RC had reported and expired. Fairlie would soon retire from his teaching post. The League of Women Voters was still conducting its studies, but not condemning us to the scrap heap. Williamson's rejection of townships in 1932 was, though unfortunate, apparently an isolated incident and not, as some of my colleagues believed, a bad sign of what might happen closer to home. All the other attempts to abandon the township for the commission system, including one in Ogle County in 1938, had failed, despite the turbulence and tragedy of this crazy decade.

On the other hand, our little organization of township officials had grown and prospered. Lake County was the first Illinois county, in 1930, to have every one of its townships join ISAHCTC. We hired, also in 1930, our first lobbyist; and, at his insistence, we now understand better "the need for officials to maintain close contact with their legislators." After 1932, the ISAHCTC included supervisors, assistant supervisors, and assessors in its membership. We retained our first attorney in 1933, a former legislator. By 1934, we were attracting state officers and sitting legislators to our statewide meetings. Our member townships topped two hundred by 1936, and we showed our unity by protesting, en masse, state legislation affecting poor relief. We changed our name that year to the Illinois Association of Township Officials (IATO), to reflect our inclusion of other offices. Hugh Cross, Speaker of the Illinois House of Representatives, addressed our 1939 convention; and 1939 was the year township tax collectors joined IATO.[121]

I might have laid aside my 40 years of township record books wondering what the next four (or five) decades might bring. Would our critics ever let us rest and do the people's work? Would other counties consider reversion to county commissioners? Should IATO cozy up so obviously to senators and representatives? Did the township tide really die in 1923?

Notes

1. Since Clark County, Illinois, voted in June of 1879 to retain its townships, all countywide referenda on the township issue have occurred in November. A new referendum wrinkle in this era, however, did involve a non-November vote. In April 1902, Chicago residents voted to disable their governmental townships. It would be 70 more years before another partial county referendum on township organization occurred in Illinois.
2. That the other 90 counties did not also participate in countywide decision-making about whether to have commissioners or supervisors, however, does not mean change agitators were absent in some of those counties. In Carmi, for instance, A. W. Charles, the editor

of the *White County Democrat,* pushed for an abolition referendum there in 1906 and 1907. Charles ran a newspaper in a county seat in a county that had not voted on the idea of township abolition since 1876. He used his newspaper as a platform from which to attack the board of supervisors repeatedly as he tried to make a case for a referendum to redress a situation that he said had become "burdensome and expensive." Editor Charles called the board to task because of how long they met and how much the supervisors earned for their service: "The county board met…3 hours…at two and half per day. This is 83½ cents per hour, which is considerably in excess of salaries of county officers." Railroad companies, he said on 22 November 1906, should refuse to pay their property taxes because of illegal township levies. On into March of 1907, Charles continued to jab at the supervisors: "If there is anything more powerful than kings and potentates, than law and gospel, it is the supervisors." He called the town meeting concept into question as "meaningless and useless," and concluded that just because several counties had revoked the commission format for townships meant nothing other than the fact that they had "jumped out of the frying pan into the fire." Even though Charles labeled his paper the "Official Democratic Organ of White County," he left little doubt that supervisor incompetence and gall did not prefer one party to another. A couple of weeks before the 1907 general election, when four counties were to vote on the issue of shifting to townships, Charles issued one last pronouncement in his year of fury against supervisors: "If White County could be cleaned of saloons and township organization, its county orders would be negotiating at par in five years. In other words, White County' s promises to pay were meeting with less than perfect confidence on the part of potential creditors. Carmi Township and the other White County townships survived the Charles assault, and they remain intact today. *Carmi White County Democrat,* 22 November 1906, 4; *Carmi White County Democrat,* 13 December 1906, 4; "The Big We," "Investigate the Supervisors," "Township Organization," and other untitled items, *Carmi White County Democrat,* 21 March 1907; *Carmi White County Democrat,* 24 October 1907, 4. Two counties to the west of White County, in Franklin County, in 1932, voters almost had a chance to vote out townships. Petitioners gathered more than four thousand signatures asking the board of supervisors to certify an abolition referendum for the November ballot. Dr. Joe E. Moore, the man who led the abolition campaign, believed he and his compatriots had enough signatures to exceed 20 percent of the total residents who had voted in Franklin County at the 1928 presidential election. Moore, however, had submitted the petitions eight days beyond the official filing date. When the board, citing that shortcoming, failed to certify for the ballot, Moore' s attorney' s filed a writ of mandamus that demanded the board give voters a chance to decide the question. Circuit Judge C. H. Miller ruled on 24 October that the question of abolition would not appear on the ballot that year. Given the fact that neighboring Williamson County did vote in 1932 to discontinue its township format, one can speculate that Franklin might very well have reverted to commissioners if Moore had filed in a timely fashion. "Supervisors Hold Meeting Yesterday," *Benton Evening News,* 14 September 1932, 1; "Mandamus Writ Aimed at Supervisors, *Benton Evening News,* 19 October 1932, 1; "Will Not Vote on Change of Government," *Benton Evening News,* 25 October 1932, 1. Hicks said in September 1932 that he knew of abolition efforts cooking then in Wayne, Kane, DuPage, and "other counties" in addition to Winnebago, Williamson, and Greene where petitioners were successful at ballot access, and Franklin, where they failed. "7 Counties Vote on Commission Plan Nov. 8th," *Rockford Register-Republic,* 20 September 1932, 14. A week before the Rockford piece ran, Illinois Attorney General Oscar E. Carlstrom (on 13 September) had issued an opinion concerning ballot access on "the question of the continuance or discontinuance of township organization" in DuPage County that fall. Assistant State' s Attorney Russell W. Keeney wanted to know the deadline for filing petitions and whether the proposition could go to the voters "as a matter of public policy under the Act of 1901." Carlstrom told Keeney that the county clerk needed to give the voters 30 days notice prior to the referendum and that there were special rules pertaining to discontinuance of townships, so DuPage could not take it to the voters as a matter of public policy under the 1901 Act. Illinois, Attorney General, *Illinois Attorney General's Report and Opinions* (Springfield, 1932), 540–41. DuPage petitioners would have needed fewer signatures than agitators collected in Winnebago County that year.

3. Howard, *Illinois: A History of the Prairie State,* 456. The state and county superintendents were constitutional officers and therefore still off limits after 1891 to women voters.
4. Illinois, General Assembly, *Laws of the State of Illinois Enacted by the Forty-eighth General Assembly at the Regular Biennial Session* (Springfield, 1913), 333.
5. Later in this chapter, I will be discussing the League of Women Voters, an organization that has done much to educate women and all voters about governmental matters, using tools such as voter handbooks. For a pre-League handbook, see Alice Greenacre, *A Handbook for the Women Voters of Illinois* (Chicago: Chicago School of Philanthropy, 1913). Greenacre included a section on county government and in it discussed the differences between township and commission organization.
6. "Women Use Vote to Deal Saloon Blow in Illinois," *Chicago Tribune,* 5 November 1913, 1.
7. "Sample Official Woman' s Ballot," *DuQuoin Evening Call,* 3 November 1913, 4; "Tuesday' s Election," *Pinckneyville Advocate,* 7 November 1913, 1; "Pinckneyville and Willisville Are Both Arid Districts; DuQuoin is the Only Oasis Left in Perry County for the Thirsty," *Pinckneyville Advocate,* 7 November 1913, 4. Women also voted on the issue of adopting township organization for Perry County in 1915, but the local papers did not separate female votes from male on that issue. See, "Official Sample Ballot," *Pinckneyville Advocate,* 29 October 1915, 4; "The ' Drys' Win at Last Tuesday' s Election," *Pinckneyville Advocate,* 5 November 1915, 1.
8. "Township Organization," *Marion Egyptian Press,* 10 September 1914, 1; "Why Township Organization?" *Herrin News,* 22 October 1914, 6.
9. George W. Young, "Special to the Women Voters," *Marion Egyptian Press,* 28 October 1914, 4; "Women Voters: Do You Want to Be Disfranchised?" *Marion Egyptian Press,* 2 November 1914, 1. This latter item, a small ad measuring perhaps two inches by three inches, may have been the first display advertising about the township issue to appear in an Illinois newspaper. Such display ads had not appeared, to my knowledge, before 1900; and they became quite common after World War I. Attribution of such ads to a group or individual was sporadic in the 1910s.
10. "Women Invited to Vote Over the Township Board," *Marion Leader,* 3 November 1914, 1; "Total Vote of County Given Out," *Marion Evening Post,* 1.
11. "Women' s Specimen Ballot" and "Men' s Specimen Ballot," *Virginia Enquirer,* 31 October 1917, 8. Even though Perry County allowed women to vote on township organization in 1913 and 1915, women apparently did not enjoy that privilege there in the 1916 Perry referendum. "Sample Official Ballot," *Pinckneyville Advocate,* 3 November 1916, 3; "Official Vote Perry County, Illinois, November 7, 1916, *DuQuoin Evening Call,* 18 November 1916, 1. Here is what one editor suggested the women might do, however.

"The women we understand will not be entitled to vote on this proposition [township organization] but if they should be handed the small ballot, they, too, should be sure to vote against it. "Vote ' Against' Township Organization Tomorrow," *DuQuoin Evening Call,* 6 November 1916, 1.

12. Howard, *Illinois: A History of the Prairie State,* 455.
13. A BUSINESS MAN [pseud.], "For Its Adoption," *Virginia Republican-Gazette,* 19 October 1923, 1. Beginning in the early 1920s, women could vote on all such referenda; and some even occupied township offices. The editor of the *Republican-Gazette* told readers to ignore political parties "and vote for the man (or woman) instead of the party." Elsewhere in that 1924 editorial, however, he used masculine pronouns, in referring to candidates, five more times without referring again to the feminine. "Time to Select Your Candidates," *Virginia Republican-Gazette,* 8 February 1924, 4. A full-page advertisement in that same issue, from the Township Organization Committee, asked for consideration of its candidates for township offices in Cass, saying, "Let' s give the plan a fair trial by electing *men* who have favored Township Organization from the beginning" and "You are the only one to pay the bills, *Mr.* Voter" (italics mine). Township Organization Committee, "Lest We Forget What We Started Out to Do," *Virginia Republican-Gazette,* 28 March 1924, 12. A similar reference, but to "Mr. Taxpayer," appeared in the *Ogle County Republican* 14 years later, when Ogle County was considering discontinuance. Ogle County League for Commission Government, "The Question, Mr. Taxpayer," *Oregon Ogle County Republican,* 20 October 1938, 1.
14. F. D. Bramhall, "Cook County and Chicago," in vol. II, *A Report of the Joint Legislative Committee of the 47th General Assembly,* by John A. Fairlie (Springfield, 1913), 180. Bramhall, a professor at The University of Chicago, had given this material as part of a paper before the American Political Science Association in 1911.
15. Illinois, General Assembly, *Laws of the State of Illinois Enacted by the Forty-first General Assembly at the Regular Biennial Session* (Springfield, 1899), 91–92
16. "Fight to End Townships," *Chicago Tribune,* 20 April 1901, 5. In addition to North Chicago and West Chicago townships, the city still had functioning, in 1901, South Chicago, Hyde Park, Lake, Lake View, and Jefferson townships. The legislation making possible the (apparent) abolition of Chicago townships, by eliminating the township offices or the duties of the offices, passed and became law 11 May 1901. Illinois, General Assembly, *Laws of the State of Illinois Enacted by the Forty-second General Assembly at the Regular Biennial Session* (Springfield, 1901), 314–15.
17. "Town Meetings Prove Funereal," *Chicago Tribune,* 3 April 1901, 13.
18. "Vote Abolishes Town Offices," *Chicago Tribune,* 2 April 1902, 2; "The City Election," *Chicago Tribune,* 2 April 1902, 4.
19. "Official Count is Complete" and "Edict Ousts the Towns," *Chicago Tribune,* 6 April 1902, 8. As Chicago annexed adjacent areas, subsequent to the 1902 demise of its townships, surrounding civil-political townships lost control of portions of their territories to the city. For instance, the southeastern corner of Maine Township became the Edison Park neighborhood of Chicago.
20. My main source for the foregoing convention background is Cornelius, *A History of Constitution Making in Illinois,* 67–77.
21. They also created a Committee on Chicago and Cook County. Illinois, Constitutional Convention, *Proceedings of the Constitutional Convention of the State of Illinois Convened January 6, 1920,* vol. I (Springfield, 1920), 27.

22. Illinois, Constitutional Convention, *Proceedings of the Constitutional Convention of the State of Illinois Convened January 6, 1920,* vol. II (Springfield, 1921), 1271.
23. *Proceedings of the Convention,* vol. II, 1272.
24. The Smith committee proposed in the next section that board members in commission counties begin serving concurrent four-year terms rather than staggered three-year terms, as had long been the rule. Jarman explained that the expense of holding off-year elections was the reason for this suggested change. He erred when, in the next breath, he said there were then 16 commission counties. The number was really 17, leaving 84 using the board of supervisors, plus the hybrid county, Cook. *Proceedings of the Convention,* vol. II, 1273 and 1285. A positive outcome of the 1920–1922 Constitutional Convention seems to have been the precedent it set for having delegates receive, in advance, the results of research by the Legislative Reference Bureau. In October 1919, the Bureau, whose secretary was E. J. Verlie and whose chief data collector was W. F. Dodd, issued a set of suggestions for delegates to consider. Among the ideas for the section on county government was that the new constitution make uniform the "language as to the popular votes required for the adoption and abolition of the township system." Illinois, Legislative Reference Bureau, *List of Suggestions for Constitutional Change Together with Texts of Constitutions of Illinois* (Springfield, 1919), 24.
25. *Proceedings of the Convention,* vol. II, 1274. Here again the Legislative Reference Bureau had a hand. Among its suggestions was the following: "The township system has been largely outgrown, at least as it is now organized, and the town meeting has in many cases become a farce. In view of this, the question presents itself as to whether greater power with respect to county organization should not be left to the General Assembly, subject perhaps to local referendum."
26. *Proceedings of the Convention,* vol. II, 1327–28.
27. *Proceedings of the Convention,* vol. II, 1328–29.
28. *Proceedings of the Convention,* vol. II, 1329–30.
29. *Proceedings of the Convention,* vol. II, 1327 and 1331.
30. Illinois, Constitutional Convention, *Proceedings of the Constitutional Convention of the State of Illinois Convened January 6, 1920,* vol. IV (Springfield, 1922), 3964–65.
31. Cornelius, *A History of Constitution Making in Illinois,* 80.
32. Illinois, Secretary of State, *Counties of Illinois: Their Origin and Evolution* (1906), 44, 58, and 64–65; Sublett, *Paper Counties,* 15. A move was afoot in 1905 to reunite Henderson with Warren. Proponents sought the merger to add to the two counties' prestige and to reduce taxes by dropping one set of county buildings and officers. As political scientists and other consolidation advocates would learn later in the century, even though consolidation appears to be an obvious improvement strategy, the step is something few localities are willing to take. "Counties May Unite," *Oquawka Spectator,* 20 December 1905, 2.
33. Robert P. Sutton, *Rivers, Railways, and Roads: A History of Henderson County* (Raritan, Illinois: Henderson County Historical Society, 1988), 46–47. Sutton was then a history professor at Western Illinois University, in McDonough County, which lies adjacent to Henderson on the southeast. In the introduction to his book, he acknowledged that university historians had ignored local history—to the detriment of the subject. While choosing not to tackle, in depth, the township tale in Henderson, Sutton did refer to areas that political townships would occupy, long before they became townships. For instance, on page 26, he introduced Biggsville, Olena, and Raritan townships by name, more than five decades before they came into existence.

34. Meyer, *Making the Heartland Quilt,* 294.
35. "Township Organization," *Oquawka Spectator,* 15 August 1906, 1. For many years, going back at least to the 1880s, communities like Raritan, an upland prairie area in the southeastern corner of Henderson, had felt Oquawka and other topographically challenged precincts were "dragging their feet," as Sutton put it, "on the push to modernize the county by turning it over to a township organization." The *Raritan Reporter* noted that Oquawka residents, if under township organization, would have to build their own bridges, as well as care for their own paupers. Sutton, *Rivers, Railways, and Roads,* 105.
36. "Township Organization Cumbersome and Expensive," *Warsaw Bulletin,* as reprinted in the *Oquawka Spectator,* 3 October 1906, 1.
37. "Township Organization," *Stronghurst Graphic,* 18 October 1906, 1; "Took a Straw Vote," *Stronghurst Graphic,* 1 November 1906, 1. Stronghurst had, in 1904, narrowly lost its bid to become the Henderson county seat. Township organization was a way to help reduce the power of a longstanding seat of justice—like Oquawka. "Oquawka Still the Capital," *Oquawka Spectator,* 16 November 1904, 1.
38. The *Graphic* crowed about the Republican wave that had swept Henderson that fall and about the fundamental change in county government to occur, saying, "For the first time since the dawn of the Christian era, Henderson county voters have given a verdict in favor of township organization." "A Republican Wave," *Stronghurst Graphic,* 8 November 1906, 1.
39. "Township Organization Opposed," *Oquawka Spectator,* 21 November 1906, 1; "Township Organization," *Oquawka Spectator,* 19 December 1906, 1; Newton Bateman and Paul Selby, gen. eds. *Historical Encyclopedia of Illinois and History of Henderson County,* ed. James W. Gordon (Chicago: Munsell Publishing Company, 1911), 620; "The County Divided," *Oquawka Spectator,* 13 February 1907, 1.
40. "Circulating a Petition," *Stronghurst Graphic,* 20 August 1908, 1; R. A. Lomax, "Some Facts as to Township Organization," *Stronghurst Graphic,* 22 October 1908, 10.
41. W. A. M. Crough, "About Township Organization," and COMMITTEE [pseud.], "Some Other Facts Concerning Township Organization," *Stronghurst Graphic,* 29 October 1908, 1.
42. "Vote on Township Organization," *Stronghurst Graphic,* 12 November 1908, 8.
43. The *News* editor said Williamson was one of just seven "has beens," whereas the number of nonadopters actually was then 19 of 102. "The Approaching Election," *Herrin News,* 20 October 1900, 1; "Official Vote of Williamson County," *Marion Leader,* 8 November 1900, 5.
44. "No Township Organization," *Herrin News,* 10 November 1905, 1; "It Was a Mistake," *Herrin News,* 17 November 1905, 4.
45. "Two Propositions," *Herrin News,* 1 November 1907, 4; "Dry, Drier, Dryest," *Herrin News,* 8 November 1907, 1; "Making County into Townships," *Herrin News,* 3 January 1908, 1; "Chance for Candidates," *Herrin News,* 7 February 1908, 1; "Many are Called," *Herrin News,* 3 April 1908, 4.
46. "Vote Township Organization," *Herrin News,* 30 October 1908, 4; "Official County Vote Compiled," *Herrin News,* 13 November 1908, 3. Discontinuance, according to the 1870 Constitution, required only a majority of votes cast on the issue, whereas supporters, in order to secure adoption, had to tally more than half the total voters who voted that day in the county.

47. "Township Organization," *Marion Egyptian Press,* 10 September 1914, 1; "County Organization and Road Taxes, *Herrin News,* 22 October 1914, 1; "Total Vote of County Given Out," *Marion Evening Post,* 5 November 1914, 1.
48. "Supervisors Take Paper for Pay," *Marion Republican,* 6 November 1929, 1.
49. "Tax Board to Hear Taxpayers," *Marion Republican,* 25 June 1931, 1; "Plan Reforms in Operating the County," *Marion Republican,* 5 August 1931, 1; "Fix Blame for Increase in County Cost," *Marion Republican,* 8 September 1931, 1; "Board Demands Collection of Delinquent Tax," *Marion Republican,* 7 October 1931, 1; "Sheriff Demands Cash for Work for County," *Marion Republican,* 22 October 1931, 1; "County Board Hears Reports on Finances," *Marion Republican,* 3 November 1931, 1.
50. "Auditor Shows Condition of County Finance," *Marion Republican,* 5 April 1932, 1; "County to Get $8800 Relief from State," *Marion Republican,* 23 April 1932, 1; "Petitions Ask Vote on County Government," *Marion Republican,* 11 June 1932, 1; "Board Hears of Petitions for Vote on Change," *Marion Republican,* 13 June 1932, 1; "County to Name Road Superintendent," *Marion Republican,* 14 June 1932, 1; "Vote on County Change Likely," *Marion Republican,* 17 June 1932, 1; "Board Calls for Vote on Change of Government," *Marion Republican,* 13 September 1932, 1.
51. "County Will Vote on Commissioners," *Carterville Herald,* 16 September 1932, 1; *Carterville Herald,* 21 October 1932, 2; "With the Editor," *Johnston City Progress,* 20 October 1932, 1.
52. A. M. Townsend, "Economy in Commission Form of Government," *Carterville Herald,* 4 November 1932, 4; A. M. Tow[n]send, *Johnston City Progress,* 3 November 1932, 4; A. M. Townsend, "Special Ballots for Commission Plan Votes," *Marion Republican,* 1 November 1932, 2.
53. "Think Township System is Gone," *Marion Republican,* 8 November 1932, 1; "Next Election Will be Held on January 2," *Marion Republican,* 11 November 1932, 1; "Change of County Government in Feb.," *Carterville Herald,* 18 November 1932, 1.
54. "Petitions Ask for Election on Townships," *Marion Republican,* 15 August 1935, 1.
55. "Refer Petitions on Election to State Attorney," *Marion Republican,* 10 September 1935, 1; "Discuss Forms of Government," *Marion Republican,* 25 October 1935, 5; "Parties Look at Nov. 5 Vote," *Marion Republican,* 9 October 1935, 1; "Trades Council for County Government," *Herrin News,* 24 October 1935, 1; "An Election Next Tuesday," *Marion Republican,* 2 November 1935, 2; "County Election on Two Issues on Tuesday," *Marion Republican,* 4 November 1935, 2; "Atwood Wins Re-election to County Board," *Marion Republican,* 6 November 1935, 1.
56. "With the Editor," *Johnston City Progress,* 23 September 1938, 1; "With the Editor," *Johnston City Progress,* 30 September 1938, 1.
57. "With the Editor," *Johnston City Progress,* 4 November 1938, 1; "Township or Commission Form of Government," *Marion Republican,* 5 November 1938, 2; "County Change Needs Majority of Votes Cast," *Marion Republican,* 5 November 1938, 1.
58. "Complete Tax Assessment is in the Making," *Marion Republican,* 27 August 1935, 1; "The Carterville Twp. Vote," *Carterville Herald,* 11 November 1938, 1; Barbara Hubbs, *Pioneer Folks and Places: An Historic Gazetteer of Williamson County, Illinois* (Herrin: By the Author, 1939), 23; *Atlas and Plat Book of Williamson County, Illinois* (Rockford, Illinois: Rockford Map Publishers, 1975), 3 and 5; "Election Districts," (Williamson County, Marion, 26 July 1983, Mimeographed). I obtained a copy of "Election Districts" at the Williamson County courthouse in January of 1993, at which time the 1983 districts were still in effect.

59. Lovett, *Notes on the Location of Illinois County Seats,* Cass County.
60. "An Unusual Stand," *Virginia Enquirer,* 31 October 1901, 1; J. J. Paschal, "Township Organization," *Virginia Enquirer,* 31 October 1901, 1.
61. *Virginia Gazette,* 1 November 1901, 4; 25 October 1901, 4.
62. "Tuesday' s Election," *Virginia Enquirer,* 8 November 1901, 1.
63. "Organized for Campaign," *Virginia Enquirer,* 17 October 1917, 8; "85 Out of 102 Counties," *Chandlerville Times,* 5 October 1917, 1.
64. "A Plain Business Proposition," *Virginia Enquirer,* 17 October 1917, 8. This ad called on "Republicans, Democrats, Progressives, Prohibitionists, and all other voters" to stop the township movement. George R. Kuhlman, "Believes Present Government Good," *Virginia Enquirer,* 24 October 1917, 1.
65. "Why Cass County Should Come Under Township Organization," *Virginia Enquirer,* 24 October 1917, 5; "Why Cass County Should Come Under Township Organization," *Chandlerville Times,* 26 October 1917, 1; "Answer to Kuhlman' s Joke," *Virginia Enquirer,* 31 October 1917, 1; "Answer to Kuhlman" Joke," *Chandlerville Times,* 2 November 1917, 8. The committee' s letter to the *Enquirer* may actually have been a paid item, but it looks like a letter.
66. "Palmer, Republican Elected by the Majority of 1,266," *Virginia Gazette,* 9 November 1917, 8; "For the Best Interests of Cass County Vote for Township Organization," *Virginia Enquirer,* 30 October 1918, 8; "Township Organization Defeated by 18 Votes," *Chandlerville Times,* 8 November 1918, 4.
67. Illinois, Auditor of Public Accounts, Township Organization File, Cass County, Canvass of Voting Precincts of the Votes Cast at an Election Held in the County of Cass and State of Illinois on Tuesday November Sixth A. D. 1923. Originals are in the Illinois State Archives.
68. "Executive Committee, Cass County Farm Bureau, "A Petition for Lower Taxes," *Virginia Republican-Gazette,* 30 March 1923, 1; "Township Form of County Rule is Being Asked," *Virginia Republican-Gazette,* 31 August 1923, 1; "State Urges an Adjustment of Cass Land Tax," *Virginia Republican-Gazette,* 14 September 1923, 1; "New County Rule to be Submitted to Vote on Nov. 6," *Virginia Republican-Gazette,* 21 September 1923, 1.
69. "Commissioner Rule Means High Taxes: Township Form Only Salvation for Tax-Payers," *Virginia Republican-Gazette,* 26 October 1923, 1; A BUSINESS MAN [pseud.], "13 Supervisors in Schuyler Co. Cost Only $874," *Virginia Republican-Gazette,* 2 November 23, 1; J. E. Hansell, "That the People May Know," *Virginia Republican-Gazette,* 2 November 1923, 1; "Corrected Ballot and How to Mark It," *Virginia Republican-Gazette,* 3 November 1923, 1.
70. "Township Organization," *Chandlerville Times,* 26 October 1923, 8; "Specimen Ballot," *Chandlerville Times,* 2 November 1923, 8; "Vote ' No' Next Tuesday," *Beardstown Illinoisan-Star,* 31 October 1923, 1; "Vote for Beardstown, Vote No," *Beardstown Illinoisan-Star,* 3 November 1923, 1; "Tomorrow, ' Vote' Against," *Beardstown Illinoian-Star,* 5 November 1923, 1; Illinois, Auditor of Public Accounts, Township Organization File, Cass County, Canvass of Voting Precincts of the Votes Cast at an Election Held in the County of Cass and State of Illinois on Tuesday November Sixth A. D. 1923; "New Era is at Hand," *Virginia Republican-Gazette,* 9 November 1923, 7.
71. "Illinoian-Star Platform for 1934–1935," *Beardstown Illinoisan-Star,* 17 September 1934, 2.

72. "Next November," *Beardstown Illinoisan-Star,* 29 September 1934, 2; "Some Questions Answered," *Beardstown Illinoian-Star,* 22 September 1934, 2; "Township Organization Costs in Beardstown Township," *Beardstown Illinoian-Star,* 27 September 1934, 1; "Morgan—Cass—Expenses," *Beardstown Illinoisan-Star,* 27 September 1934, 2; "Tax Rates," *Beardstown Illinoisan-Star,* 28 September 1934, 2; "County Government," *Beardstown Illinoian-Star,* 29 September 1934, 2; "County Government," *Beardstown Illinoian-Star,* 31 October 1934, 2; "How to Vote on Township Organization," *Beardstown Illinoian-Star,* 5 November 1934, 2.
73. "Not in Favor of Commission Form," *Ashland Sentinel,* 7 September 1934, 1; A. T. Harrison, "A. T. Harrison Asks Where is Saving?" *Ashland Sentinel,* 28 September 1934, 1; T. L. Barber, "Barber' s Salary $44.80 for 9 Months," *Ashland Sentinel,* 12 October 1934, 1; W. H. DeSollar and others, "Comparing Counties Under Different Forms of Government," *Ashland Sentinel,* 19 October 1934, 1; M. E. Coker, "Cass Auditor Says Commission Form is More Expensive," *Ashland Sentinel,* 26 October 1934, 1.
74. T. L. Barber, "T. L. Barber Says: Don' t Pick A Lemon," *Chandlerville Times,* 14 September 1934, 1.
75. "Change in Form of Government to be Voted," *Virginia Gazette,* 17 August 1934, 1. The *Republican-Gazette* became the *Gazette* in 1930.
76. "Vote to Retain Supervisors by 2 to 1 Majority," *Virginia Gazette,* 9 November 1934, 1; "Township Organization," *Beardstown Illinoisan-Star,* 7 November 1934, 2. By the way, this Beardstown editorial may have been the last newspaper item ever in Illinois to carry the title "Township Organization."
77. This group was the forerunner of the Township Officials of Illinois. Frederick Mercer Van Sickle, in his authorized history of TOI, said officials had perceived certain actions by the general assembly as threatening to their well-being, and they felt the need for "fraternity and unity." A half dozen officials met in Watseka, the seat of Iroquois County, in November 1907 to establish the organization, linking highway commissioners and township clerks. In 1909, the ISAHCTC held its first annual meeting. Thus began, in an unobtrusive manner, what would become a formidable foe for anyone who would challenge the right of the people to enjoy the benefits and pay the price of having townships. Van Sickle, *Grassroots Democracy,* 42.
78. Newell, "Township Government in Illinois," 501–3.
79. John A. Fairlie Papers, University of Illinois Archives, Urbana; "Previous Editors of the American Political Science Review" and "Former Presidents [of the American Political Science Association]," *American Political Science Online,* 2001, <www.apsanet.org> (23 May 2001); Kelly Guthrie, email correspondence with author, 24 May 2001.
80. Illinois, General Assembly, Joint Legislative Committee, *A Report of the Joint Legislative Committee of the 47th General Assembly* (Springfield, 1913), II:1; John A. Fairlie, "Governmental Reorganization in Illinois," *American Political Science Review* 9 (May 1915): 252–57.
81. *Report of the Joint Legislative Committee,* II:20–22; John A. Fairlie, "County and Town Government in Illinois," *Annals of the American Academy of Political and Social Science* 47 (May 1913): 62–78.
82. *Report of the Joint Legislative Committee,* II:10. For a compilation of survey results, see pages 176–79.
83. The 1913 meeting, most likely in Decatur, focused on educating the 64 attending clerks and highway commissioners. Educating officials is still an important function of TOI annual conferences today. Joliet was the site for the 1914 ISAHCTC meeting; and, for the

first time, clerks and commissioners decided it was in their interest to lobby the general assembly for higher pay and road law improvements. In 1915, the group expressed concern for the Tice Law (after Homer J. Tice of Greenview), which, in 1913, had switched responsibility for main roads from townships to counties, among other changes. By 1919, association president Kendall was talking merger of his group's clerks and highway commissioners with the "county commissioners association" because of mutual interests in road laws and statewide government reforms. ISAHCTC, at least as far as Van Sickle reported, was just beginning to flex its muscle in matters that the general assembly controlled. Van Sickle, *Grassroots Democracy,* 42.

84. Illinois State Normal University, *The Index* (Normal: 1892): 1; "Alumni Register, Illinois State Normal University, Classes 1892–1922," *The Normal School Quarterly* 83 (April 1922): 8; "Ex-Senator Hicks Dies in Crash," *Rockford Register-Republic,* 3 February 1936, 1; "Head-on Collision Sunday is Fatal to Rockford Attorney," *Decatur Review,* 3 February 1936, 12.
85. Illinois, *Bluebook of the State of Illinois: 1915–1916* (Springfield, 1915), 87; *Bluebook of the State of Illinois: 1917–1918* (Springfield, 1917), 146; *Bluebook of the State of Illinois: 1919–1920* (Springfield, 1919), 190. When Hicks was in the general assembly, three representatives served each district.
86. Illinois, *Bluebook of the State of Illinois: 1917–1918,* 256.
87. Township clerks and highway commissioners continued with their annual meetings, at various downstate venues, in cities like Springfield, Peoria, and Kankakee. They continued also to pass resolutions urging changes in state laws affecting townships. Memberships still only cost $10.00 per township in 1929. Van Sickle, *Grassroots Democracy,* 43.
88. Illinois, General Assembly, Bills, Petitions, etc., 1923–1924, "S. B. No. 125," State Archives, Springfield.
89. "S. B. No. 125"; *Senate Journal,* 53rd G. A., 1923, 493 and 1135–36.
90. Illinois, General Assembly, Bills, Petitions, etc., 1923–1924, "Committee Blank 'B' House: Senate Bill No. 125," State Archives, Springfield; *House Journal,* 53rd G. A., 1923, 1093, 1217, 1302, 1328, and 1337.
91. Illinois, State Tax Commission, Township Budgets and Appropriations, Correspondence Files, Township Receipts and Expenditure Files, Commission Meeting Minutes, State Archives, Springfield. The General Assembly created the STC in 1919 as an element of the Department of Finance, giving the governor the responsibility of appointing the three commission members. In 1943, the state dissolved the STC and turned its duties over to the Department of Revenue.
92. Victor S. Yarros, "County and Township Government," *Bulletin* (of the) *Illinois League of Women Voters* 2 (June 1922): 9–10; "Know Your Local Government," 14–15. The *Bulletin* began with the May 1921 number and by 1929 had become *The Illinois Voter.*
93. Illinois, State Tax Commission, Minutes, 16 July 1931, State Archives, Springfield. Custer was the state's treasurer from January 1925 to January 1927 and from January 1929 to January 1931. Clayton, *The Illinois Fact Book and Historical Almanac, 1673–1968,* 140.
94. H. S. Hicks, "The One Room Schools of Illinois," (Illinois, State Tax Commission, Springfield, [1932], Mimeographed), 13; H. S. Hicks, *Why?* compiled for the Illinois State Tax Commission (Springfield, [1932]).
95. The 10 township counties, in descending order of 1931 taxes levied were Winnebago, Lake, LaSalle, Rock Island, Peoria, Madison, Kane, Macon, St. Clair, and Sangamon.

Hicks might have improved his argument if he had used a set of 10 counties that were more comparable in population and culture to the 16. H. S. Hicks, "Is the Township an Obsolete Form of Government?" *Illinois Journal of Commerce* 14 (October 1932): 14 and 25. The Illinois Chamber of Commerce, headquartered on North Wacker Drive in Chicago at the time, was an active township critic through its journal, *Illinois Journal of Commerce,* and no doubt in other ways through its lobbying efforts. Besides sponsoring several Hicks publication projects, the Chamber conducted an advisory referendum of its own among "business men in all parts of the State" in 1934 to obtain answers to and commentary about four tax-related questions. Question number 4 dealt with the "Eventual Elimination of the Township." As a way to provide the referendum voters with background, the Chamber sent along a foreword to the question that read in part: "It is generally recognized that the first approach to cutting the costs of government must lie along entirely practical and simple lines. The elimination of many of the functions of the township is now generally recognized as being possible. Some changes have been effected in the State of Illinois; that is to say, about 17 counties are operating today under the so-called commission form of government....This proposition is timely because at the present time a number of counties in Illinois are actively engaged in securing signatures to petitions which will permit the voters to determine whether or not they desire to change from the township form of government to the county commission form." Not surprisingly, the "business men" strongly endorsed the Hicks/Chamber position on abolition of townships. J. T. Meek, "What the Referendum Brought Out," *Illinois Journal of Commerce* 16 (November 1934): 5–6, 16, and 29.

96. "7 Counties Vote on Commission Plan Nov. 8th," *Rockford Register-Republic,* 20 September 1932, 14; "Seven Counties to Vote on New Government," *Marion Republican,* 21 September 1932, 1; "7 Counties Seek Relief in Change of Government," *Johnston City Progress,* 13 October 1932, 1. Hicks' s careful research and condemnation of townships caught the attention of the League of Women Voters that fall and winter. In her report to the Illinois League' s December 1932 convention, about research she and her committee had done on local government, Marietta Busey Tawney cited Hicks along with various state-funded works. Tawney wrote, "Former Senator Hicks' studies of comparative costs of local governmental units have evoked widespread interest." "Report in County Surveys to Illinois State Convention," 17 December 1932, Marietta Busey Tawney Papers, Illinois Historical Survey, Urbana. Hereafter, I cite as Tawney.
97. In the League of Women Voters' Township Kit of 1933, there is a reference to Hicks as a "Former Member of the State Tax Commission." Barbara Morgan to Mrs. Guy A. Tawney, "County Organization vs. Township Organization," 6 April 1933, Tawney.
98. H. S. Hicks, *Waste in Local Government: An Analysis of Local Expenditures of the Minor Governmental Units of Winnebago County, Illinois* ([Rockford]: Rockford Chamber of Commerce, 1934).
99. H. S. Hicks, "The Costliness of Local Government," *Illinois Journal of Commerce* 16 (September 1934): 12–13 and 27; H. S. Hicks, "Railroad Taxes in Illinois," *Illinois Journal of Commerce* 16 (October 1934): 21–23.
100. H. S. Hicks, *Wasting Tax Money by the Millions: A Study in Township Road Building Costs in 12 Illinois Counties,* 2d ed., ([Chicago]: Illinois Chamber of Commerce, 1935). From the contents of the second edition, it seems likely that the first edition also carried a 1935 imprint.
101. H. S. Hicks, *Waste in Local Government: An Analysis of the Expenditures of the Minor Governmental Units of St. Clair County, Illinois* (n.p.: Illinois Chamber of Commerce

and East St. Louis Chamber of Commerce, 1935); H. S. Hicks, *Waste in Local Government: An Analysis of Local Government Costs in Sangamon County, Illinois* ([Springfield]: Springfield Chamber of Commerce, 1935); H. S. Hicks, *The Wasteful System in Local Government: An Analysis of the Expenditures of the Minor Governmental Units of Kane County, Illinois* (n.p.: Illinois Chamber of Commerce, Aurora Chamber of Commerce, and Elgin Association of Commerce, 1936); H. S. Hicks, *How Many Taxing Bodies, How Much Government Do Taxpayers Need?: An Analysis of Expenditures by Minor Governmental Units of Macon County, Illinois* ([Decatur]: Association of Commerce of Decatur and Illinois Chamber of Commerce, 1936).

102. "Ex-Senator Hicks Dies in Crash," *Rockford Register-Republic,* 3 February 1936, 1; "Head-on Collision Sunday is Fatal to Rockford Attorney," *Decatur Review,* 3 February 1936, 12; Association of Commerce of Decatur, Illinois, Foreword to *How Many Taxing Bodies, How Much Government* by H. S. Hicks; Clayton, Foreword to *The Wasteful System in Local Government: Kane County* by H. S. Hicks; "Herbert S. Hicks," *Rockford Register-Republic,* 5 February 1936, 12. In October of 1937, the Illinois Chamber of Commerce and the Pekin Association of Commerce jointly published the last in the series of booklets Hicks had begun. Clayton again wrote the foreword. In it he called Illinois "the most over-governed state in the United States" and urged voters "to take advantage of the legal means which now exist to secure a lesser number of more efficient bodies." Clayton, Foreword to *How Many Taxing Bodies, How Much Government Do Taxpayers Need?: An Analysis of the Expenditures of Minor Governmental Units of Tazewell County, Illinois* ([Pekin]: Illinois Chamber of Commerce and Pekin Association of Commerce, 1937).

103. T. B. Manny, "The Unit Requirement Idea for Local Government in Rural Areas," in *Standards of Living: Proceedings of the Thirteenth American Country Life Conference, Madison, Wisconsin, October 7–10, 1930,* by the American Country Life Association (Chicago: The University of Chicago Press, 1931), 131–34. See also Manny' s "Government Structure, Powers and Problems in the New Rural Municipality," in *Rural Government: Proceedings of the Fourteenth American Country Life Conference, Ithaca, New York, August 17–20, 1931,* by the American Country Life Association (Chicago: The University of Chicago Press, 1932), 154–57.

104. Frank O. Lowden, "The Problems of Rural Government," *Proceedings of the Fourteenth American Country Life Conference,* 1–9. See also Lowden' s "Scrapping Useless Governments," *National Municipal Review* 22 (March 1933): 122–24 and 136.

105. Paul W. Wager, "Can Local Self-Government Be Preserved in Our Rural Areas?" in *Rural Government: Proceedings of the Fourteenth American Country Life Conference, Ithaca, New York, August 17–20, 1931,* by the American Country Life Association (Chicago: The University of Chicago Press, 1932), 54–60.

106. Arthur W. Bromage, "Functions and Powers of Village and Township Governments," *Proceedings of the Fourteenth American Country Life Conference,* 158–61. See also Arthur W. Bromage and Thomas H. Reed, *Organization and Cost of County and Township Government,* Michigan Local Government Series (n.p: Michigan Commission of Inquiry into County, Township and School District Government, 1933); and Arthur W. Bromage, "Shall We Save the Township?" *National Municipal Review* 25 (October 1936): 585–88.

107. M. H. Hunter, *Costs of Township and County Government in Illinois*, bul. 45 (Urbana: University of Illinois, Bureau of Business Research, 1933), 30–31; H. K. Allen, "Relative Merits of Township and Non-Township Organization," *Illinois Journal of Commerce* 17 (January 1935): 8. See also, Harry Kenneth Allen, "Efficiency and Economy in

Local Government—With Particular Reference to Illinois" (Ph.D. diss., University of Illinois, 1936); and H. K. Allen, *Control of Expenditures in the Local Governmental Units of Illinois,* bul. 61 (Urbana: University of Illinois, Bureau of Business Research, 1940).

108. L. D. Melton, "The Elimination of Township Government in Oklahoma," *National Municipal Review* 27 (August 1938): 405–7. Even though the Oklahoma constitution provided, as did that of Illinois, for county-by-county abolition by a vote of the people, Melton did not know of any instance when such local initiative had occurred. The Depression-inspired Oklahoma solution was a constitutional amendment to remove the authority of townships to levy taxes. Melton wrote (page 406), "Thus were townships eliminated as active subdivisions of government in Oklahoma by the indirect process of taking away the necessary authority to levy taxes for township purposes. Townships were finally put out of business in fact as well as in name by an exercise of the power of the purse."

109. "Townships: Excerpt from 'The Existing Structure of Local Government in Illinois';" "County Organization v. Township Organization: Excerpt from Analysis by H. S. Hicks, Former Member State Tax Commission;" "Report in County Surveys to Illinois State Convention," 17 December 1932;" "Program of Work for the Illinois League of Women Voters, 1934–35;" "Department of Government and Its Operation: Explanation of the Program of Work," August 1938. Tawney. Katharine Croan Greenough, *Know Your County* (Washington: National League of Women Voters, 1937). "Women Voters' League Hears Talks on County," *Rockford Register-Republic,* 22 September 1932, 1. The Illinois League would become much more negatively critical of townships in the next half century (chapter 6).

110. Illinois, General Assembly, *Laws of the State of Illinois Enacted by the Fifty-seventh General Assembly as the Regular Biennial Session* (Springfield, 1931), 47; Illinois, Commission on Taxation and Expenditures, *Report and Recommendations of the Illinois Commission on Taxation and Expenditures* (Springfield, 1933).

111. "Proposals for the Reorganization of Local Government in Illinois: Part 1—The State Outside of Cook County," (Chicago: Griffenhagen & Associates, 30 November 1932, Mimeographed), i–iv.

112. "Proposals for the Reorganization of Local Government," I/21–22, I/24–25, and X/14.

113. Illinois, Commission on Taxation and Expenditures, *Report and Recommendations,* 12–20 and 77.

114. *House Journal,* 58th G. A., 1933, 194, 234, 254, 284–85, 323, 376–77, 416, 417, 1339, and 1508; *Senate Journal,* 58th G. A., 1933, 574, 727, 737, 772, and 1,182; Illinois, General Assembly, *Laws of the State of Illinois Enacted by the Fifty-eighth General Assembly at the Regular Biennial Session* (Springfield, 1933), 1108–9; Illinois, General Assembly, Bills, Petitions, etc., 1933–1934, State Archives, Springfield. This odd method for determining a county's population became an issue 15 years later in the call for an abolition referendum in Macon County. "Written Opinion Backs Balloting on County Rule," *Decatur Herald and Review,* 12 September 1948, 3.

115. Department of Commerce, Bureau of the Census, *Fifteenth Census of the United States: 1930,* vol. 3, pt. 1 (Washington: 1932), 614–20. The 13 Illinois counties exceeding 75,000 residents in 1930 were, in descending order: Cook, St. Clair, Madison, Peoria, Winnebago, Sangamon, Will, Lake, Rock Island, LaSalle, DuPage, Vermilion, and Macon. For St. Clair, the second-most populated county of the decade, 72,571 voted in the 1932 general election. "93 Percent of Registered Vote in County Cast," *Belleville Advocate,* 14 November 1932, 1. "Democrats Carry Every Office in Cass County Vote," *Springfield Illinois State Journal,* 10 November 1932, 5.

116. *House Journal,* 58th G. A., 1933, 948; *Laws Fifty-eighth General Assembly,* 155–56.
117. *House Journal,* 58th G. A. , 1933, 2063; James T. Burns to Mrs. G. A. Tawney, 1 February 1932, Tawney.
118. "State, County and City ' Dead Wood' Is Found by Solons in Thrift Study," *Springfield Illinois State Register,* 16 September 1934, 6; "Next November," *Beardstown Illinoisan-Star,* 17 September 1934, 2.
119. Illinois, Retrenchment Commission, *Report of the Retrenchment Commission* (Kankakee, 1935), 1, 3, 11, and 14.
120. For an idea of how long some of the officials, who were retiring after the 3 April 2001 election, had served, see Erin Brothers' s "Retiring Officials Have Pieced Quite a Quilt," *Township Perspective* (July–August 2001): 17–23.
121. Van Sickle, *Grassroots Democracy,* 43–44.

6 *Fortress Astoria: 1940–1989*

ILLINOIS'S ORGANIZATION OF TOWNSHIP officials enjoyed only modest success negating the criticism about which I spoke in chapter 5; but, after the Depression and World War II, the Township Officials of Illinois (TOI), which had succeeded the IATO during the war, came into its own as a scrappy defender of local government. At the local level, besieged township officials could turn to TOI for comfort, advice, and a timely visit from a TOI road warrior. In Springfield, legislators and others came to realize just how much TOI cared about keeping township organization alive and healthy. During the half century that chapter 6 covers, TOI exceeded ten thousand members, employed a pair of effective leaders, enjoyed the benefits of a monthly magazine, purchased a building to serve as its headquarters, instituted a self-insurance setup for townships that would later yield to TOI useful monetary surpluses, and became adept at working the Illinois political system and in squelching efforts to reduce the number or influence of the fourteen hundred plus townships. All this TOI accomplished from the small Fulton County farming/mining community of Astoria, a place that its leaders rendered almost fortress-like in the sense of repelling from there the un-American (to TOI) notion that citizens in Illinois's 85 township counties would be better off with a board of three at-large commissioners and without civil-political townships.

Neither the seven countywide referenda during the period nor the two single-township referenda altered the township landscape. Overall, the score was eight victories for Fortress Astoria and one for the other side. Each decade saw at least one countywide referendum (figure 11); and many of the campaigns, TOI played a significant supporting role.

Figure 17 features a pattern that has become quite commonplace in this monograph: a southeast-to-northwest arrangement of voting counties. While all seven of these countywide referenda seemed to spring from locally inspired (independent) efforts, the fact that they lay on the map in a pattern so similar to patterns I earlier presented leads me to conclude that the tide that pushed southwestward across the state still had a glimmer of life left in it.

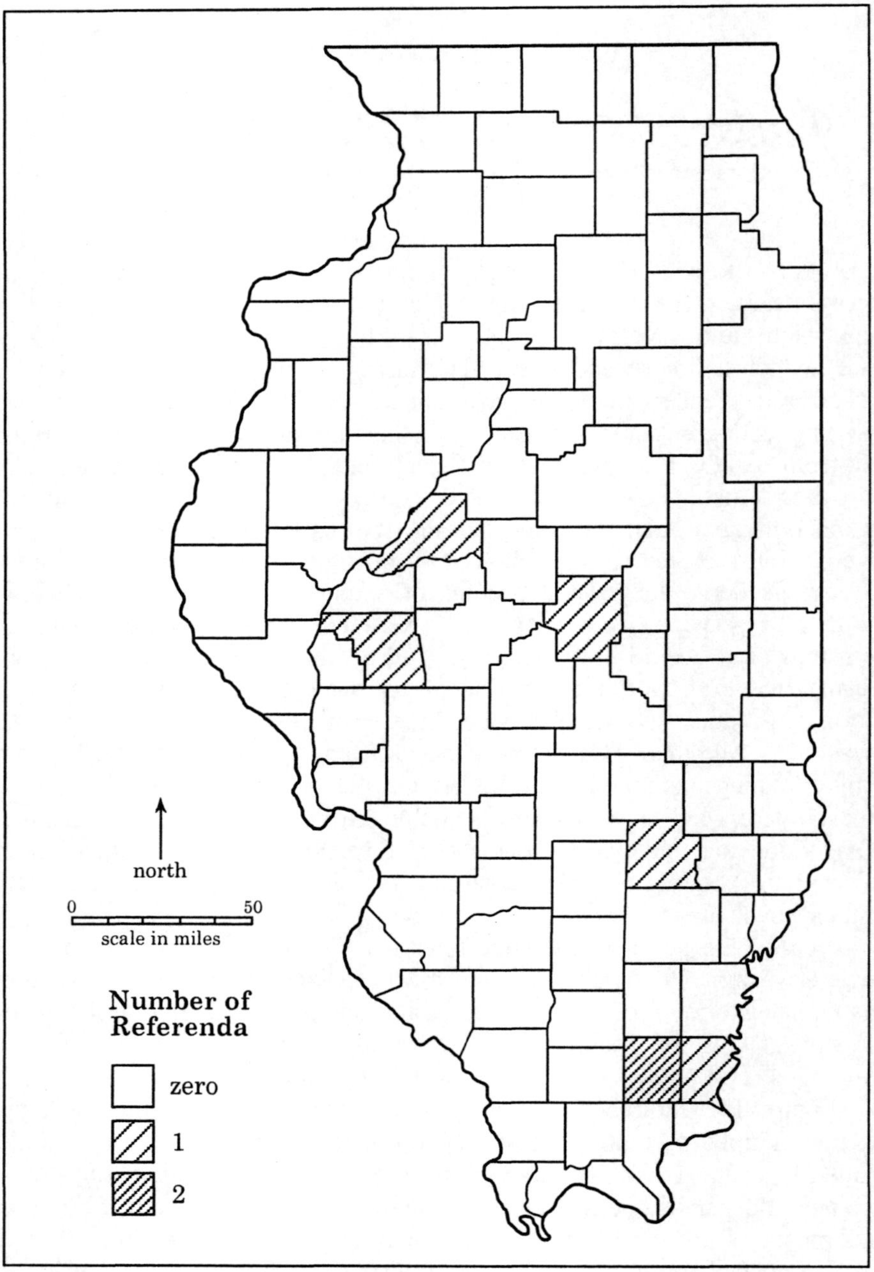

Figure 17 Countywide Referenda on Township Organization, by County: 1940–1989.

The Kost Years

Troy Augustus Kost spent much of his life in Astoria Township, the six-mile square minor civil division that occupies the southwestern corner of Fulton County. When Kost died in 1993 at the age of 90, George Miller praised him as being not only the only TOI secretary-treasurer for nearly forty years but also its "CEO." The organization, in Kost's honor, had earlier retired the position of secretary-treasurer, so successor Miller became the first executive director.[1]

Kost, born 1 November 1903 in Astoria, grew up on a farm nearby and graduated from Astoria High School in 1922. After his high school years, he repaired and sold automobiles, but soon displayed a knack for highway matters, eventually devoting 23 years of his life to the service of Astoria Township Road District as its highway commissioner. For several years, he held that elective post along with his paid position as the TOI secretary-treasurer. In the 1950s, he gave up the road district job to take a Springfield appointment as assistant director of the Illinois Department of Public Works. By that time, he had represented township interests as a member of the Farm-To-Market Road Commission (1945–1946) and Highway Traffic and Problems Commission (1947–1948). His Springfield experiences served him and TOI well after he returned full-time to Fortress Astoria, especially in the critical months of the 1969–1970 Constitutional Convention. In politics Kost was a Republican and in his personal life active in the Astoria Christian Church, where he was deacon, elder, and trustee. He belonged to the Astoria Masonic Lodge, Knights of Pythias, Peoria Consistory, and Mohammed Shrine of Peoria. His steadfast Christian convictions often reverberated through the items he wrote for the TOI magazine and no doubt permeated his unwritten record.[2]

As a young highway commissioner, Kost found himself in a good location from which to access the state township officials' conventions, because many were either in Peoria or Springfield. It was in Joliet, however, at the 1934 convention of ISAHCTC, that he won a seat on the association's governing board as, according to Van Sickle, its "tenth vice-president." By 1941, he had moved up to be first vice-president of IATO. A couple years later he became acting secretary, and finally, by 1945, secretary-treasurer. He served in that slot until failing eyesight and advancing age led to his resignation in 1979.[3]

The Magazine that Kost Bought

TOI uses *Township Perspective* to keep in touch, 11 times a year, with the eleven thousand or so township officials in its member townships. Nearly every Illinois township now belongs to TOI; and all township officials in those townships receive a free subscription, courtesy of the dues the townships pay, generally with tax dollars.[4] Gratis copies go also to state legislators, other state officials, Illinois's congressional delegation, and anyone else who Executive Director Bryan Smith believes ought to be absorbing the township story.

Township Perspective is the direct descendent of a trade magazine that began publication in February 1941, a few months before Troy Kost became IATO's first vice-president. H. L. Cecil founded *The Illinois Supervisor and County Commissioner,* as a commercial venture, in Springfield. His chief audience was to have been the membership of the Illinois Association of County Officials, an umbrella that covered eight state associations, including those for sheriffs, auditors, and county board members. By May of 1941, IATO leadership had decided to adopt the magazine as its "official organ;" and, after only three issues, the publisher changed the name to *Illinois County and Township Official (ICTO),* the name it retained until April 1999, when Smith changed it to *Township Perspective.*[5]

Over the six decades plus that the magazine has appeared the message has remained consistent. Its editors have kept subscribers apprised of upcoming and completed conventions, seminars, and the like; legal matters that might impact official duties; potentially useful new products and services; and personnel changes among officials (including obituaries of the rank and file). It has also provided warnings about threats to township offices and township government itself. Publisher Cecil made it clear early on that he would do his best to "serve, protect and perpetuate home rule," by which he meant local government—in the face of increasing "centralization of power and usurpation of local unit authority by state and nation."[6] Other individual owners followed Cecil, all of whom provided Kost several pages of each issue, which he ably used to promote the TOI agenda. When the opportunity came in 1971 for TOI to purchase the magazine, Kost saw the chance for TOI to own its print forum. On the first of August that year, the transfer came true; and, in the September 1971 issue, Kost announced the appointment of a new editor, Lee Ahlswede. That arrangement held fast for eight years, until the TOI board of directors chose, because of financial exigency, not to reappoint Ahlswede. Kost served briefly as editor in 1980; and, since then, the executive director has been the ex officio editor.[7]

Centrists at the Gate

County and township advocates placed home rule on a pedestal in the years after World War II. Centrists, in the minds of *ICTO* editors and Kost, were demons determined to wreck a way of life that Americans treasured and for which tens of thousands died between 1941 and 1945; and centrists were a menace to home rule. "This is not flag waving or patriotic bellyrash," said editor Richard Cecil in June of 1941, "nor is it Fourth of July oratory or campaign balderdash—the root of the American system is based in the official close to the people." His editorial target was not some foreign enemy but rather the Illinois general assembly, which was threatening yet another probe of township government. Editor Cecil urged, in return, an investigation of the legislature. Kost, in his first bylined contribution to *ICTO,* three years later, continued to hammer home the need for vigilance on the part of township officials, prompt payment of TOI dues, and attendance at the upcoming Peoria annual meeting. "We must fight," said Kost, "every step of the way because as you know there are within this great state of ours forces at work all the time in an effort to do away with home rule solely for personal gain." He noted also how some of Illinois's "sister states" had, through inattention, seen home rule die.[8]

Centrism, socialism, and communism somehow came together, in TOI rhetoric of the postwar years, as anathematic to home rule. TOI went on record at its 1945 meeting in Peoria as an opponent of federal health insurance because members felt it "would contribute to the further socialization of the...Federal government which in turn...[would] violate the principles and purpose of home rule of the local communities of America." At the 1946 meeting, also in Peoria, Kost warned of forces present in Illinois that would place the township under county rule and would put counties at the mercy of the state. Later in that conference, Joseph R. Thomas, president of the Township Officials Association of Ohio, spoke to the Illinois delegates, saying, "Centralization of government is no new thing....That's not going ahead—that's going back." Addressing the Revenue Laws Commission in Springfield on 16 March 1948, Kost left little doubt where those who advocated the downfall of township government ranked on his list of public enemies, when he said,

> Yes, we know that some of the bigshots of the state, who happen to be professors and so forth, have advocated that the townships be done away with, that local government could be administered much more efficiently and more cheaply from a centralized source. This is an erroneous statement and I might even add an UnAmerican [*sic*] statement for as long as townships stay with local people administering local affairs, we need have no fear of Communism and Bolshevism taking over.

Good Americans, he went on to say, would resist communism and "never give up home rule without a struggle." He closed by suggesting professors and other "super experts" who were critical of local government's home rule prerogative should "emigrate to Russia but leave our form of government alone."[9]

Later in 1948, at another Peoria TOI meeting, Secretary-Treasurer Kost was taking no prisoners as he rallied his troops against township government's enemies, people who he felt were "sincere in their conviction that home rule must go and that government be centralized." Where, he asked his loyal audience, did foes of township government "get their information"? A lot of it was coming from "higher brackets of government," which Kost said "Reds" had infiltrated. "These ideas are promoted from Washington and from some of our college professors whose records are extremely shady as to whether they are red, white and blue, or just plain red." Township officials, and their county counterparts, had to campaign vigorously against such sedition. Placing an area's government in the hands of three commissioners "was Hitler's idea for running Germany a few years ago—to put it under a selected few whereby it can be dictated to by centralized sources." Big companies and other pressure groups wanted, he said, three or four officials they could influence instead of a county board with a dozen or two dozen supervisors watching out for the public's welfare. Kost went on to cite enemy efforts to chip away at home rule through centralization of road functions or assessment duties. "We, the township officials," he concluded, "are the last of that great fundamental form of government which is really close to the people."[10]

Working together, however, officials could preserve that precious way of life. One tactic was for them to do their "jobs a little bit better." Another was to seek a change in the 1933 law that governed the number of signatures centrists needed to get the abolition of township organization on the general election ballot. Petition circulators had to find 5 percent of the electorate to sign on for an abolition referendum in counties "having a population of less that 75,000" compared to 20 percent in counties coming in above that number. Election officials determined the number of signatures necessary by looking at how many in the county voted at the previous presidential election. A TOI resolution from the 1948 annual meeting complained, "determination of population by the vote cast at a presidential election is not a proper or accurate method of determining the real, true, actual population of a county." TOI suggested the general assembly amend the statute by dropping the section about presidential elections so that the most recent census figure for the county would suffice. Many more Illinois counties, therefore, would fall into the higher category; and petitioners in those counties that moved up a category henceforth would need four times as many signatures.[11] Probably not coincidently, that very fall, there was an Illinois county involved in an abolition referendum struggle where a different method (the wished-for TOI method) would have posed a much bigger petition task for the centrists who wanted to convert that county from township organization to commissioner government.

Centrists did not, however, actually need to abolish townships in order to lessen township status. Coextensive township road districts were also vulnerable to elimination or consolidation. In 1951, Kost, playing up a rural-versus-urban motif for the conflict, attacked "the County Road Unit propaganda." He blamed the road unit idea (district consolidation) on a centrist group sitting "around a large table with a highly polished top in some committee room deep in the heart of one of our large metropolitan cities." No doubt "high-powered engineers" were present at the meeting, those who dreamed of "super duper highways." Big contractors, eager for "those large, juicy state and federal contracts," were surely at the gleaming table. "Of course, the play-boys" that made a living on stocks and bonds were there, beside a bureaucrat or two from Springfield and the nation's capital. There might have even been a "high-powered" professor present, helping cook up the unit road notion. TOI took its case against road district consolidation to the people and flooded the general assembly with hundreds of petitions opposing the unit idea. County road units did not, perhaps as a result of the TOI campaign, become the norm through legislation; but individual counties could vote themselves a unit district if a majority of a county's voters and two-thirds of the townships voted that way. Saline County, for instance, held a referendum in April 1953 on the notion but rejected it by a sizable margin. Kost went to Harrisburg, Saline's seat, before that election to mobilize opposition to change.[12]

County Contests before Con Con

To a certain degree, the war that Kost, TOI, and *ICTO* waged with the centrist enemies of home rule and township government prior to the 1969–1970 Constitutional Convention (Con Con) was shadow boxing. There were, however, between World War II and the late 1960s, real contests in four Illinois counties, Macon, Clay, Morgan, and Saline, between local forces that embraced township government and those that denigrated it.

Even though Macon County had been an easy conquest for those urging the township innovation in 1859, with no unsuccessful adoption attempts before that year or abolition votes afterward, this county, with its overwhelmingly dominant county seat, was by the 1940s a prime candidate for governmental reform. Many of the blue-collar families that had swelled Decatur's ranks and made it one of the state's largest cities knew little or nothing of township government or the rural bias that a board of supervisors, because of its makeup, could sometimes exhibit. Indeed, there had been discussion of such a referendum in 1943; but the war focus plus the erroneous idea that 20 percent of the voters would have had to sign petitions left the issue hanging. After the war ended, however, the two daily newspapers in Decatur, under common ownership, stood solidly behind reversal of the 1859 adoption and gave the reform topic thorough coverage, coverage that *ICTO* editor Robert L. Morris called "propaganda."[13]

Macon County's auditor, Ellis B. Wheeler, acting as a private citizen, spearheaded the countywide petition drive, which culminated with the submission of petitions containing 2,628 signatures on the first of September. Wheeler estimated he had 482 more signatures than 5 percent of the number who voted in Macon County in the November 1944 election. County board members, many of whom were township supervisors, began immediately to cast doubt as to whether Wheeler had collected enough names; but State's Attorney Ivan J. Hutchens assured the supervisors and the public that the law was clear, that not registered voters but the number voting in 1944 was the correct base from which to determine the signature target. Several board members reviewed the 149 pages of petitions in the hope of finding fault, and some asked for a second legal opinion from Illinois Attorney General

George F. Barrett. A few days later, Barrett noted that 42,917 had voted on 7 November 1944. Therefore, Macon County was considerably below the 75,000-vote cutoff. Wheeler had exceeded 5 percent of 42,917, and the referendum was to occur that fall.[14]

If readers or TOI had any doubt where the Decatur press stood on the abolition matter, a lengthy Sunday editorial, entitled "It Cost $18 to Pay $36 Bill," cleared up the uncertainty. *Herald and Review* staff warned the public that change opponents would claim, "change is 'communist' or 'nazi'." These accusations, said the editors, would be the work of "people who are trying to get our minds off the facts." The editorial staff chose to tell a little story, supposedly true, to illustrate the facts and solidify the call for governmental overhaul. A local printer, probably the job press side of the *Herald and Review,* took an order to prepare a thousand tax complaint forms for the Macon County Board of Review. The printing bill came to $36.72. Because three members of the board of supervisors had to meet and approve, at a cost of $18 in salary ($6 for each of three committee members) plus mileage, Macon County spent as much to approve the bill as it paid for the forms. Having three commissioners and no committees, said the paper, would save a "substantial" amount of public funds.[15]

The public did not have to depend on local newspapers for coverage of the Macon referendum. Both sides eagerly sought public occasions to present their cases. Decatur Township's highway commissioner defended townships at an American Business Club luncheon and challenged Wheeler's claim that Macon County would save money under the commissioners. Father Victor C. Goyke, a local Catholic priest, debated Wheeler in an event the local League of Women Voters co-sponsored with the Community Forum. Auditor Wheeler "used the cold logic of efficiency and money saving," said the *Review,* while Father Goyke, a "skilled and heartwarming speaker" appealed to the emotions, saying the township form was "'tested and true'." Goyke did note, however, in an apparently unemotional aside, that the 17 commission counties were much smaller in population than Macon. He said, "'I doubt the effectiveness of the commission form in a county as heavily populated as Macon.'" Goyke's premise for turning down reform will appear again, in the 1990s, when voters in another heavily populated Illinois county will have to compare their township circumstance with that of the much less populated commission counties. The local League sponsored a speaker from Morgan County, Mrs. Orville Foreman, of Jacksonville, who came eastward to Decatur to heap praise on the commission format. All of Morgan's three commissioners were, she told her audience, from rural portions of the county, even though more than half the county's population resided in Jacksonville.[16]

Decatur's editors kept the matter before the readers with news items during the week and with investigative reports and editorials over the weekend. The 26 September *Herald and Review* carried a long article by Ellis D. Roberts about waste among the several township road districts. "With 17 of these unique divisions in the county," said Roberts, "one finds an astounding duplication of road services." All but three of Macon's districts had their own tractors; all had their own trucks; and most, in his opinion, were "making a thin use of mechanical equipment." His editors, in a companion piece, noted correctly that dissolving townships would not automatically dissolve their road districts. That related reform would, and probably should, deserve a separate vote. A week later, the *Herald and Review* attacked another township-level responsibility, property tax assessment, and found far too much diversity in assessing philosophy and methodology. "The commission form of county government offers a distinct and immediate advantage to the people of the county in giving them uniform assessments for tax purposes." The next weekend, *Herald and Review* readers learned that the editors had found no adverse effects (of making the switch to commissioners) on local option (liquor sales), libraries, cemeteries, or justices of the peace. Ten days before the referendum, the editors summarized their position in a lengthy editorial that concluded with this thought: "A vote against the continuance of township organization in Macon county gives us our best chance to get efficient, competent county government." Finally, on the weekend before the second of November, the editors provided readers with a facsimile ballot, marked "Against," and ran an editorial in which they cited some of the politics playing out as the big day approached. "There are times," said the *Herald and Review,* "when it is necessary to show the little political fixers that they cannot lead an independent electorate around by the nose.... Adopting commission government is a blow at machine politics in Macon county."[17] Someone had actually linked township government with a political machine.

On Tuesday night, 26 October, one week before the election that also sent Harry Truman back to the White House, TOI mobilized its out-of-town cavalry to aid in the Macon County confrontation. The incursion had been in the works for some time because *ICTO* carried a piece in its October issue promoting the appearance of Joseph R. Thomas, the Ohio township executive who sometimes addressed TOI conventions. Township officials from other Illinois counties on hand that Tuesday night to support their Macon brethren included Troy Kost, who spoke briefly about the obvious, to him, benefits of having 41 board members rather than just three. Thomas, who had presided over the Township Officials Association of Ohio for 21 years, said commission government was centralization and that centralization was the "path communism takes." Old Joe Stalin, warned Thomas, "'can't get this country through Washington. But communism can come in when the rule is taken from the people and placed in the hands of a few.'" Mixed up in the centralization movement, he believed, were "'professors and others'" who used "'paper rather than experience'" to recommend abolition of township organization. Thomas implored Macon voters to stop centralization "'before it's too late.'" Introducing Kost and Thomas that night at the Masonic Temple was attorney E. Wayne Schroeder, the man who had chaired the Decatur Association of Commerce's Taxation Committee. Schroeder's committee had just released a report that showed taxes would increase by at least $65,000 annually if Macon County dropped townships.[18]

In the end, it was not even close. Auditor Wheeler lost his bid for reelection, and private citizen Wheeler saw his abolition effort go down to defeat by a margin of more than two to one. Rural areas overwhelmingly opposed giving up their townships and their guaranteed representation on the Macon County Board. Decatur's Association of Commerce did recommend, however, after the outcome was safe, reduction in the number of assistant supervisors (then at 24) and enactment of a law to help equalize assessments across the county.[19]

I have to wonder what H. S. Hicks would have thought about this 1948 contest and the protownship stance the Decatur Association of Commerce adopted. After all, Hicks had died in Macon County, in the employ of the Decatur Association, gathering data for a study to condemn township government, a study the Association would publish along with posthumous praise for his efforts. Not once, in all the coverage of the 1948 referendum, did anyone mention the local work of this Rockford reformer.

Four years after Macon County reaffirmed allegiance to township organization, TOI dispatched its road warriors several counties south-southeast of Decatur to help townshippers in Clay County battle the commissioner menace. At the courthouse in Louisville, Kost and long-time TOI president, William J. Vieregge of Gilman in Iroquois County, assured an audience consisting mainly of township officials that the township, compared to the alternative, "was more indicative of home rule, constituting the basis for all representative government."[20]

Like Macon County, Clay had adopted township organization at the outset of the Civil War and had not voted again on the issue. "Fairly consistent in ability and efficiency" is how the *Flora News-Record* described the operation of Clay's townships and board of supervisors. For a rural county like Clay, road conditions, however, were a major matter. Resurfacing dirt roads with gravel, widening and straightening roads for higher speeds, and replacing aging bridges were improvements that farmers wanted and needed. From Clay County, citizens could and did travel regularly to nearby commission counties Edwards and Wabash, where, according to one reformer, rural roads were "very nearly all gravel." That same change advocate claimed another advantage of having three commissioners versus the 14 then forming the Clay County Board of Supervisors: the ability of voters "to get to the seat of their trouble" much easier. "Buckpassing" would diminish considerably if the referendum succeeded. Another correspondent, an opponent of change, cautioned *News-Record* readers that the three commissioners would take control "away from the voters" of the townships and centralize it in Louisville. And roads would be no better.[21]

Voters in Flora, Louisville, and the rest of Clay County liked their townships and board of supervisors by a margin of 5,174 to 1,800.[22] Even without support from Fortress Astoria, Clay County townshippers would likely have prevailed; but the tradition of going on the road in defense of local townships and road districts had proven successful. Losing even one county to the forces of centralization might encourage centrists elsewhere to try their luck.

"I do not think I risk anything in predicting," wrote Illinois College's President J. M. Sturtevant, in 1879, "that if the people of this [Morgan] county do not adopt it [township organization] at this election, the time is not distant when they will."[23] Here was at least one professor that Troy Kost could have embraced, but Sturtevant was wrong about the direction voters in his adopted county would go. Not only did Morgan County folk disagree with the president in 1879, they eschewed townships in 1880, 1882, 1884, 1886, 1888, 1892, 1894, 1895, 1902, and 1907. Then Joe Boyle tried. Actually, he tried twice, 1956 and 1957, both times apparently without benefit of advice or intervention from Astoria.

Joe T. Boyle was a Jacksonville businessman and salesman. City directories placed him at the Morgan County seat from the late 1930s, when he would have been about thirty years of age, until the middle 1960s. For several years, including the 1950s, he owned and operated a vending-machine business. His wife, Virginia, was an art teacher in the local public schools.[24]

It only took the signatures of 50 registered voters to fill the petition requirement for a referendum on adopting township organization. Boyle and 51 others submitted such a petition to the county commissioners toward the end of September 1956. Most signers resided in Jacksonville, but a few were from Chapin or Concord. Commissioners Hobart Riggs and Ralph Heaton immediately requested an opinion from State's Attorney Albert W. Hall as to the "sufficiency" of the petition. Hall responded by noting the late submission did not provide time enough for the proper preparation of ballots. Absentee ballots, for instance, had already gone out to voters who had requested them. Hall, in advocating ballot denial for November, left open the question of whether the petitioners might have access to the ballot at the next general election.[25]

That next general election was to occur in November 1957, and Boyle made certain he had plenty of signatures so election officials would have no problem fulfilling their obligations to the voters again. When the *Waverly Journal* asked Boyle why he was heading another petition drive, he said he represented no organization but rather "'a large number of people who would like to see the matter brought to a vote.'" Dissatisfaction festered, said Boyle, because of the existing tax structure and ongoing reappraisal of real estate and commercial and industrial personal property. Boyle knew, and told the *Waverly Journal,* that most Illinois counties were township counties; and he believed "'the township form of government [was] closer to the people.'"[26] If

Boyle was not in league with TOI, he somehow had picked up one of its buzz phrases: closer to the people.

Neither Troy Kost nor any other major TOI player ventured into Morgan County for what may turn out to have been the final time that adoption of township organization will reach an Illinois county's electorate. No mention of the locally well-publicized prelude to the 1957 referendum appeared in *ICTO*. Maybe it slipped under the radar at Fortress Astoria. Maybe Kost chose to ignore Morgan because he feared having to report an unfavorable outcome at a time when politicians and academics were glaring at township government. On the other hand, Astoria may have provided aid and comfort of a clandestine nature.

Several outsiders did find their way to Jacksonville to share views on the competing governmental systems; and a couple visitors had township ties. Arthur Witte, clerk of Adams County (Quincy), spoke at length, during a League of Women Voters debate, about his experience with township organization. Opposing him was a local League member, Mrs. Francis Berry, who iiked what Morgan had depended on since becoming a county in 1823—three county commissioners.[27] About a month later, the League and the Morgan County Farm Bureau sponsored a public meeting at the courthouse during which Kevin Burke, the supervisor of Woodside Township in neighboring Sangamon County, was to share the stage with a former Morgan County commissioner, Virgil Wegehoft; Southern Illinois University professor Irving Howards, an expert on governmental matters; and Maurice Scott, from the Illinois Taxpayers' Federation of Illinois, who had already stated his opposition to the change in Morgan. State Senator George Drach of Springfield covered both sides of the issue when he spoke to the Waverly Lions Club a couple weeks before the referendum. Drach noted how the people seemed to prefer the township approach while "most political economists" liked the streamlined nature of the commission approach. He praised those who showed enough interest in local government to raise the question of reform.[28]

Editors of the *Jacksonville Journal* took a dim view of what Boyle and collaborators were suggesting. The *Journal* carried Maurice Scott's unanswered comments about 1957 being a "'mighty poor time'" to switch, a time when Illinois might just be doing "away with the township supervisor form." *Journal* editors resurrected comparisons that the State Tax Commission had made, of township and commissioner counties in the 1930s, to help bolster contemporary observations about the newspaper's fear that townships would further divide Morgan into factions. In a later item, half news and half editorial, the *Journal* pecked at the township idea, dropping hints about how confusion would reign if Morgan chose to switch. Township *proponents* have also made, of course, good use of the same strategy in trying to convince voters to hold the course and keep townships in place. The *Journal* completed its case on 3 November with a short editorial telling voters how to vote against change and warning that the election's outcome might "vitally affect" Morgan County's future.[29]

Boyle and friends paid for a few paragraphs in the last Sunday edition of the Jacksonville paper before the election. The commissioners were a connection to England and Virginia, said their ad, and totally unsuitable for the modern era. "What would you rather ride in," they asked, "an oxcart or an automobile?" They wanted "Mr. And Mrs. Jacksonville" to ponder the question as to whether they could remember when they ever had a commissioner who was something other than "a retired farmer who moved into town and got himself elected County Commissioner just to pass the time away." Could they do worse? Chances were good, said the ad, they could do a lot better.[30]

Tradition trumped townships that fall in Morgan, thus confirming the fact that the township tide really had died 34 years earlier in Cass, the county immediately to the north of Morgan. Roughly four thousand agreed with the *Jacksonville Journal* while about two thousand took Boyle's side. The reformers won not a single one of Jacksonville's 26 precincts but did take Waverly's three and a few scattered others. Boyle thought he might try again, but never did; and, after 17 failed attempts to put it in the township column, Morgan remains among the 17 Illinois counties today taking a different governmental route and among 15 that never sampled what Fortress Astoria espoused.[31]

Kost moved no troops to Harrisburg in 1966 when Saline County faced abolition of its townships, nor did *ICTO* carry any notice of the debate and referendum. By late 1966, Kost and TOI had bigger concerns to consider, such as the recent ruling of the United States Supreme Court over equal representation and the increasingly strident call in Illinois for another constitutional convention. Saline County has always had at least one commission county on its doorstep. Pope County, to the south, remains one of three Illinois counties never to have voted on township organization. Gallatin, Saline's eastern neighbor, and Saline were the last of the nineteenth century, late-majority counties to slough off commissioners and embrace supervisors, in 1890 and 1889, respectively. Williamson, to the west of Saline, had reverted to commissioners following a referendum in 1932, and seemed to be doing well without townships. Saline County was, the proponents of a return to the old ways believed, an opportune target.

Events played out over a period of approximately sixty days, between the filing of nine hundred petition signatures in the middle of September and the November election. The supervisors seemed surprised that so many of their constituents would want a chance to vote them out of office. Board Chairman C. L. Blackman invited any of the petition signers out to Stonefort, which straddled the Saline-Williamson line and where residents would gladly, he said, tell visitors about taxes being three times as high on the Williamson side as those in Saline County. Blackman and his board asked attorney L. M. Hancock to provide an explanation as to exactly what all this uproar could mean to the way Saline conducted governmental affairs. Hancock seemed to favor the retention of supervisors and townships as he produced figures on the prevailing tax rates for all county/township services in Saline and Williamson: $.971 per $100 of property valuation in Williamson versus $.401 in Saline. No wonder the west (Williamson) side of Stonefort might have looked longingly eastward to the Saline side of the village.[32]

Opposing forces placed sizable display ads in the *Harrisburg Register* during the week before the referendum. On the side of keeping things as they had been since the 1890s was the Bipartisan Committee to Continue Good Government. The BCCGG advised readers of the *Register* that the little yellow ballot was "one of the most important slips of paper you'll see in many years." With it in hand, voters had the chance to preserve "a democratic representation form based on the American ideal." As matters stood then, the board of supervisors consisted of the 13 township supervisors and five assistant supervisors from Harrisburg, but "under the commission form of government, it [was] possible that as many as 12 of the 13 townships in the county would no longer be represented on the board." Surely no one could stampede, pressure, or coerce a board with 18 members. What about three? A couple days later, another BCCGG ad told of the threat to "a vestige of our democracy" that the founding fathers had set up and the nation had fought wars to preserve. How could there be anything wrong with a form of government that was the norm "in 85 Illinois counties?" After all, it was "still the American way—the best way!" BCCGG included in this second display a nugget that probably reveals the underlying cause for this change movement in Saline: plans to build a new courthouse. Reformers will tell you, said the ad, "that by voting out township government you will be voting out the proposed new courthouse. THAT IS OUTRIGHT DECEIT!" The truth, said BCCGG, was that only the Saline County Building Commission could alter the plan to construct a new county building, "no matter what the outcome of the election may be."[33]

The Committee for a Better Type of County Government (CBTCG) chose the fourth of November for its advertising splash. Cost of governing under township organization was one big CBTCG complaint. From 96, at the time, the number of elected officials (and costs of keeping them) would decline sharply, if voters would use the little yellow ballot appropriately. "We say the cost of the Township form of government is over $30,000 per year compared to $13,400 for the Commission type of county government." A board of county commissioners could, said the CBTCG, still choose to go ahead with a new courthouse "or repair the old one." As to concentration of all commissioners from one place, the record showed that Saline's officials (clerk, sheriff, etc.) then serving countywide lived in several different localities, not just Harrisburg. Besides, the supreme court of the land had decreed as inappropriate the tendency for rural, lightly populated areas to enjoy excessive representation on a governing board.[34]

Register editors sided with the BCCGG. "In our opinion," said an editorial, "no one has yet shown clearly and unmistakably that either form of government would effect any great economics in comparison with the other." Retain township government, the editorial staff argued, because of its genuinely representative nature. And retain it they did on 8 November, 7,103 to 2,665.[35]

Professorate, Commission, and League

Troy Kost had no formal education beyond Astoria High School, yet he often had to contend with professorial criticism in his role as chief defender of the township faith and sometimes countered academic calls for the demise of townships with acerbic commentary about the patriotism of professors and their tendency to affiliate with those who wished to profit financially from township dissolution. Also posing potential threats to the township way of life in the late 1950s were a state commission focusing on township problems and the nagging negativity of the League of Women Voters.

A favorite outlet for academic criticism of townships in the postwar years was, as it had been in the Depression era, the *National Municipal Review.* In short articles, which authors and editors toned down to serve a broader audience, professors hammered at this layer of local government that had somehow escaped the reforming zealots of the 1930s. Professor Clyde F. Snider, University of Illinois, began his "The Twilight of the Township" by saying that "An early landmark in the development of American political institutions is today approaching oblivion." He ended with the observation that "From whatever angle considered, the conclusion seems inescapable that the sunset of township government even now is past." The *Rock Island Argus* found Snider's article to its liking, building an antitownship editorial around his comment that "Both reason and experience assert the civil township of the Midwest is a costly relic of other days." Are "Iowa Townships Still Here?" asked Richard C. Spencer of Coe College. In Iowa, where townships had lost already a good deal of their power and did not even make the list of governmental bodies in the United States Census Bureau's recent count, they somehow managed to persist. "Local sentiment" tended, in Iowa at least, to guard jealously the few remaining duties that townships officials could perform. James W. Drury, University of Kansas, based his article, "Townships Lose Ground," on a study he had been doing of the fifteen hundred Kansas townships. Roads, cemetery maintenance, fire protection, and extermination of prairie dogs were some of the functions he found townships still performing. Kansans, in the 1950s, seemed prepared to let their townships slip away—once the legislature solved the problem of how to dissolve them and transfer their assets, debts, and duties. In "Townships on Way Out" Paul W. Wager, University of North Carolina, ran through a list of states that had been reducing township responsibilities, saying "Every argument advanced twenty-odd years ago for their elimination has become even more cogent with the passage of time." Tradition, home rule claims, and "political advantage," were, according to Professor Wager, their salvation thus far. He noted how township officials in several states had united "into powerful state associations" to protect jobs and political influence.[36]

I am guessing that Fortress Astoria did not subscribe to the *National Municipal Review* (which became the *National Civic Review* in January 1959). TOI leaders had known for years, for decades, where the township stood in the eyes of professors who studied and pontificated on governmental efficiency. As long as Kost could paint ivory-tower professors as slightly unpatriotic, for even suggesting townships were unnecessary, and largely disconnected from the reality of providing services at the local level, he could confidently ignore their longstanding lamentations about the persistence of townships.

Townships in Illinois had come under commission scrutiny both before and during the Great Depression, but not until 1957 did the general assembly drag out its microscope for another close-up look. The Township Government Problems Commission resulted from the Seventieth General Assembly's desire to improve local government. One creation of that general assembly was a commission to examine counties; and, according to H. B. 892, signed into law 8 July 1957 by Governor William G. Stratton, the other was to look at their smaller brethren. The township report was to reach a subsequent general assembly by 1 March 1959; and, unlike its sister commission on counties, the township commission was not to extend beyond its report date.[37]

TOI was a supporter of the Township Government Problems Commission (TGPG) as was the Illinois Agricultural Association (parent of the Farm Bureau). *ICTO* carried several news articles about the progress of the commission and a number of informative columns by the commission's executive secretary, Professor Irving Howards. Merle K. Anderson of Durand (Winnebago County) and Joseph Rittenhouse of Bement (Piatt County) were among the five "public" members Governor Stratton appointed to the TGPG. Anderson and Rittenhouse served then also as members of the TOI board of directors. Six members of the general assembly, three from each chamber, rounded out the commission. Apparent to Howards, and probably to Kost as well, was the fact that the legislators had "not organized [the TGPG] for the purpose of recommending the elimination of township government in Illinois." Nevertheless, vigilance on the part of TOI was the price of persistence in the political climate of the late 1950s.[38]

The TGPC met several times in Springfield, beginning in October 1957, and also held public meetings around the state. Rep. Rollo R. Robbins, the Republican from Augusta (Hancock County) who had been the chief sponsor of H. B. 892, won the uncontested right to chair the commission. Robbins reported that Southern Illinois University had offered to conduct research for the commission at no charge. TOI's Rittenhouse revealed to the members at the first session that his organization had long favored a careful study of the township situation.[39]

At its second gathering, TGPC members heard spokesmen from the Illinois Agricultural Association and the Taxpayers' Federation of Illinois. Among the list of suggestions that Maurice Scott of the Taxpayers' Federation offered was that a "higher number of voters should be required to change from commission form of government to Township form of government." He made that suggestion two weeks after Morgan County had turned down the township model, and what he wanted increased was the number of signatures needed to get such an issue on the ballot. Maybe 5 percent of the legal voters of the county would provide a better indication of community sentiment, he said. "Such an election is costly and should not be effected by a petition of merely 50 signatures."[40]

Troy Kost got his chance to speak at the third session of the TGPC and asked for the commission's help in "strengthen[ing] the present township government." Justices of the peace ought not serve, as they then did, on township boards. Townships ought to have budgets, like the counties; having budgets would help reduce the amount of "tax problems" at the town meetings. Kost told commissioners that townships were often too small and that he welcomed "a study of this problem." Perhaps some township boundary lines were ripe for change, he admitted.[41]

After a series of intelligence-seeking local hearings later that year in Olympia Fields, Rockford, Bloomington, Edwardsville, Macomb, Carbondale, and Peoria, the TGPC fulfilled its mandate by reporting to the Seventy-first General Assembly and went out of existence. Some of its "hardly…startling" recommendations, as Howards wrote in 1960, gained legislative blessing. Among the favored ideas of TGPC was Kost's suggestion to remove justices of the peace from township boards. A new law replaced justices with three township citizens (first auditors, later trustees), elected at-large, who would serve with the township supervisor and town clerk as board members. Highway commissioners received more freedom in contracting for purchases up to $1,000. On many other TGPC recommendations, the legislators resisted change; but they did create a new commission to study "township laws for the purpose of codification and revision." If that new commission was to have more impact than the one that just folded, said Professor Howards in the *National Civic Review,* its recommendations would need more "energetic support…from significant organizations around the state."[42]

Kost faced not a single female adversary among the professors who were advocating or documenting the decline and fall of the township in the 1950s. He faced not a single female when he testified before the TGPC. The great majority of the general assembly members in those days was male. Toward the middle of the 1960s, however, Secretary-Treasurer Kost had to contend with some mighty determined females, those who belonged to the Illinois League of Women Voters. That confrontation would continue for Kost beyond the adoption of a fourth Illinois constitution in 1970; and, even today, the League can prove a challenge for TOI leadership or local township officials.

Townships were on the League's watch list year after year in the 1960s. By 1963, 160 governmental studies were underway at the local level of the Illinois League. Cook County chapters, in particular, were giving townships a lot of thought. "The leagues involved in these studies felt strongly that township government should be given a great deal more attention," said *The Illinois Voter.* Some leagues were hosting workshops, prior to the annual township meetings, to prepare study personnel for what to expect when they attended the meetings. In 1966, as the Illinois League looked at its program of study plans for 1967–1969, township government was a leading contender for even more scrutiny. Then, by October of 1969, on the eve of Con Con, the state League began asking its members to consider the possibility of townships disappearing altogether from the Illinois political landscape.[43]

The Illinois League of Women Voters was the "most active interest group" at the 1969–1970 Constitutional Convention, according to Ian D. Burman, who conducted a study of lobbying at the convention; and squarely in the League's sights was township government. On 18 February 1970, the Chicago headquarters of the League issued a lengthy statement for convention delegates in general and for the Committee on Local Government in particular. Mrs. Andrew J. Koller, chair of the League's Committee on Structure of Local Government noted that statement represented the culmination of a three-year study by local leagues in Illinois and that a prime concern was the need to reduce "the number of units of local government." She and her fellow leaguers hoped the new constitution would name only the county and municipality and leave subsequent general assemblies the task of creating other forms of local government, as the need arose. "There should be no mention in the Constitution of township government. The League of Women Voters favors the elimination of township government." Koller said Illinois needed to get busy and provide a means to transfer each township's assets and functions to the appropriate county or municipality.[44]

Troy Kost might downplay his townships' professorial critics as being out of touch with governmental reality. He might influence the makeup and recommendations of legislative commissions. When it came to the League of Women Voters, however, he would continue to face a determined and effective adversary through much of the balance of his TOI leadership years.

One Man, One Vote

During his TOI watch, Kost witnessed the demise of all the boards of supervisors. After the elections of April 1972, the supervisors (primary and assistant) in counties like Kost's Fulton lost their board seats. The United States Supreme Court had ruled four years earlier that having one supervisor represent a township of, say, three hundred residents and another represent three thousand on the same county board was unfair to those in the township where the larger number lived. Commentators began referring to the court's 1 April 1968 ruling on equal representation as "one man, one vote."[45]

Illinois had tried over the years to compensate the more populous parts of counties, beginning in the 1850s with the assistant-supervisor solution; but the number of citizens per urban assistant never got down to that ratio in the lightly populated rural townships. The *Bloomington Pantagraph* sarcastically calculated that for McLean County to recast its board of supervisors so that all segments of the county would have one board member per 387 people, as did Cropsey Township in the 1960s, the board would have required 240 members.[46]

Legislating into life the one man, one vote principle fell to the Seventy-sixth General Assembly; the pertinent laws came from house bills 2806, 2808, 2809, 2810, 2811, and 2812. A township's supervisor could still serve on the county board, but she or he needed to win a separate election for the privilege. County boards were to determine whether their members would run at large, as in commission counties, or in districts of roughly equal population. If in districts, was it to be a single member or multiple members? The county had to decide. A single township might constitute a board district, if the population were just right. Multiple-township districts were possible, as were districts that split a township. No board from any of the 84 affected counties could have fewer than five nor more than 29 members, so counties like Sangamon and LaSalle, with 51 members each, were going to need fewer chairs in their boardrooms. No county could choose to have more board members after April 1972 than it had right before that transitional election. Assistant supervisors were to disappear.[47]

TOI sided not with a few recalcitrant defenders of the status quo but rather with the great majority of townshippers who were embracing the change. An unsigned November 1971 article in *ICTO* said the forthcoming divorce of county boards and township supervisors "should strengthen both county and township government." In turn, the people would see better delivery of governmental services at the local level because county board members would concern themselves "with the policies, operation and procedures of regional county matters" and township officials would worry about the "affairs of their districts [townships]." The writer, probably editor Lee Ahlswede, cited the separation of county board and township supervisor that had been the mode in Cook County since the 1870 Constitution. So the boards of supervisors, first convened in 24 counties in 1850, had lasted 122 years; but, on the first Monday in May 1972, they vanished.[48]

A Fourth Constitution

By the 1960s, a fresh look at the 1870 Constitution was long overdue. Patching the state's charter was difficult at best and often impossible even after passage in 1950 of the so-called Gateway Amendment, which was, according to Cornelius, to make other amendments easier "by largely eliminating the negative impact of those failing to vote on the amendment." Between Gateway and the late 1960s, only 6 of 15 constitutional amendments had succeeded; and of those 6, only 2 were the result of the change that Gateway provided. A diverse coalition led the charge for another convention to consider overhaul of the 1870 document. Both major political parties, the League of Women Voters, newspapers, most politicians who mattered, the Independent Voters of Illinois, the Chicago Bar Association, and "countless business, labor and civic groups supported the convention call," wrote Joseph P. Pisciotte, who was to become the convention's executive director. On the other side, opposing a call for convention were the AFL-CIO, SOS (Save Our Suburbs), and TOI.[49] Once Troy Kost saw that a new constitution were inevitable, however, he was able to make TOI one of the most effective interest groups at the convention and to come away with a huge win, the biggest win of his long TOI service.

Kost viewed the possibility of a new constitution with great concern, given the battles that he had been fighting with academics, the League, and others over the role of townships in the local-government mix. He knew that township government had a rightful seat at the governmental table as a result of the 1870 Constitution and careful guardianship of the privilege since then. Whether that place at the table would exist after a 1970 Constitution was highly uncertain.[50]

Each of the state's 58 senatorial districts would send two elected delegates to Springfield in December of 1969 to deliberate Illinois's future. Kost, though an avowed opponent of overhauling the constitution, decided he might just as well be one of the overhaul crew and threw his name into the pot. Approximately five hundred men and women filed the necessary petition signatures, and a lottery determined ballot positions for both primary and final elections, one in September 1969 and the other two months later.[51]

Fulton County was part of the Forty-second District, which also included Warren, Knox, and western Peoria counties on the west side of the Illinois and Cass, Morgan, and Menard to the east. Kost was one of 10 from the district on the primary ballot and, with the endorsement of area Farm Bureau executives, used his good name to carry himself to a fourth-place finish in the September primary. John Knuppel, a Menard County attorney led the primary, with James Thompson of Galesburg and Ralph Killey of Monmouth claiming second and third places. Kost had only one-third the votes of the well-known Knuppel, but he had made the cut for the final ballot.[52]

According to the *Galesburg Register-Mail,* Kost "increased the momentum of his campaign through October and early November." He picked up at least one newspaper endorsement before the final election, from the *Lewistown Fulton County News,* and ran an advertisement in his hometown paper, the *Argus-Searchlight.* In his ad, he said the 1870 Constitution was a "good one, but some changes [were] necessary to bring it up-to-date with the present times and present problems." He favored an income tax (which Illinois did not then have), elected judges, elected state representatives, and elected statewide officers. Candidate Kost promised to devote all his time to his delegate duties. On 18 November, the voters ranked the four candidates in exactly the same order as they had come out of the primary. Even though Kost had closed the gap on the third-place finisher, Knuppel and Thompson easily took the two the top spots and went to Springfield for their district.[53]

Troy Kost went, too, but to represent several thousand township officials, not the people of the Forty-second; and he was one of only three defeated candidates who worked the convention as a lobbyist, according to Burman. "Troy practically lived with us," said John D. Wenum, then a Lake County professor, who was a delegate from the Thirty-first District and a member of the Committee on Local Government (CLG). Thomas D. Wilson singled out Kost as the most vigorous lobbyist of the entire convention. As Kost told Burman later, Kost had heard that the CLG had "'got the idea of doing away with townships. I figured we were sunk.'" Eventually, the CLG, in Kost's words to Burman, "'began to see the light.'" To help Wenum and other committee members see things the TOI way, Kost formally testified before them twice and was constantly present, or so it seemed to Wenum, talking to delegates or just listening.[54]

TOI was among a handful of advocacy associations that the CLG slated for testimonial time. Kost testified on the afternoon of 22 January 1970, along with four other township officials. He and his delegation were amenable to constructive changes that would permit townships to do their jobs more economically and efficiently. They even allowed that the annual meeting might have passed its prime. "All the speakers felt [, however,] there should be provision for townships written into the constitution." A week later, Chairperson John C. Parkhurst of Peoria sent Kost a routine note of thanks, in which he urged Kost to "keep the issues of the constitution alive in your area." Parkhurst had no idea then how lively Kost was going to become or how much the CLG would be hearing from him and his townshippers.[55]

What turned Kost the affable lobbyist into Kost the Con Con activist was Member Proposal Number 42, which came not from Wenum, Parkhurst, or any of the other CLG members, but from Delegate Robert R. Canfield of Rockford. An attorney and former state representative and state senator from some of the same area that H. S. Hicks once served, Canfield, according to Wenum, sent up several trial-balloon proposals. Most "did not go very far," but Number 42 got Kost's attention because Canfield proposed abolition of township organization and the establishment of commission government in all 85 counties having townships. Of course, township organization, as a way of manning the county boards in 84 of those counties, was already set to expire in 1972; and Canfield's proposal did not actually say he wanted to rid the state of its civil-political townships; but that is the way Kost interpreted the suggestion.[56]

Kost was taking no chances. He wanted the CLG and the convention as a whole to understand that he could bring disfavor upon the whole constitution. He rallied the township militia with a letter to the membership that began, "This letter is one of the most important you will ever receive from this office." Canfield's proposal, if adopted, he warned, would "abolish township government in Illinois." That proposal and others were "receiving favorable consideration" because delegates were under the impression that "the people back home want it." He told his people that "dreamers and do-gooders" were at work and that TOI folk needed to get busy and let the framers know "the people still want township government in our State constitution." He urged them to "write, telephone, telegraph or see your delegates in person" and to bring into the fray friends, neighbors, and even companies doing business with townships. "We must flood this convention with our wishes to keep township government in our constitution." The letter came with enclosures containing delegate names and contact information, plus a map so township officials could determine whom they should contact. "Home rule in Illinois," Kost told them, was in dire jeopardy that spring.[57]

Canfield and Kost appeared before the CLG on the morning of 11 March 1970. Delegate Canfield confirmed that, indeed, abolition of townships was his ultimate goal. He told how township government, active in only 22 states, was diminishing in importance elsewhere, too. Townships, although "once useful," were now almost "meaningless." He urged the CLG to purge "all reference to township governments [from the constitution] and leave the decision to the people," even in Cook County. Local option was the best way to determine the fate of townships in all 85 township counties, not, as apparently Kost had inferred, with blanket extinction via the new constitution. In answer to a question from Delegate Joan G. Anderson of Western Springs, a Chicago suburb, Canfield admitted that, in those places where townships were performing functions "very efficiently...the people will continue this form of government." Delegate Ted A. Borek, a Chicago automobile dealer, sparked discussion by citing figures comparing Illinois and Wisconsin that indicated to him fewer governmental units meant lower governmental costs. Parkhurst noted that the "real problem of township government...[was] whether the township exists as an autonomous unit" not whether the people could make a choice to have them or not have them.[58]

Then Troy Kost got his chance. For his second appearance before this critical committee, he came alone. Sure, there were probably more efficient ways for townships to deliver their services. Perhaps "a redrawing of township lines" (consolidation) would help. Maybe consolidation ought to be mandatory, but "as to Constitutional revision," he felt, "strongly that the Constitution should remain as is." Townships would, of course, be gladly willing to accept "special district functions" and attendant tax revenues. Finally, Kost made a comment that the recording secretary for the CLG (or perhaps Chair Parkhurst in reviewing the day's events) thought ought to be on the record verbatim—a very unusual occurrence in these minutes. Troy Kost closed his testimony by saying, "'Township officials will be very unhappy if they are not referred to in the Constitution.'" The antecedent of "they" was not the officials but rather their townships. Kost wanted, above all else, to have constitutional standing for his beloved townships.[59]

As winter edged into spring that year, the response to Kost's 28 February letter to his soldiers became evident to Parkhurst, his committee, and the other hundred delegates to the convention. Letters poured into Springfield demanding retention of the township. Burman reported that Delegate Anderson got 257 letters about township government, mostly in favor and most "generated" by "The Township Officials of Illinois." Her next highest category for incoming correspondence, branch banking, brought in only 69 pieces.[60]

TOI troops also responded to Kost's alert with petitions, resolutions, and (even) snapshots. Several, virtually identical, petitions containing hundreds of signatures came in from Woodford and Marshall counties urging retention of townships and road districts. "Proposal No. 42," said point 3 of the Metamora petition, "is just another step in the erosion of 'Home Rule' and denying our citizens the direct right of self-government guaranteed by our Constitution." F. H. Engelmen, the road commissioner of Harlem Township, sent Parkhurst petitions from each of "17 rural townships of Stephenson County," which Engelmen said totaled 2,359 signatures, urging the delegates "to do all you can to retain township government." Annual meetings for townships occurred, as usual, in early April. Some officials, like those in Jackson Township of Will County, took the opportunity to pass appropriate resolutions supporting the Kost cause. That resolution, in part, said electors opposed "changes in the constitution which would in any way limit, weaken or remove local and/or township government from the people of Jackson Township." Harry Meinert, the supervisor of Rock Grove Township in Stephenson County, chose a graphic approach. He sent to Parkhurst photographs purporting to demonstrate how efficiently his township's road commissioner removed huge quantities of snow from the township roads.[61]

By the time most of the reactions reached Springfield, the CLG had decided townships were perhaps a necessary evil, if delegates hoped to get a modern constitution past the voters. Approval did occur on 15 December 1970. Section 5 of the local government article (VII) reads as follows.

> The General Assembly shall provide by law for the formation of townships in any county when approved by county-wide referendum. Townships may be consolidated or merged, and one or more townships may be dissolved or divided, when approved by referendum in each township affected. All townships in a county may be dissolved when approved by a referendum in the total area in which township officers are elected.

Townships, like school districts, special districts, and certain other units of local government would be able to exercise only those powers that state law granted (section 8, article VII). Home rule authority, a privilege of counties and municipalities, did not extend to these lesser units. The 1970 Constitution mandated that no dissolution of townships or any other unit of local government was to occur unless the general assembly had passed a law governing the transfer to another unit of "assets, powers and functions" and "payment of outstanding debt" for the disappearing unit. This last section (section 12) was to prove extremely useful in thwarting township abolition efforts in the years that lay ahead. Although Fortress Astoria had seen Con Con as a life-threatening (for townships) event, in the end, the township kept its chair at the governmental table. Lobbyist and activist Troy Kost had much to do with that outcome, of course; but here are some of the other ingredients in the Con Con win.[62]

Perhaps because so many convention delegates had experience with local government and probably because the issue of home rule was so important to the Chicago delegates (and others), delegates listed CLG as their top choice for Con Con committee assignment more frequently than they did any other; 62 of the 116 chose CLG. President Samuel W. Witwer had to narrow the 62 to a workable number, say 15, while maintaining good balance among the (technically) nonpartisan factions at the convention. Witwer chose four Daley Democrats, including Richard M. Daley, the son of Chicago's powerful boss and now the city's mayor; three Cook County Republicans; three Downstate Democrats; five Downstate Republicans, including Wenum and Chair Parkhurst; and no Independents. Anderson, herself a committee member, described the CLG 15 as "colorful, complex, [and] often unpredictable." The chair, John C. "Parky" Parkhurst, had been a township official, a county official, and a state representative (for ten years). Anderson and Betty Keegan were League veterans but did not take on township abolition as a committee cause. Of the 15, the chair, Wenum, Anderson, Keegan, and four others came from areas with active township government. The remaining seven members, Daley, four others from Chicago, and the two from Little Egypt, were from nontownship areas of the state—a nice balance after all.[63]

The CLG enjoyed a good reputation among members and observers. Wilson described Parkhurst's committee as "diligent, slow, and cautious." Parkhurst had an informal style, members were on a first-name basis, and they were comfortable injecting humor into the mix. Even three decades later, Wenum still considered Parkhurst "a good friend." Wilson believed that this committee achieved "a high level of respect among" the overall delegation, and David Miller called its home rule provision "the most important innovation in the 1970 Constitution."[64]

Besides Troy Kost and Robert Canfield, many others presented their views in person to the CLG, both in Springfield at regular meetings of the committee and away from the capital at many locations as part of the plan of Con Con leadership to begin selling the constitution to the electorate. Several men from the "County Officials Association" were on the 22 January agenda.

Among them was a Mr. Briggs from Madison County who praised the annual meeting of townships as "the only place in which people have a right to vote and voice their opinions." A Mr. Harness, from St. Clair County, told delegates he believed people were proud of their townships; and, if someone were to try abolishing township government, "people would resent it and would probably not vote for the proposed new constitution. Rep. Clarence Neff, from Henderson County and a member of the County Problems Commission, told Parkhurst and other committee members on 28 January that they needed to tread carefully on the township and road district topic because "if township government were abolished, they would vote against the Constitution." Richard Friedman, executive director of the Better Government Association, reported to the CLG on the 29th about his organization's six-month investigation of townships in four counties—Cook, DuPage, Lake, and Will—and emphasized his remarks were only speculative vis-à-vis rural counties. As for Cook, and these three Collar Counties, the Better Government Association concluded that townships had become unnecessary. Herbert Klynstra of the Illinois Agricultural Association, among other topics, mentioned on 3 February his employer's view that abolishment of townships, if necessary, ought to be a legislative matter. When the CLC convened in Arlington Heights, on 11 February, Senator John Graham, a Cook County Republican, said home rule and township government were essential; and Rep. Eugene Schlickman, also a Cook Republican, agreed. Nearly a dozen township officials, including Ethel Kolerus, the supervisor of Wheeling Township in Cook, lined up to praise townships. Warren Wilson, a certified public accountant, favored townships if for no other reason than the fact that township tax collectors were more accessible to the public than their county counterparts. Amy Kummerow, from the Champaign-Urbana Democratic Central Committee, argued when the CLG met in Champaign on 6 March that, while township government was "undoubtedly valuable in 1870," there was in 1970 "little to justify its continuation." Frank J. Kopecky, director of the Legal Services Agency (in Champaign County), lambasted townships from every angle but especially from his own perspective as a welfare facilitator. Townships, he said, are an "archaic reminder of the past" and "an expensive and dysfunctional remnant of local government." With the township, "expertise, professional management, and the efficient resolution of citizens' problems have all taken the back seat to political patronage and the continuation of a grossly inefficient governmental system.[65]

Wenum told me he had gone to Springfield in December 1969 ready to "write townships out of the new constitution." He had grown up in rural Wisconsin and seen township government failing to do its jobs adequately. As far as he knew going into Con Con, counties could plow and otherwise maintain rural roads better than the typical township. But Wenum and his fellow committee members came to believe that so many other improvements were possible with a modern charter for Illinois that they could not risk failure at the ballot box by riling thousands of townships officials, their relatives and friends, politicians who favored townships, and the countless Illinoisans who looked with great favor on their little slice of the local government pie. So the CLG and the other delegates took what Elmer Gertz and Joseph P. Pisciotte called a "populist" solution. Give townships constitutional standing and let the people later, county-by-county or township-by-township, decide. Parkhurst told his committee in late March that any mention of townships in the constitution would "elevate" their status; and, of course, it would make Troy Kost extremely happy. Wenum said he felt their committee did not capitulate completely in that the provisions for intergovernmental cooperation (section 10, article VII) and special services area taxation to give residents of those areas access to a higher level of services (section 6, article VII) could later make townships less popular. He and other silent township critics hoped these provisions would "undercut" the townships' roles.[66]

On 12 May, Parkhurst assigned committee members to prepare the final draft-language for the local government article. To delegates Donald D. Zeglis of Momence and Ralph Dunn of DuQuoin he gave the task of writing the township segment. Parkhurst wanted the pair to emphasize the fact that down through Illinois history, since the 1848 Constitution, townships had always been "creatures of statute, not of the constitution." In other words, Zeglis and Dunn were to give townships constitutional standing but leave them dependent on future general assemblies for their existence.[67]

Parkhurst, though a former township official, had major reservations about the continued role of this governmental unit. While speaking in Bloomington at a public forum about the progress his committee was making, he lamented the fact that Illinois would still have far too many government units after Con Con ended its labors. The answer, he said, was to "'beef up county government.'" When an audience member asked about abolition of townships, he replied, "'We buried it in our committee so as not to arouse the natives. It's dead and has been for a long time.'" Parkhurst wrote responses to letters that had come to him from citizens who felt strongly that townships were unneeded. He told Mary Ellen Harkrader of Peoria that he had appreciated a thoughtful note from her, someone who realized "township government may not be the most efficient and most desirable system of local government." Rosemary Trump, of Peoria, had written Parkhurst about the fact that only three spectators attended the 1970 annual meeting of the City of Peoria Township. Thanking her for the *Journal-Star* clipping about the meeting, he said he was inclined to think the "Local Government Committee could take on the supporters of Township Government and win the issue, if we were brave enough." The committee had decided discretion was "the better part of valor" and opted for "local referendum to determine the fate of a township." Alice DeViney of Palatine had written him about several issues that concerned her, among which was "elimination of Township government." He noted in his letter to her how he and his committee had also heard, from sources other than her, about authoritarian handling of annual township meetings and lack of accounting of townships for tax money. Knowing by that May date that townships were going to survive if Illinois adopted a new constitution, he thanked DeViney for her "personalized and thoughtful" letter and said, the CLG had "had literally thousands of letters from proponents of the township form of government, but only a few from opponents."[68]

Parkhurst and committee were also receiving advice from the state's newspapers as they heard testimony, read and responded to letters, devised their stance, and wrote article VII. "Let New Constitution End Outmoded Townships" ran the headline of a *Danville Commercial-News* editorial on 15 February 1970. A *Moline Dispatch* editorial reminded delegates and readers, "Several of the reasons for the existence of townships in Illinois have evaporated." The *Bloomington Pantagraph*'s editors noted how far townships had fallen from the days "this New England import...overwhelmed the commission style of county government that prevailed during the early days of Illinois statehood." Court cases (like one man, one vote) and legislative action had "sliced away at the powers and offices of township government." As a result, the *Pantagraph* saw the actions of Troy Kost as those of a man begging for enough scraps to ensure life after the new constitution. Only by finding a new source of political muscle, "in modern language, muscle means money," could township government survive.[69]

What later became section 5 of article VII, arrived at the Committee of the Whole on Thursday, 30 July 1970, as section 8 of the Committee on Local Government's "Proposed Local Government Article," dated 9 July 1970. Delegate Anderson served as the committee's chief spokesperson on section 8, leading off with prepared comments about each of the three proposed segments of the section. She reminded the delegates on the floor that the role of townships as election districts for county board members was about to change drastically due to the recent reapportionment legislation, so that "townships...as government entities [would] no longer have any connection with the county." Her committee recognized, however, the "great long history [of townships] as subdivisions of the county" and felt townships needed "a little special treatment to get them from there to here to the future." Those words must have pleased Troy Kost greatly on that July afternoon, as did the wording of section 8.1: "The General Assembly shall provide by law for the formation of townships in any county when approved by county-wide referendum." Townships had made it to the floor of the convention. Section 8.2, a necessary part of the forward-looking package, allowed for consolidation, merger, division, and dissolution of one or more townships, following voter approval. Finally, 8.3 permitted voters, as those in Johnson County had done in 1876 and in Williamson in 1932, to vote to dissolve all their county's townships. Anderson noted this last segment of the section allowed dissolution only "in the total area in which township officers were elected," in order to cover the Cook County situation of inactive townships in Chicago and future situations in any township county where one or more townships might have become extinct due to the dissolution of individual townships.[70]

Delegates asked whether existing townships would receive a grandfather exemption and remain in place, because 8.1 said the general assembly "shall" permit townships. Anderson, in reply, noted the transition schedule, the route the state would take between 1870 and 1970 constitutions, would allow existing townships to "remain in force." When Delegate David Davis of Bloomington asked to see the fine print about that allowance, Anderson deferred to Parkhurst amid laughter from the chamber. Parkhurst told Davis he had "the boys working on it, and when we get the thing in language, we'll present it to you." Davis seemed to derive some satisfaction but reiterated the importance of them being very clear about this transition element. Anderson promised Davis she would deliver the provision to him personally.[71]

Delegate Paul E. Mathias of Bloomington, however, still harbored doubts about whether residents of townships then in existence would truly feel safe if the new constitution won approval. Mathias, who had for many years worked for the Illinois Agricultural Association as general counsel, was looking after the interests of the rural folks whom the IAA served. Instead of depending on Parkhurst and his boys to come up with a suitable transition provision to cover townships, he had an amendment to propose: "Townships as they exist on the date of the adoption of this article shall continue until consolidated, merged, divided or dissolved in the manner provided by law." In explanation of his concern, he reminded delegates of the Canfield proposal to abolish township organization and of the "correspondence" he and fellow "downstate township-governed" counties had received in its wake. Unless they did something before the local government article went to the Committee on Style, Drafting, and Submission, Mathias was afraid they were "going to just reactivate all that opposition." Clearly, Kost and Fortress Astoria had left a strong impression on some delegates.[72]

Later in the session, when his amendment came up for a vote, Mathias restated his fear about leaving the appearance that all was not well with existing townships. He offered to withdraw the amendment once the Parkhurst transition provision was in place, at the second reading of the article. His Bloomington colleague, Davis, rose in to say no one Davis knew at the convention was "afraid of the instant abolition of counties," in response to Anderson's question as to whether they were singling out townships for special protection. Davis said there were "a great many township...supervisors...who [were] scared to death that this Convention is going to eliminate all township government." Supervisors, having endured indoctrination, feared the worst; and it would take little more than a newspaper article implying the same, for them to "panic." With Parkhurst's permission, Anderson then polled the CLG; and the committee accepted a slightly revised version of the Mathias amendment. By heeding the Mathias-Davis warning, delegates protected turf until the transition provision took effect; supervisors and other townshippers could rest easier, as could delegates from "township-governed counties" and Troy Kost.[73]

Delegate Robert L. Butler, a member of the CLG, raised an issue that in the 1990s would prove crucial in efforts of a new generation of township critics: Could one "heavily populated township" determine the fate of all the other townships in a county? According to 8.3, said Anderson, the answer was yes. She felt 8.2 gave the behemoth township an alternative, because, henceforth, people in a single township could decide to dissolve their unit, leaving the rest in the county (supposedly) unaffected. In the 1990s, township critics often gathered petition signatures, for a countywide referendum, from population centers and not from far-flung outposts of a county. TOI officials thought that unfair to outlying townships and prevailed on the general assembly to provide statutory relief, which it did in 1997 (chapter 8). Delegate Betty Ann Keegan supported Anderson in defense of the flexibility factor they had built into 8.2; but Butler persisted in trying to convey to the convention the conflict he saw between 8.2 and 8.3. He said he liked 8.2 because each township would thereby control its fate, regardless of what neighboring townships might do, and continued to emphasize the inequity of 8.3, which could allow a populous township to dissolve all townships in that county. Butler, who represented six of the state's 17 commission counties, as well as three township counties, had not gotten much satisfaction by the time Anderson and the presiding officer on the floor suggested they move on to other matters. In essence, when it comes to dissolution, 8.3 (countywide referendum) seemed to take precedence over 8.2 (individual township choice) because several townships might vote a particular way on dissolution but still lose when a majority of voters chose the other alternative.[74]

When the township section of the local government article came up for a vote at that first reading, the support was overwhelming: 74-0 on 8.1, 79-1 on 8.2, and 102-0 on the amended 8.3. Applause followed.[75]

Among the last of the delegates to speak on section 8 (later to become section 5 of article VII) was Robert R. Canfield, the author of the member proposal that had caused Kost consternation. He rose in support of the Mathias amendment but more so to express his disappointment in failing "to achieve the goal that [he] set out to achieve by terminating townships." Canfield still felt "it would have been a great thing," and his community had supported him on the matter. Despite what he called "a substantial defeat," he could still express faith in the people of township counties to do the right thing, even though the article did not turn out the way he "would have done it."[76]

On second reading, 13 August, the local government article passed on roll call, 84-4. Butler voted for it, as did all the other CLG members present. Mathias voted aye, but Canfield was absent. Prior to that vote, Parkhurst had a chance to elaborate for the delegates his strong support for the article and to share some misgivings. One misgiving involved the failure of Con Con to "present the people some alternatives for the proliferation and the Balkanization...in [the] local governmental jungle which now surrounds us in the state of Illinois."[77]

Minor changes continued to the end. Third reading of the article and vote (99 in favor, 9 against, 3 abstaining) took place late in the evening of 30 August. None of the late opposition to the article concerned townships. Parkhurst warned delegates they were likely to get questions back home about the article's "lots of little innovations." He suggested they consult the final edition of *The Weekly Illinois Constitutional Convention Survey,* for which he and other committee chairs had provided summaries of their committees' various articles. About townships, Parkhurst had said therein the people would have "a chance to decide the future role of township government." Of course, he meant, as he continued, it was to be their first chance to decide township-by-township, if they so chose. Finally, he wrote that "a dominant city" not electing township officials could no longer "control the political destiny of others." On the last day of August, delegates approved article VII for final enrollment in the constitution, 83-0.[78]

All that remained to do in Springfield was for delegates to craft a statement about the constitution for the voters. On 2 September, Parkhurst and other delegates labored over the draft of the local government segment of the "Address to the People." Much of that segment dealt, perforce, with the innovative home rule allowances in article VII. Townships were to get one sentence: "Townships may be dissolved following popular referendum." Parkhurst thought it might seem less harsh to the people (especially, I suppose, Kost and crew) if it were to read, "Townships may be more easily combined or dissolved following popular referendum." He felt consolidation by referendum was a reform that deserved recognition in the address.[79]

Delegate Harlan H. Rigney, a farmer from Red Oak in Stephenson County, urged Parkhurst to leave the township sentence as first written. "I suggest to you that out our way this [consolidation] is a subject better left unsaid. I recall the tremendous outpouring of letters that we had on this subject." Many had told him that Con Con meant "the end of our township government," and they opposed calling the convention "on that basis." Rigney said he thought they had not really made a "significant change in this area." For his part, it would be much better to "leave it unsaid."[80]

Parkhurst again advised that he was merely trying to soften the wording and would not object if the sentence disappeared altogether from the address. Delegate David Kenney, a professor of political science at Southern Illinois University-Carbondale, told Rigney if there were people out Rigney's way concerned about abolition, there were people down Kenney's way who expressed the concern that townships "won't be abolished." So the old Upland-South versus Yankee division had popped up again on the township issue. Rigney's constituents feared losing their townships; Kenney's could not understand why counties deep in Little Egypt still allowed them to exist. The meeting chair turned to Delegate Anderson, who was still shepherding the township section of article VII. She said there ought to be more, not less, about townships in the address for the people, noting townships had gotten recognition as "another form of local government." As the meeting moved on to other matters, Parkhurst and Anderson conferred about the disputed wording and came back with a recommendation to drop the township sentence. Parkhurst brought a laugh when he noted, in suggesting deletion, that townships did not "even have a sentence in the Old Testament." The sentence disappeared as, to the very end of the convention process, framers were looking over their shoulders to see who was watching and who would be in their corner as the document went to the people.[81]

In a voting booth in Astoria on the fifteenth of December 1970, Troy Kost surely chuckled as he marked his blue ballot in favor of the new state charter. His yes vote would have helped Fulton County be one of the 29 counties that gave the main package of the constitution an affirmative majority. Another was Cook County, which did more than any other county to push the 1970 Constitution over the top, by roughly four hundred thousand votes.[82] Yes, Troy Kost must have been thinking that 1970 had been for him and TOI a mighty good year.

County Contests in the Decade after Con Con

Kost was approaching 67 years of age when the Sixth Constitutional Convention adjourned sine die on 3 September 1970 (figure 18). Some might have considered such a successful conclusion to a long struggle for inclusion in the new constitution as a good occasion for turning the fortress over to another, younger defender; but Kost had in him several more years of solid service. Looming later in 1970, as he probably knew on the day Parkhurst, Wenum, and the rest of the CLG packed it for home, an Illinois county's townships were going on the referendum balance, right across the river from Fortress Astoria. There, in Mason County, the Farm Bureau was championing a countywide vote on retention of township government. Later in the 1970s, abolitionists would be active far to the south, trying to place a countywide referendum on the ballot and in suburban Chicagoland trying out the new constitution's provision for individual township dissolution.

The last time residents of Mason County got to vote on township organization was April of 1863, a couple months before the Battle of Gettysburg. And they would not have voted in November of 1970 had it not been for the specter of one man, one vote. Reapportionment was in the wind that fall, and an equal-representation plan was to be in place for Mason and each of the other township counties (save Cook) by 1 July 1971. Delbert Hackman, a farmer from near Havana, the county seat, and president of the Mason County Farm Bureau filed the referendum petitions. He explained to the local press that the Farm Bureau wanted citizens, not the board of supervisors, to determine the cut of their next ruling body. "'Through the years,'" he said, "'we have slowly watched our power as voters be preempted by the legislature and the courts.'" The time had "'come to reverse this trend.'"[83]

Mason County, with only 14 members on its board of supervisors (13 township supervisors and one assistant from Havana), was well below the limit of 29 members that H. B. 2806 had specified for boards beyond 1972. If Mason retained township organization, the new board could not be any larger than 14 but had to have at least 5 members. Those members, elected in April of 1972, would each have to represent a population-driven district within Mason or run at-large. Either way, district or at-large, there would be at least two more board members than the three commissioners that a defeat of township organization would bring.[84]

Figure 18 Troy A. Kost, circa 1970. Kost guided the Township Officials of Illinois for nearly forty years. (Courtesy of the Township Officials of Illinois)

Although Hackman delivered the goods to the county clerk, the real change agent in Mason County that fall was Farm Bureau Executive Secretary Jerry Erwin. And, while ostensibly the impetus was concern over erosion of citizen control, I get the feeling that he just preferred commissioners to township supervisors and the larger county board. Erwin's first contribution to the debate appeared on 19 October. In what was essentially a guest column in the *Mason County Democrat,* he summarized Hackman's comments of six weeks earlier and then ticked off some advantages of commission government: reducing the number of board members from 14 to 3, saving taxpayers at least $30,000 annually, replacing township welfare with a countywide system of assistance, establishing the county commissioners as a new board of tax review, and allowing road districts and road commissioners to remain as they were. Erwin closed his opinion piece by promising the Farm Bureau's objective was "to present the issue to the voters and to inform them of the choices." He was back next with a letter to the editor that appeared the day before the vote. His major premises in that letter involved economics; and the conclusion seemed obvious to him: commission government for a county like Mason meant lower administrative costs. Some saving would occur when county officers took on new duties, at no additional cost to the county, e.g., the county clerk taking over for township clerks. Standing committees, a necessity with an "unwieldy" 14 on the current board, would disappear, and produce savings. Township auditors (costing $4,000 annually) would no longer serve. Commissioners would earn less in total salary and claim fewer compensated miles if the referendum succeeded. His comparisons, of course, as he noted, were with the existing board of supervisors. No comparisons were possible with the replacement setup that the board would have to make if the referendum failed.[85]

Uncertainty about the post-township arrangement gave the commission alternative in Mason County an unusual advantage, because township advocates could not urge retention of the status quo. But they could and did fight back by forming a committee and buying advertising space in the newspaper. The Committee for Responsive Government, featuring the slogan "Don't Lose Your Local Government," ran ads on the same days as the Erwin items appeared. Whereas Erwin had opted for lots of words, his opposition went with a few "Facts," as they put it. The committee's pro-township 2 November ad noted that a "majority" of Illinois counties were under township organization, implying that there was something good about being with the majority. Without providing documentation, the committee claimed less road maintenance money would be available under a board of county commissioners. The committee tried a common scare tactic by pointing out that any two commissioners could "rule the entire County," and threw in some misinformation by claiming that neighbor Cass County had been the last county to change "from Township to Commission form" and then to revert some years later.[86]

Troy Kost must have taken some enjoyment in subtitling a section of his December *ICTO* column "Do-Goodness Bites the Dust." He also pointed to the potential mischief when just two commissioners can team up to decide on a course of action for an entire county. Mason voters kept their townships and allowed the board of supervisors to determine the makeup of the board that would succeed it in 1972. The tally was an overwhelming 5,153 to 1,237; but I could find no evidence to suggest that Kost ventured into Mason for the 1970 fray, even though it lay along one of the principal overland routes between Astoria and Springfield.[87]

John Crawford Wilson, Harrisburg resident and charter member of the Saline County Taxpayers Association, told me that the real impetus in his county for the 1976 campaign to reform his county's government came from "the quarry owners south of Harrisburg." Companies extracting bituminous coal from strip mines (quarries) had been quietly agitating the question of commission-style management of county affairs since the late 1960s. According to Wilson, whom attorney Robert Rath of Harrisburg described for me as "the archivist of county secrets," Saline's quarrymen envied the quality of "oil and chip surfaces" that county commissioners in neighboring Williamson County had authorized. Better roads meant easier ingress and egress for coal-haulage trucks, and perhaps more profit from the mines. Replacing the 13 at-large (since 1972) county board members with 3 commissioners, even though road districts and their road commissioners remained, might yield better county roads for southern Saline County.[88]

Never well funded and lacking solid leadership, the Saline County Taxpayers Association (SCTA) longed for reversion to commission government for the county but struggled through the first few years after its founding "about 1970," according to Wilson. Into the SCTA leadership void stepped Harold James, a retired local man "with time on his hands" and extensive real estate holdings. Wilson described James as "aggressive and civically active," a man who had worked hard on another occasion to secure a better airport for his county. Rath remembered him as "extremely frugal and conservative," and as one who had opposed local school bond issues. James always "wanted something to fill his time," said Wilson. In 1976, he filled some of that spare time by taking up the twin tax causes of tuberculosis and townships.[89]

SCTA began circulating petitions in early summer, one to eliminate all vestiges of township government, the other to eliminate the county's longstanding tuberculosis tax. Many counties had already dropped the tax on property to provide funds to treat those that had TB or that health authorities suspected of having it. No county, of course, had dropped townships since 1932.

Petitions for both causes reached the Saline County Board the evening of 5 August 1976. Rath, on behalf of James, "an officer in the Saline County Taxpayers Association," made the formal presentations; but he told me later

James had "conned" him into the task and, as a result, Rath "drew the wrath of the county board" for so doing. Wilson, then a member of the board, moved that they "acknowledge receipt of the petitions." The SCTA had collected eighteen hundred signatures, which represented perhaps 10 percent of those voting at "the last general election." There was confusion apparently as to whether Saline needed 10 percent or just 5, so they opted for the higher number to be safe. Someone had already gotten State's Attorney Mike Henshaw to vet the petitions, and he told the board he believed the township petitions to be "legal." According to Henshaw, the board really had no choice but to "place the question on the ballot," a pronouncement abolitionists twenty years later would loved to have heard from state's attorneys of their counties. Someone at the board meeting then produced a letter from the State Board of Elections to County Clerk Maynard Cannon informing him that a special election would be necessary in January if dissolution prevailed that November.[90]

Most of the debate and coverage by the press occurred just before the election. Opponents of dissolution formed the Taxpayers for Township Government (TFTG) and found a suitable spokesman in Vernon Moss. TOI assisted in ways that Kost did not specify, although there is no evidence to suggest that he or any of his lieutenants personally ventured into the county. Harold James remained in the forefront for the SCTA.

Vernon Moss served then as supervisor of East Eldorado Township and had been chair of the county board until the prospect of conflict of interest over a new job led him to resign his board seat on the same night Rath presented James's petitions. Both the *Harrisburg Register* and the *Eldorado Journal* gave Moss space to air his argument for retention of townships. He pointed out, in a 29 October item that each commissioner would have to represent 8,400 Saline subjects, compared to only 1,938 for each of the 13 current board members. Democracy would suffer if Saline County unwisely joined the trend toward "more and more control of local affairs" by outsiders in Springfield or the District of Columbia. Moss quoted an Illinois Department of Local Government Affairs official, Edward H. Levin, who had said recently, according to Moss, that townships were "'just about the least expensive form of government around.'" Appointees, which someone other than the voters had chosen, were a commission-county trademark, he noted in a 1 November contribution to the *Journal.* He also warned then of the concentration of power in "a very limited part of the county," if commissioners came on the scene, and passed on a suggestion from an unnamed "state legislator from a commission county near Saline," who said that any county already having township government had "'better keep it.'"[91]

The TFTG used both papers to tell its story, running ads that emphasized a few key points and kept text to a minimum. Ads urged voters not to allow the proponents of change to mislead them. Saline County had employed the commission-style government "from 1848 to 1890." Why should the county "go backwards?" Continuance of township government proclaimed, "Let's Go Forward." After all, townships were working "for 85 Illinois counties." Even more important, perhaps, for the local situation was the uncertainty issue. "Don't vote to abolish Townships until you really know who takes over."[92] Transition uncertainty played well in 1976, and TOI would (and should) continue to use it as a defensive strategy until the general assembly fulfills its constitutional responsibility to provide for a smooth transition. Troy Kost may not have taken an autumn sojourn to the Shawnee Hills in 1976, but the Saline outpost of Fortress Astoria probably benefited from TOI advice and counsel.

James and the SCTA countered with a display ad in the *Journal* of 29 October, a text-rich ad emphasizing waste in local government and other factors that had prompted the petition drive that year. Nowhere, of course, did James or his group mention "quarry owners" lusting for better roads. Instead, the SCTA ad asked taxpayers if they wanted to save "up to 25% in Our County and Township Taxes," and promised their vote on the township issue would "Save [them] more tax dollars" than any vote they would cast that November day. Voters had a chance, said SCTA, to cut the number of government officials from 104 (13 board members and 91 township officials) to just 3 (commissioners). SCTA then offered favorable cost comparisons with two commission counties, implying that both of those counties, Massac and Williamson, had recently "gone to a 3 man board of elected commissioners." Saline, said the SCTA ad, was going with a tax levy of $27.10 per person, compared to $18.70 for Massac and $15.80 for Williamson.[93]

The "idea of a three-man county commission" exerted quite an appeal for the *Register*'s editors. Tax savings seemed to be a foregone conclusion, but one of the TFTG premises had hit home with the editors. Would "it not be difficult," said the *Register,* "to find qualified, high caliber individuals" to assume the full-time jobs as county commissioners "for what amounts to starvation wages?" Furthermore, county affairs would be in the hands of just two commissioners, because getting three to agree would be difficult. Editors challenged readers to think who could possibly fill such positions and gave the governmental reform a resounding "maybe." Voters decided they could not think of such commissioner candidates and voted down discontinuance, 7,684 to 4,037. They also voted to keep the tuberculosis tax.[94]

ICTO said, "The right side won." Moss got kudos for leading the "pro-township forces," as did voters for not buying "the ideas put forth by the proponents of township abolition." TOI and the magazine took credit for providing "assistance."[95]

Consolidation

A strange thing happened to Troy Kost between his first appearance before the CLG at Con Con, in January 1970, and his second, in March. He apparently caught the consolidation bug. Or maybe he just offered up consolidation, "a redrawing township lines to reach an optimal scale of economy" (as the official CLG minutes paraphrased him), in order to avoid even worse possibilities for townships during that critical period of constitutional maneuvering and posturing.[96] What had Kost been saying about consolidation of small governmental units before his March testimony? What did he and TOI think in 1974, when the consolidation rubber hit the township road—from one end of Illinois to the other?

Consolidation or merging, to achieve economies of scale for any small unit of government, usually met with immediate disdain from the pugnacious TOI secretary-treasurer in the 1940s and 1950s. To Kost, no matter what you called the process, the result was more centralization of power and less access for the individual citizen. With equal vigor, he opposed unwise consolidation of schools, road districts, and townships.

A modest amount of school consolidation, he did admit, was "advisable"; but much consolidation of rural schools occurred before local roads were ready to carry school buses properly. Children were sure to suffer, he complained, when weather made school access impossible for them. "Enemies of home rule," he wrote in 1948, advise school consolidation "because the children [are] no longer in the rural districts." Citing as an example one district that resulted from the consolidation of six one-room school districts, Kost worried that little children there and elsewhere would be "standing in the rain waiting for their buses" or walking even farther to classrooms than ever before. If they did somehow have bus access, then they would be leaving home early and returning late.[97]

Merging road districts or merging townships "for road purposes" did not win favor from Kost. Roads that Indiana townships had built to "A-1 condition" had in 12 years, under county control, he said in a 1948 legislative summary, deteriorated to the point where the "good folks in the smaller communities [were] clamoring for help." As for Illinois, he confessed in December 1950, "in a very few cases" consolidation of township-size districts into larger road districts might "be advisable, but generally speaking, it is not." Small districts, like the small farmer, could do the job better, he felt, because they could time advantageously their work against the weather. "If the grain would ripen as he wished and the weather would be ideal all year," a farmer could harvest all the grain in the county. That miracle would not happen for the farmer or for the highway commissioner, however. Big farms and big road districts required lots of equipment and personnel. Kost had to watch mergers of road districts win approval in several of Illinois's commissioner-run counties: Union, where the voters decided to drop the number of districts from 13 to 4; in Randolph, which went from 17 to 4 (Kaskaskia Island plus 3 east of the Mississippi); and Massac and Pulaski, where the decision was to go to the unit district, 1 per county. Not all road district mergers in Illinois succeeded, as I pointed out above in the case of Saline County in 1953 and as was the case in Johnson County a couple years earlier. But

Troy Kost saw the glimmer of an unwelcome (to him) trend toward making bigger governmental units at the expense of the little ones.[98]

Township consolidation, of any stripe, was evidence of what he called in a 1950 magazine column, "creeping socialized bewitchment." Several years later, speaking before the Pennsylvania Township Officials, he chastised "reformers and do-gooders" who favored merging or abolishing townships. When he and other TOI representatives testified before Parkhurst's committee on 22 January 1970, as the minutes reported, no one that day suggested solving township ills by making bigger townships out of sets of two or more existing townships. Then came the Canfield proposal to abolish township organization and with it the townships themselves.[99]

At his second turn before the CLG, Kost countered Canfield's testimony from earlier in the morning of 11 March by "strongly" urging township retention, by favoring "the allocation of special district functions" to his townships (or maybe "to municipalities), and by suggesting the possibility of a careful "redrawing of township lines" in order to assist small, under-funded townships cope better with future conditions.[100] I doubt, however, if he expected the "redrawing" of boundaries to put over three hundred Illinois townships on the consolidation block four years later.

Senators Jack T. Knuepfer of Elmhurst (DuPage County) and Daniel Dougherty of Chicago sponsored the legislation that brought boundary redrawing (consolidation) to local Illinois ballots in November 1974. S. B. 485, which became law 1 September 1972, mandated that county boards in township counties review the assessed valuation of all townships as of 31 December of that year and prepare a plan to permit those townships having less than six million dollars of assessed property the chance to merge with one or more adjacent townships to produce wealthier and presumably more viable units of government. Voters in all affected townships would have to approve; and townships coterminous with municipalities, like the the City of Peoria Township, were not eligible for such consolidation. *ICTO* carried a summary of and commentary on an early version of the Knuepfer/Dougherty bill in its January 1972 issue. The writer, presumably Editor Lee Ahlswede, projected 434 of the 1,433 Illinois townships "would be 'eligible' for consolidation," and admitted some consolidation "was desirable, so long as it [was] not baseless or arbitrary." The stronger townships could thereby become, the less vulnerable they would then be to the criticism of "those who talk about eliminating townships entirely." By the way, asked the writer, where was a comparable call for merger of small special taxing districts? Their merger "into townships" would enhance the gaining townships and simultaneously reduce the total number of governmental units.[101]

Macon County got into the consolidation action first, holding a referendum for Milam and Mount Zion townships on 8 January 1974. *ICTO* had questioned "the legality of such an early election" but reassured township officials and reminded township opponents that consolidation (really a form of abolition) still required all affected townships to approve. In this case, Milam Township, comprising 24 square miles of agricultural land in southeastern Macon, would have merged with Mount Zion and lost its identity. Voters in both, however, turned down the proposal by a significant difference, roughly seven to one. A few days before the Macon referendum date, Senator Knuepfer, according to a February 1974 *ICTO* story, had expressed the sentiment that maybe nothing would come of his consolidation legislation after all. In particular, Knuepfer cited the Illinois League of Women Voters as "death on Townships" and averse to any change that might serve to strengthen townships—like merger into a stronger unit—and leave them less vulnerable to outright dissolution.[102]

The Milam-Mount Zion vote was the first of 93 such 1974 referenda, offering the opportunity for voters to abolish financially vulnerable townships and merge them with one or more adjacent townships. All failed, and the *ICTO* rightfully claimed a "big" win for the friends of township government. The *Bloomington Pantagraph* said, in a 1976 special on townships, the overall margin, across the state in these tests of the people's township loyalty, was two to one. A few scattered townships, like Lynn in Knox County and Fayette in Livingston County, did vote for merger; but one or more of their prospective partner townships nixed the deals. Township government ended up getting quite favorable press because of its big 1974 statewide sweep in the consolidation matter. Kost and Ahlswede were so happy with a post-election *Alton Telegraph* editorial that they reprinted it in the magazine, despite the fact *Telegraph* writer Bill McFadin also therein reminded readers that township government was "in many cases, inefficient" and an "additional level of government in a state" with too many governments trying to govern.[103]

A Plague of Women Vipers

Con Con Delegate John Wenum did not coin the expression, "a plague of women vipers," nor did he seem to subscribe to such a negative view of the League of Women Voters; but Wenum is the person who called it to my attention.[104] Troy Kost and Lee Ahlswede never employed, to my knowledge, that sobriquet for the League; but my guess would be that the thought might have crossed their minds as TOI and the Illinois League confronted one another in the 1970s. In the end, the gentlemen from TOI prevailed over the ladies from the League. Cook County was the principal field of battle; and revenue sharing proved to be an ally of Fortress Astoria in this critical period, as township government tested new monetary muscle—the manna from Washington, D.C.

The Illinois League of Women Voters had backed away at the 1970 Constitutional Convention from a test of will against TOI and all that it represented. Leaguers "testified…and worked for a Local Government Article," said a 1974 item in *The Illinois Voter,* "that would permit an orderly reorganization of the structure and functions of local governments in Illinois." Though the League had really wanted the framers of the new constitution to write townships completely out of the system, the League "accepted the spelling out of provisions for referendum to dissolve townships."[105] League leaders might not have so gracefully ignored their longstanding insistence on township superfluity had they known what President Richard M. Nixon and Congress then had in the works.

A month after Illinois voters, with the League at the vanguard in the campaign to adopt, approved a new constitution, President Nixon suggested in his 1971 State of the Union Address that the federal government share its bounty with local governments in a new way. Within guidelines, local officials would be able to spend federally provided dollars as they saw fit. On 12 September 1972, the senate, with both Illinois senators (Charles Percy and Adlai Stevenson) voting nay, agreed to a $33.6 billion revenue-sharing bill making "39,000 states, cities, counties and townships" eligible for a super infusion of cash. Senator Stevenson based his negative vote, according to an *ICTO* piece in October of that year, on his belief that the federal government could do a better job itself dispensing funds for welfare and general aid to education. Proponents of sharing the wealth, like Illinois Governor Richard Ogilvie and Troy Kost, wanted the federal money in Illinois but could do without the federal government's "purse-string controls over state and local governments." Not every aspect of the final version of the Fiscal Assistance Act of 1972 pleased Editor Ahlswede, who, in his November 1972 column, said that states would have far fewer restrictions on their ability to spend the bonanza of bucks than would townships. On the other hand, townships (cash-poor and otherwise) would gain the "muscle" that the March 1970 *Pantagraph* editorial said they needed to survive after Con Con.[106]

Townships had no trouble finding ways to flex their new monetary muscle. Federal dollars could supplement property taxes and motor fuel taxes for road and bridge maintenance and repair. Human services could now go beyond the basics of life maintenance to support drug abuse and delinquency programs, expanded health care, gathering places for seniors and youth, garbage collection, and "even ecology," as a *Chicago Sun-Times* writer put it. Ahlswede cautioned township officials to think carefully before turning the federal funds over to their county, as some had by October 1973 already done. "It would seem," said the magazine's editor, "that any township anywhere in the state, could readily find projects which would be helpful to its citizenry and which would be a credit to the township itself." Beyond the citizen benefits, however, Ahlswede cautioned that Congressional renewal of the sharing program would depend largely on how local governments used what they had received. Once they did use these dollars, of course, townships needed to publicize "the good deeds" to their constituents.[107]

Despite League protests, townships continued to receive federal funds well into the 1980s and to use them for all sorts of purposes that Troy Kost could have only imagined when he took office during World War II. The *Daily Herald,* out of Arlington Heights (Cook County) but circulating through many of the Chicago suburbs, accused the League, in a 12 August 1974 editorial, of turning its opposition to township sharing of federal dollars "into a political tool" with the result being "a serious threat to critically needed health services." Townships, particularly those in the *Herald*'s immediate area, had a chance with the revenue-sharing program to once again become "a vital, meaningful local instrument"; and the League ought to reconsider its antitownship stance and "reaffirm its basic principles" so that it could "bring the same kind of objective, reasoned and impartial judgment to this issue as it provided for so many others in the past." Kost and Ahlswede published the *Herald* editorial and its companion cartoon in October 1974.[108]

With Illinois Public Act 77-1655 in force after 17 September 1971, the Illinois League set out to make an example of several Cook County townships and one DuPage County township. The law that Governor Ogilvie signed for the Seventy-seventh General Assembly, the first general assembly to convene after Con Con, provided the specifics on how a referendum would take place if "10% of the number who voted in the last general election" wanted to test the viability of a single township.[109]

Ahlswede kept a close watch on the League's offensive. He noted the League had had "a gleam" in its collective eye since Con Con and had long before that expressed an interest in township abolition. Where to take the single-township effort was the question League leaders had first to answer. Ahlswede reported, in an October 1972 article, he had heard the women first looked downstate to places where local League chapters were "active and strong" and where there existed (perhaps vulnerable) coterminous townships and where townships lay adjacent to sizable cities. "Obviously," mused Ahlswede, "townships in Cook County came under closest scrutiny." The League had its headquarters in Chicago, and aggressive League chapters peppered the area. Among the League's staunchest allies, in the opinion of Editor Ahlswede, were Chicago's mayor, Richard E. Daley, and the president of the Cook County Board, George Dunn. Cook County had, incidentally, just opened branch offices to bring county functions like tax collecting closer to the people; and, to Ahlswede, that meant one more effort to "run townships out of business." Daley and Dunn were both Democrats; most Chicagoland township officials were Republicans. Philip Raffe, Maine Township (Cook County) clerk, according to Ahlswede, had said, "either most of the League of Women Voters are Democrats or they have been duped by the Democrats." The editor, and no doubt the publisher of *ICTO* back in Astoria, Troy Kost, knew that any League success, no matter where the vanquished township, could "quickly spread to neighboring counties" and maybe throughout the state. There was definitely sufficient reason for concern among TOI members, but Ahlswede was still able to allay somewhat their fears by referring, in an editorial that accompanied the October article, to the "legal technicalities" that could save townships for the time being. The main technicality was the fact there existed then no "statutory implementation" procedure for replacing townships with some other form of local government.[110]

The League chose as its best bet for individual abolition, four townships in northern Cook County, Maine, Niles, Northfield, and Palatine, plus Milton Township in DuPage County. Concurrently, as an alternative to outright abolition of individual townships, abolitionists planned to run for township offices on the promise of downsizing or disabling their township if elected. Petition drives to collect signatures in all five test townships unfolded during the summer of 1972, in advance of the general election on 7 November that year. One source said the League gathered about twelve thousand signatures in the five abolition-targeted townships. Immediately, however, township officials in those five began the process of trying "to give the request for referendums a legalistic heave-ho," as *Chicago Tribune* writer William Sluis reported. Northfield's supervisor, Mark Clayton, told Sluis they had "'no intention of placing this issue on the ballot.'" Besides, continued Clayton, state law said "'nothing about what would happen to a township's assets or duties'" if the unthinkable occurred. Somebody would have to clean up "'an unholy mess.'" Richard Troy, the League's attorney who filed the petitions, countered in the same *Tribune* article with the best argument one could make under the circumstances, which was a promise to work out the transition subsequently "'in the courts or in the legislature.'" A quarter century later, abolitionist Bob Anderson would be resorting to the same lame vagueness when townshippers would challenge him in McHenry Township (McHenry County) in a similar drive to get rid of a single township (chapter 8). Sluis also checked for his story with John J. Nimrod, the Niles supervisor. Nimrod said he feared for the people of his jurisdiction, saying, "'elimination of township government would relegate the suburbs to much the same status as wards in Chicago.'" Somehow, that was a bad thing, I guess. He also called the attention of Sluis to the uncertainty over transitioning out of townships.[111]

Both Northfield and Niles voters did have a chance to pass judgment on their townships, but only one set got to do so in 1972. Milton Township (Wheaton, Glen Ellyn) resolved its dissolution dilemma by calling a special town meeting, whereat, what *ICTO* called, "attendee-delegates" voted to ignore the whole thing. Maine (Park Ridge), Niles (Skokie, Morton Grove), and Palatine (Palatine) townships deferred acceptance while "their legal counsels" looked for loopholes or anything else to enable them to keep the question of dissolution from going before the electorate. The League brought mandamus suits against the officials of the three lagging Cook townships, demanding that the petitioners receive their referenda as requested. Stalling tactics (what *ICTO* called "hearings, motions, counter-motions") made it impossible to post the legally required notices of the referenda in a timely fashion, so 7 November passed in those three without a vote on the future of township government.[112]

Northfield (Northfield, Northbrook) was the 1972 exception where signatures produced a referendum. Its township officials had accepted the League's petitions and given notice to prepare the ballots accordingly for 7 November, but expected a pair of court cases to foil the abolitionists. Both suits were, as it happened, *against* Northfield Townships itself. A "bond-holder company" prayed for relief from the courts on the premise that dissolution of the township would leave the plaintiff "holding the bag" on money the township owed it. A group of township citizens sued the township on behalf of residents in unincorporated areas who would supposedly have lost access to a variety of services, which the township ordinarily provided.[113]

By early October, when the Northfield lawsuits had failed to bear the fruit township officials desired, they organized a campaign to "spread [the] township gospel," as the *ICTO* phrased it. Brochures, posters, and fact sheets appeared, some with the assistance of publicist Ahlswede and TOI. Speakers dispersed to make solo presentations or to debate the leaguers. Women, who favored retention of Northfield, played a prominent role as door-to-door canvassers. Letters helped convince wavering newspaper editors that 1972 was "the wrong time to vote township government out of existence."[114]

This grassroots effort turned out to be useful because the Cook County Circuit Court tossed both lawsuits, allowing the referendum to proceed. Voters agreed with the township's message and confirmed Northfield 17,661 to 11,426. As League members waited for results that November night at their election headquarters in Northbrook and saw George McGovern concede early to Richard Nixon, one of the women commented, "'He sounds kinda like us.'" At least that is what *ICTO* reported. On 8 November, the *Tribune* noted that the day before had "marked the first time in Cook County history that township residents were able to decide whether to retain that level of government." Of course, the writer might have mentioned that Chicago voters had had a similar chance in 1902 (chapter 5).[115]

Following their Northfield defeat, dissolution diehards looked forward to the spring 1973 quadrennial election of all township officers. Diehards had two strategies. First, their preferred plan was still to abolish one or more townships altogether via the constitutionally and legislatively approved referendum. Second, as a backup, they tried to put as many antitownship candidates on the ballots for township offices as they could find, with the intent, as the *ICTO* editor saw it, to "dismantle the township unit from within."[116] I begin with the backup.

Township residents enjoyed, for over a century, the privilege of approving their township's budget at the annual spring meeting. Following a 1970 taxpayer revolt in Cook County's Proviso Township (Broadview, Maywood, Westchester), however, at which disgruntled voters "packed" the audience "and slashed the budget to near zero," the general assembly "stripped citizens of their right to vote on the town[ship] budget." Since then, township boards have had the final budgetary say. "'It kind of makes a joke out of the whole "grass roots democracy line,'" one critic told James Kloss, as Kloss was preparing an article prior to the 1973 township elections. Lacking, in 1973, the Proviso possibility, League strategists could still run qualified township residents (even League members) for any of the township offices, because all incumbents must run again every four years (1969, 1973, 1977, etc.).[117]

From several sources, I compiled a list of townships where at least one candidate who pledged to dismember from within made it to the ballot in 1973: Bloom Township (Chicago Heights), Elk Grove Township (Elk Grove Village), City of Evanston Township, Lyons Township (Countryside, Lyons, Western Springs), Niles Township, Palatine Township, and Rich Township (Park Forest). All but one abolition candidate lost; TOI celebrated. Ahlswede and Kost viewed the Grace Lee margin of victory in the supervisor race in Northfield as an improvement over the margin by which Northfield survived the November 1972 abolition scrape. Lee beat Ann Reason, running on the Citizens to Improve Township Government label, 4,732 to 1,905, whereas the earlier referendum had produced a 60/40 split of votes. In Palatine Township, John J. Scollay, "who had promised to schedule a referendum on whether to abolish township government," lost by a four to one factor. SCOTS (Some Citizens Opposed to Township Spending) ran Verna H. Beaver, a League leader, against incumbent supervisor Charles N. Anderson in Niles Township. His was the Township Improvement Party (TIP). SCOTS candidates, like Beaver, claimed municipality residents in the township paid for the cost of two governments. True, countered the TIP folks, but there existed "no guidelines for the disposal of assets and liabilities of township government." According to *ICTO,* Beaver, after losing by several thousand votes to Anderson, said the League still had "to do a lot of educating." The only reform candidate to succeed in April 1973 was an Evanston alderman, Quaife M. Ward, who defeated the City of Evanston Township's incumbent supervisor, Coleman Miller, by fewer than a thousand votes. Ward had campaigned on the promise of abolishing his township, but it is still in business three decades later.[118]

The League succeeded in putting dissolution on the ballot in only one township that spring. Leaguers had hoped holdover petitions from 1972 in Palatine Township would provide ballot access there or that new petition drives in Maine Township and Worth Township (Alsip, Oak Lawn) would do the same, but all failed. They did make the grade in Niles, however, thanks to hard work and a ruling by Cook County Circuit Judge Harry G. Comerford. Early in 1973, the Niles Township Board had accepted the petitions from 1972, but decided to hold the referendum on 13 March instead of at the regular township election in April and to open only 50 of the township's 123 polling places because the clerk had limited time between the board's decision and the March date. *ICTO* said another reason for the schedule adjustment was to keep the referendum "free from the campaign" and to avoid confusing voters who later in April would also have village officers to elect. The League objected, as one might expect, because township opponents wanted as much publicity and as many thoughtful voters as they could muster. The April date fulfilled those criteria better than 13 March. Judge Comerford bought the League's argument, ordered all Niles polling places open, and told the clerk to prepare ballots for the first Tuesday in April—township election day.[119]

SCOTS, Beaver, and the League lost to TIP, Anderson, and TOI badly in the Niles referendum, 9,155 to 19,620, according to the *ICTO* summary. Again, as in Northfield five months earlier, local townshippers combined forces with Astoria assets to provide fact sheets and other advertising as to the value of keeping the status quo. Proponents publicized cost comparisons for welfare delivery showing $88.00 per case under the township and a projected $230.00 under Cook County. The Niles campaign attracted media coverage across Chicagoland and, according to *ICTO* "dominated the local news media, taking play from village elections." Anderson issued a victory statement, covering his own election as supervisor and the defeat of the abolitionists, in which he said (said *ICTO*) TIP had won "because people want local government and wouldn't like what the would have gotten had we lost."[120] I thought Cook County qualified as local government, too.

Farewell, Old Soldier

While Lee Ahlswede did the heavy lifting for the magazine during the 1970s and as George Miller edged closer to becoming the next garrison commander at Fortress Astoria, Troy Kost continued to manage TOI affairs and watch his organization count victory after victory in his last decade of full-time service. In addition to the wins at the ballot box, there were a number of little advancements. Newsletters (to the residents) began to appear in townships like Maine and Palatine in 1974. At least 17 counties had active associations of township officials by 1976 that embraced more than one TOI division (clerks, road commissioners, etc.). Kost must have taken pride in the way Ahlswede transformed a condemning editorial from the *Decatur Review* on 26 August 1975 into an opportunity to re-educate the *Review*'s editor and put the township side before the readers of that newspaper and others in the Lindsay-Schaub chain. TOI's Secretary surely took a major dose of pleasure in 1975 from the failure of several bills in Springfield that would have weakened the state's townships. Both H. B. 1484 and S. B. 801 offered the magic that abolitionists had been seeking—methods of transitioning from townships to their replacements. Another 1975 legislative proposition would have removed townships from the "general purpose government" category under revenue sharing and thus killed the township government's golden federal goose. Kost surely approved of the 1978 report from the Township

Government Laws Commission, which recommended *increasing* township powers.[121]

Troy Kost remained a TOI asset to the end of his official duties as Secretary, late in 1979, when George Miller moved from the TOI presidency to become executive director. Kost had watched over TOI for nearly four decades, leaving it stronger and more proactive than it had been at the outset of his paid service in the 1940s. He saved township government during the 1969–1970 Constitutional Convention, when many observers predicted, many opponents prayed, and many township officials feared it would disappear from the Illinois scene. Fortress Astoria is no more; but the memory of this dedicated man and what he did for TOI should persist, just as Astoria Township persists, along with the other fourteen hundred townships that his colleagues trusted to his loving care for those many years.

Miller Time

Like the man who preceded him and like Bryan Smith, who would follow, George H. Miller prepped for his TOI command as a local township official (figure 19). Unlike Kost and Smith, however, Miller did not spring from rural roots in central Illinois, but rather took his initial township steps in the near western suburbs of Chicago. Born on 18 April 1928 in Chicago, Miller grew up in the city and began there a career with Bresler's Ice Cream, eventually rising to the position of driver-salesman. In 1960, Miller and his family bought a house in Melrose Park, where he would make his official residence for over forty years. As a route driver for Bresler's, he joined the Teamsters Union; became a union activist; and, in 1965, worked to gain (and gained) his union's support for a man running for political office in Miller's Leyden Township. The Miller candidate, a Democrat, lost; but Miller, working so effectively, caught the eye of local Republicans. He subsequently joined the Republicans, ran at the next township quadrennial election (1969) under the GOP standard for an auditor slot (trustee since 1977), and won. Miller's affability, ability, work ethic, and loyalty to the township cause soon won him respect and support from other township officials, no matter their party. By 1973, he had become president of the Auditors Division of the Township Officials of Cook County (TOCC). A year later, he won a similar post in TOI's Auditors Division; and, by 1975, he was president of TOCC. TOI had itself a shooting star.[122]

Man with a Mission

George Miller made his first significant appearance in *ICTO* in April of 1975 when Ahlswede presented a question-and-answer piece to give the new TOCC president statewide exposure. From the outset of the interview, Miller tried to assure township officials outside of northeastern Illinois that they and their Cook County brothers and sisters were "underneath it all...pretty

Figure 19 George H. Miller, circa 1979. Miller replaced Troy Kost in 1979, became the first TOI executive director, and served until 1996. (Courtesy of the Township Officials of Illinois)

much alike." Common foes, like the League; common needs, like good roads; and county boards that were no longer the dominion of township supervisors (as had long been the case in Cook) ought, Miller hoped, make for much more of a fraternal feeling across the 85 township counties. He even employed the adjective "rural" to characterize parts of his county and explained, when Ahlswede asked, that "most all of the 30 townships in Cook County have rural areas in them." His own township, in fact, still had a few small towns and farmland, and, like other townships in Cook, was "constantly battling the Big City influence and powerful machine efforts to weaken or destroy township government." In Cook County, said Miller, stood "the first line of defense [a second fortress, if you will] in the ongoing struggle to avoid annihilation. The TOCC president praised Kost and TOI president Joe Burgett and pledged that Cook County's township officials would continue to serve "as a vital contributor to this Illinois township progress" then underway.[123]

By 1976 George Miller had set his sights on higher office and was building support among the TOI rank and file. In April, he grabbed more *ICTO* ink as a result of a memorandum he had circulated that February to TOI's board of directors, warning them "of continuing dangers to township government." He had gotten word that certain "'U.S. Congressmen from Illinois'" were wavering in their support of revenue sharing for townships. Somebody in TOI had not been keeping "'Washington and Springfield'" appropriately apprised about all the great uses to which townships had been putting federal dollars. Educate the legislators, said the man from Melrose Park, or suffer the consequences. He made appearances at meetings of various county associations of township officials to explain how the TOCC conducted its affairs. Miller used these forums to urge local township leaders to eschew featured speakers and to embrace panels and discussion groups so as to involve more members in their meetings. Of course, the breakfast-and-banquet circuit also served to showcase his talents and to mark him as man to remember when the time came to elect the next TOI president.[124]

It actually took him two tries at the presidency of TOI. He failed in 1976, losing to Warren Behr of Waukegan Township (Lake County), a man who had worked his way up the vice-presidential ladder, much as Kost had done decades earlier. The breakthrough for George Miller came in 1977, at the annual meeting in Peoria. In a surprise move for many, Miller and Don Scheib of St. Charles, "two unnominated candidates," as Van Sickle characterized the pair, won the presidency and vice-presidency respectively.[125]

TOI's new president did not disappear between annual conventions, as had many of his predecessors, but instead pushed the membership with spoken and written exhortations. Early on he used an open letter to the members to reassure them that "downstate townships everywhere" would get his "sincere and full attention." He promised to fulfill as many invitations from county associations as time and commitments would permit. "'Tell the people township's story,'" he urged Rock Island County's township officials in the summer of 1978. If opponents were to eliminate townships, all other "local governmental bodies" would be vulnerable; and, perhaps, "only the Federal government" would survive. For the sake of township survival, partisan politics had no place, so they ought to park "'the donkeys and elephants...by the door.'" A presidential column soon began appearing in the magazine, and he used his space effectively to chastise big government and to plead the case for his favorite form of little government. Get out and vote for "those candidates who agree with our way of thinking on township government," he told readers in his September 1978 column. TOI and other bastions of small units of government were up against such foes as the National League of Cities, the National Association of Counties, the League of Women Voters (again), and lots of senators and congresspersons—in the ongoing battle over who should continue to collect the golden eggs from revenue sharing. His February 1979 magazine contribution, reminiscent of Kost rhetoric from the 1940s and 1950s, said their "whole form of government" might depend on what they did in the year 1979. At stake, after 1981, would be township's portion of revenue sharing, a share since 1972 that had brought Illinois's townships $172 million. The Illinois League was back to it old tricks, and its members and other abolitionists were unlikely to confine future efforts to "only one or two counties." Miller revealed in *ICTO* that he was planning, for the 1979 TOI convention in Peoria, a show of force for important state and federal politicians, as TOI kicked off its Diamond Jubilee celebration, which would culminate in 1982.[126]

Revenue sharing obviously meant much to Illinois townships, their residents, and their officials; but it also was good for TOI, which in the late 1970s, with inflation rampant, found itself in financial trouble. Miller's April 1979 presidential column warned the membership that there were soft spots in the armor of Fortress Astoria, soft spots that only more revenue would harden. Over the previous decade, while TOI income was increasing 11.3 percent, expenses were increasing far faster. Rent for the headquarters building, an Astoria storefront, had gone up 180 percent. Insurance was up 175 percent, and on and on it went. When Troy Kost could work the hallways of power at Springfield, TOI needed no legislative director. Since Kost had become unable to perform that function, TOI had resorted to hired lobbyists. They cost money. Dues to the National Association of Towns and Townships, auditing fees, and legal fees were draining fortress funds in ways they had not just 10 years earlier. Cost cutting, to include TOI board members taking less reimbursement for expenses than Miller felt right, had ceased to be a useful option. TOI dues, promised the president, were almost certain to increase when the directors again convened. Those dues, he reminded readers, could come from local taxes or they could come from a township's "Federal Revenue Sharing entitlement." Yes, revenue sharing, as then defined, needed all the protection Astoria and its fourteen hundred plus township outposts could muster.[127]

As the president from 1977 to 1979, Miller (along with other perceptive TOI leaders) came to realize that the organization not only had a financial crisis but also a leadership dilemma. Despite his countless and critical contributions and his great standing with the rank and file, Troy Kost had become a liability as the top man. His health, especially his eyesight, was declining. He opposed educational efforts aimed at township officials, other than those TOI delivered in breakout sessions at the annual meeting. Miller told me Kost opposed educating elsewhere for fear that members would then bypass the fall conference. Kost also nurtured, in Miller's opinion, a deep-seated mistrust for suburban Chicago's township officials. He may even have mistrusted Miller. If Kost were to step down as secretary, who could afford and who would want to replace him. With 35 years of service at Bresler's, Miller was the only one among the restive leaders that could take early retirement. Given his meteoric rise over just 10 years since his election as a Leyden auditor, Miller had the confidence and the desire to try for one more rung on the TOI ladder. He said yes, and, as he put it, "took a big chance."[128]

First Executive Director

George H. Miller advanced from the presidency to the new position of executive director at the 1979 Peoria convention. He was 51 years of age, about ten years older than the retiring Kost had been when he became secretary-treasurer in the late years of World War II and about twenty years older than Miller's successor would be when Miller retired in 1996. Middle age, however, was no deterrent to this diminutive dynamo. Fortress Astoria was his to command, but he never became an Astorian. He roomed with printing company executive Ken Stevens for a couple of years at the outset. Later Miller bought a travel trailer, parked it on a rented lot in Astoria, and lived in it during the workweek. On weekends, he went back to his home and family in Melrose Park.[129]

Change was in the air in Astoria. Lee Ahlswede left the editor's chair, which, during his tenure had actually been in Cook County. He issued a farewell editorial in the September 1979 magazine, after he learned Miller and colleagues had "declined to renew [his] contract for financial reasons." Ahlswede departed with good words for the friends he had made while serving in that capacity. Kost agreed to edit the magazine temporarily, until Miller was ready to take over that responsibility (which Miller did in October 1980). The new executive director wasted no time, in his first column, as director, telling the old guard of TOI that the 1980s was going to be, if he had his way, "a decade of growth, new stature, and renewed strength for the Township Officials of Illinois." Those who chose to neither lead nor follow his lead should "get out of the way." He vowed not to abide excuses from members who claimed activism, such as better public relations with voters or more lobbying state and federal officeholders, was "too much work." The abolitionist fire was simmering under "the township pot," and he needed help putting it out, before they all cooked.[130]

One log under the pot that first winter of Miller time came in the form of a negative editorial in the *Chicago Sun-Times.* Miller immediately sought to quench the abolitionist sparks it contained by shooting off a letter to the editor in which he praised townships for the ways they had used revenue-sharing dollars. With that instant rebuttal to the *Sun-Times,* Miller began what seems to have become standard procedure at TOI headquarters: never let the ink dry on a damning editorial before sending a reply suitable for publication in the same paper and offering to write a guest editorial to elaborate even more on the township message. He encouraged local officials to write their own responses and to keep him posted of bad press to which he needed to respond. With so many township-friendly eyes scanning the state's newspapers, in all the township counties, Miller hardly needed a clipping service to help him quench other combustibles.[131]

Next on the Miller change agenda were the very bricks and mortar of Fortress Astoria itself. Until 1980, TOI had rented various offices in Kost's hometown, the last of which was a former store on Astoria's main drag (U. S. 24). As Miller put it, in his TOI valedictory column in 1996, "In 1980 TOI owned nothing but the (used and abused) furnishings in the rented building in Astoria." That circumstance changed on 23 April 1980, when the ownership transfer for the building became official. Renovation of the first floor offices started immediately, and Miller had plans for remodeling the second floor as a meeting site for the board of directors. TOI was proud, he said, of its rural roots; and Astoria was surely rural. The next issue of *ICTO* carried a map and directions for finding the command post.[132]

In the fall of 1980, Miller replaced Kost as *ICTO* editor. The old gentlemen would, however, continue to consult and to contribute his "From the Desk of Troy Kost." Miller immediately modernized the front cover of the magazine to include topical/contemporary photographs and began negotiating with Ken Stevens for better paper to improve readability (and make it more attractive to advertisers). Editor Miller solicited letters to the editor, especially letters with ideas for improving the magazine or the organization; and he wanted photographs of township buildings and the stories to accompany the pictures. "We'll put your township story in 12,000 mailboxes across the State and Nation," he promised.[133]

Athough he had the benefit of only a few college-level courses (in dairy science at the Navy Pier campus of the University of Illinois), George Miller fervently believed in educating township officials so they could do their jobs better. Through guidebooks on their duties, dispersed "continuing education seminars," and the fall conventions (which he soon began calling the Annual Education Conference), he tried to help the 12,000 officials know how to perform to standard—and perhaps avoid the criticism that followed poor job performance. His educational ally was Paul Green, then a political science professor at Governors State University. Green had run for, in the late 1970s, and won the job of supervisor in Monee Township (Will County). After attending an annual meeting during the Miller presidency, he wrote Miller complaining about the meeting being mostly a waste of Green's time, in terms of learning how to be an effective supervisor. Miller agreed, and thus began a long and productive alliance between Fortress Astoria and a professor who knew how to organize educational events. Green ran the TOI seminar series through the Institute for Public Policy and Administration, at Governors State University, and later out of Roosevelt University, but held them all over Illinois. He relinquished the duty when TOI set up its own educational arm.[134]

The early Miller years were not entirely about change. George Miller also appreciated the value of heritage and tradition; and, as the seventy-fifth anniversary of that November 1907 meeting of W. Rufus Kendall and a handful of town clerks and highway commissioners in Watseka approached, Miller hired Frederick Mercer Van Sickle as a research assistant. Van Sickle, a Fulton County native (Canton) and then an undergraduate studying history at Lake Forest College, was to compile a manuscript from TOI archives and other sources that Miller would publish later that year. *Grassroots Democracy* is a suitable memorial to the organization that Kendall, Kost, and countless others nurtured; and it has been a useful source for me. TOI further marked its anniversary year by expanding its circuit of training seminars and by preparing a guidebook to help officials better understand their duties. TOCC marked the event, coincidently I suppose, by appointing Grace Lee, former Northfield Township supervisor, as its first executive director, and certainly the first paid administrator of any local Illinois association of township officials. Lee planned to mount a strong defense against those who would chip away at townships, but also told Kathy Rosemann that she wanted to see TOCC "'initiate legislation.'"[135]

Skirmishes Here and There

George Miller's first decade in the commander's slot at Astoria served him well as preparation for the turbulent 1990s. The press, township opponents, and even a gubernatorial candidate or two kept him vigilant, helped him refine his pithy prose style, and showed that others too were willing to help defend township government.

The 1981 quadrennial election of officials in Belleville Township, in St. Clair County, brought out the same strategy that activists had tried at the 1973 election in several Cook County townships: candidates committed to destroying the township from inside. In Miller's words, the 7 April 1981 "election proved to be a disaster" for those who dared test the mettle of Supervisor Dick Caesar and his band. By over a thousand votes, the members of the Belleville outpost had weathered the first real attack on township sovereignty in the Miller era. More were to come.[136]

A couple days after Caesar celebrated his big Belleville victory, Sally Rudolph sounded off on WJBC-AM in Bloomington about the lack of interest in the township elections just completed in McLean County. Using a series that WJBC calls *Forum,* Rudolph, one of a rotating group of local opinion leaders who take up the microphone to go public with thoughts on subjects of interest to them and perhaps the listeners, suggested on 9 April that getting rid of townships altogether was a good idea. Rudolph recalled the activist era of the late 1960s and early 1970s when "township residents with long hair, beards, and beads" invaded town meetings to vote out the traditional things and vote in radical (often progressive) notions. Unfortunately, in Rudolph's opinion, "the well-organized township officials" could only tolerate so much of this participatory democracy; and they "enticed the Illinois state legislature to change the rules of the game." Do not bother with the annual meetings next week, said Rudolph. Somebody has already made all the decisions. She went on to elaborate on how revenue sharing had put Illinois "townships in a real bind." How were they going to spend all the "extra money"? As often happens with a tasty topic like township abolition, another *Forum* participant will grab the bait. On 28 April 1981, Violet Hamilton, a long-time series regular, took her turn, saying, "Rudolph's call…must not go unchallenged." Hamilton chided the national League of Women Voters for pushing its abolition agenda down to the state and local level. She outed Sally Rudolph as being a "long-time member of the League" and reminded listeners that a prominent League officer had once said, "'we learned from the suffragettes that you have to infiltrate with sweetness often to accomplish what you want to do.'" Miller must surely have wished that he had more troops in the trenches like Hamilton, because he knew there were a lot of folks like Rudolph ready to test his defenses.[137]

Did gubernatorial candidate Adlai E. Stevenson III, son of a deceased former governor, really appear as an invited speaker before a thousand township officials at the 1981 TOI meeting and suggest consolidation of some townships might just be a good idea? George Miller could not believe he had heard correctly but soon read in the Chicago papers that the Democrat candidate was "repeating" and expanding his merger message. Miller's sources then revealed Stevenson had promised to put his running mate, an outspoken opponent of township government from Lake County, in charge of a committee to study units of local government. Miller explained these developments in a letter he sent to all 1,430-odd-township supervisors. He asked each to confirm that his or her choice for governor had made a commitment to township preservation "before the elections," and to share the letter with other township officials so they could check signals with their candidates as well. Stevenson lost to the incumbent, James R. Thompson.[138]

Freedom for the press to investigate and to criticize vigorously our governmental bodies probably gave Director Miller a fitful night or two in his travel trailer in Astoria, and elsewhere as well. I cannot imagine waking every morning and wondering which of the state's newspapers was going to be running an expose or negative editorial today. George Miller lived with that shadow for a long time, and he seemed to thrive on the challenge of such word contests.

Editor Bill Groninger of the *Champaign Weekly Broadside* likened township government to the "Notorious Widow," the lady that everybody whispered about but nobody would run out of town. Why keep her and why keep townships in town he asked? "The Illinois Township Officials Association," he replied, "is perhaps one of the single most powerful lobbies around." Groninger used examples, with dollar figures, from his area to assail the township setup. Road services of townships duplicated those of other governmental units. Township assessors, "for the most part are a joke," he continued. Supervisors were not much better in his opinion. Ray Henderson, highway commissioner of rural Pesotum Township in southern Champaign County, authored a reply that Groninger ran in the next issue. Henderson attacked the editor's "facts and figures," providing some of his own about roads. He reminded readers that six months from then they might like to see a township snowplow out their way. As for taxes, Henderson paid $156 that year to support his township and over eight hundred dollars for "consolidated school districts."[139]

"The 1,434 townships," said an editorial of 20 October 1983 *Chicago Sun-Times* "exist mainly to waste federal revenue sharing money," and "efforts to abolish or merge" them or "other governments" had met with intense and effective lobbying "by local officials desperate to keep their jobs." The Miller rebuttal machine struck back with a letter the paper printed in late October. The TOI chieftain cited a study of how townships *really* used revenue sharing funds: for maintenance of more than half the state's road mileage and almost half of the bridges and for services for seniors, youth, and the poor. To Miller, eliminating units of government no more guaranteed "improved service at lower cost" than would the elimination of newspapers improves the reporting by the press.[140]

Confusion over the myriad rules regarding township budgets and federal revenue sharing prompted the *Aurora Beacon-News,* without laying the blame on part-time, inadequately trained township personnel, to suggest counties could do a better job with welfare delivery and probably with all other township duties as well. "Perhaps it is time" said the editor(s), "for the Illinois General Assembly to take a hard look at how we transact public business and let township government gracefully go the way of" other governmental units and offices that had "similarly become obsolete at the hand of the complexity of the times." A letter left Astoria for Aurora posthaste. To George Miller, it was that very complexity that cried out for the simplicity of governmental officials who worked closest to the people. What was wrong, he wondered, with "ordinary citizens" doing the work, as they had done in places like "Lexington, Concord,…Iwo Jima, Korea, and Vietnam"? Besides, only the people, he reminded the editor(s), could eliminate townships.[141]

Federal revenue sharing lived 14 years; but, in 1986, Congress failed to authorize its renewal, for townships or any unit of local government. In an especially testy mood, Miller went on at length, when he got the bad news that fall, about all the other things that Congress did fund, like AIDS research, site planning for a new Chicago federal building, and a replacement for the space shuttle *Challenger.* An irate TOI executive director listed the good things that townships had done with their federal funds and lamented the problems that were sure to occur at the township level when property taxes and other traditional revenue sources again had to provide all the funding. Especially galling to the feisty Miller was the reaction he had gotten from some Illinois township officials, when he had tried to rally support for continuing the federal program. He felt the townships that never really needed such a budget infusion should have tried to see the picture from the standpoint of the hundreds of poorer townships around the state. He knew he was being critical of board members who could help oust him as executive director, but he had by 1986 come to identify more closely with TOI's rural roots, than some observers might have expected from a native of Cook County.[142]

Patrick Quinn and his Coalition for Political Honesty took up the cause of township abolition in 1988. Quinn, who once served as the state treasurer and in 2002 won a term as lieutenant governor, seeks populist issues that can bring him favorable publicity and perhaps provide public service at the same time. In the late 1980s, he discovered township government and found a handful of abolition-minded citizens that agreed to lend their names to a lawsuit revolving around the 1933 revision of the Township Law calling for at least 20 percent of voters to sign township abolition petitions in populous counties. Quinn, on behalf of Consuelo Zemaitis, Melvin Bishop, Thomas Ciesielski, and John Komar, in a case that became *Zemaitis v. DuPage County Board of Election Commissioners,* challenged the 20 percent figure in DuPage, Cook, and LaSalle counties.

Thomas Ciesielski was in 1988, and had been for several years, on the Coalition's board of directors. He served as the LaSalle County plaintiff in the suit and later told me he knew of John Komar, but had never previously heard of Zemaitis or Bishop before Quinn's attorney, Mathias Delort, approached Ciesielski about becoming a participant. Ciesielski was a former City of LaSalle alderman (1969–1973) and had retired in 1986 after a career as an industrial glassmaker. In May of 1987, he became disgruntled with LaSalle Township over a special election regarding a cemetery. Officials in his township decided, as the law allowed in special elections, to keep the polls open only for three hours in the morning and three in the afternoon. Ciesielski believed, by so doing, they were trying to "eliminate all the working people" from the process. As he thought more about township government in the ensuing months, he found other faults—so he decided to start a petition drive in the hope of getting abolition of all LaSalle townships on a referendum ballot in 1988. If he were to fail, at least he might "shake up" some of the "insiders," as he called them, insiders like LaSalle mayor, Paul Murphy, whose wife was LaSalle Township's supervisor. He circulated his petitions largely among his friends and their friends. Many refused to sign because of concern over who would perform the township duties if they disappeared or because some good people would lose their jobs. Ciesielski failed to come close to getting even 5 percent of the county's voters to sign his petitions.[143]

Even though Quinn and Delort included LaSalle County in their suit, the real targets were Cook and DuPage, where the numbers of voters made 20 percent seem impossible. Delort sought a preliminary injunction in the

United States District Court for the Northern District of Illinois to allow referenda if the plaintiffs could gather just the 5 percent that the bulk of Illinois counties needed, those counties whose voter numbers were less than 75,000. Defendants complained, in response, that the plaintiffs did not even have the 5 percent in hand and were unlikely to approach that plateau and asked if the Township Law of 1874 (as amended many times) was the correct basis for such a lawsuit. They pointed out that the general assembly in 1982 had modified the Election Code to make 10 percent of registered voters in all counties the target for petitioners, instead of 5 or 20 percent of those who voted at the last presidential election. Defendants also asked that the court grant intervention rights to several individuals, townships, township officials associations, and TOI. Judge Ann Claire Williams, on 19 August 1988, denied the interventions and the injunction. Voters in these three counties did not see dissolution of townships on their 1988 or any subsequent ballots. Patrick Quinn and the Coalition for Political Honesty moved on to other arenas, and Thomas Ciesielski resumed his retirement. In her final Zemaitis ruling a year later, Judge Williams noted that the general assembly had just amended the Township Law to require that at least 10 percent of any county's legal voters sign dissolution petitions. The vote total from the preceding presidential election was still to provide the basis for calculating the number of signatures that abolitionists needed. Thus Cook and DuPage would have a lower standard going into the 1990s while LaSalle and most of the other township counties would have to meet a higher standard than had previously been the case.[144]

George Miller used the Zemaitis case to prod his troops "to do more for our constituents and for our form of government." Maybe cutting taxes was not always the answer. Maybe publicity and bragging a little at the right times about how well townships were serving their public was not such a bad idea. "We're engaged in a full-time war using part-time ideas," he said. Township government needed "township officials and township government advocates with the fire in their bellies to fight and work to keep township government alive and well, and growing for future generations."[145]

Another County Contest

Carl Edwards did not need help from Pat Quinn or Mathias Delort to get abolition on the November 1988 ballot in Gallatin County. And it was Edwards, rather than Zemaitis, Bishop, Ciesielski, or Komar, who posed the first real threat, at the ballot box, to the 85/17 split of township versus commission counties that Miller had inherited from Kost. Enhancing the threat was the fact that Gallatin had commission county Hardin on its southern border; several other diehard commission counties in the neighborhood; and Saline County to the west, a county that had voted (albeit negatively) in 1966 and 1976 on this issue. For these reasons, among others, Gallatin appears to have been riper for reform in 1988 than LaSalle, Cook, or DuPage.

"The county cocklebur" is how Edwards characterized himself to me when we met at the courthouse in Shawneetown. "Troublemaker" was another moniker he proudly recalled when discussing how he thought others there perceived him. Rick Gill, Gold Hill Township's supervisor, described Edwards as a "habitual complainer." This trouble making, complaining cocklebur enjoyed his maverick reputation, but it did not help him make his case against townships. On the contrary, what a local reformer should be is not a Quinn or Edwards but rather a reluctant Cinncinatus, one who temporarily abandons the ordinary life to take up a cause for the community good and then reverts to whatever he or she had been doing beforehand. Of course, a community still needs a "busybody," as Horace Brown of Ridgway referred to Edwards, to point out, as Edwards had done, malfunctioning streetlights or grave plots that are too narrow. A reputation as nosey neighbor hurts one's credibility, credibility that is critical in an effort at upsetting something so ingrained as township organization.[146]

A farmer by trade, Edwards said the idea of gathering signatures to get rid of the 10 Gallatin townships came to him earlier in 1988 when he was dining at the senior citizen center in Ridgway with local politician Ed Hines. Both had agreed, recalled Edwards, that their township roads were pretty awful and that something needed doing. Hines, said Edwards, later denied any recollection of that conversation and certainly could not remember discussing township abolition. Edwards did find an ally in his neighbor, however, Bob Walters; and those two, with the help of County Clerk Mark Mossman, determined they would need only 250 (5 percent of 5,000) signatures to make the referendum happen that fall. Signatures collected, Edwards delivered them to the five-person county board, received no challenges, and won a ballot slot.[147]

Edwards and his neighbor never had much of a chance. The county board went on record as opposing the change, perhaps because at least two of the five would have lost their board positions if voters were to have decided to go with three commissioners. George Miller came in from Astoria for a townshipper rally at Omaha, one result of which was a large ad in the *Gallatin Democrat* a few days before the vote. In it, the Gallatin County Township Committee emphasized the local control that townships provided and warned of "higher taxes," "less representation," "less road maintenance," and "AT-LARGE" control of the county if commissioners replaced the county board. Rick Gill's name was first among the ad's listed sponsors. Edwards tried to schedule a rally to give his views some air but had to cancel when the president of the local Farm Bureau, according to Edwards, scared off a township critic who was willing to come from Farm Bureau (Illinois Agricultural Association) headquarters in Bloomington. The abolitionists had no war chest with which to fund advertising, but Walters did manage to publish a letter in the Shawneetown newspaper at the last minute. He used it to remind readers of the terrible condition of the gravel roads out his way and noted how farmers gradually acquire a collection of dislodged hubcaps from roads through their property. Fortress Astoria, since its establishment in the 1940s, had won countywide abolition fights in Clay, Macon, Mason, and Saline (twice); and in November 1988 made it a sweep, with a 2,746 to 729 drubbing of Edwards and Walters in Gallatin.[148]

Three for the Future

Miller time was off to a good start by 1989; and, given the high turnover of township officials every four years, only long-serving officials would have remembered Troy Kost. George Miller had inherited from Kost the title of "Mr. Township," and he was living up to the responsibility. TOI had a stronger Springfield presence and better access to legislators than it had had when Miller began his weekly commutes from Melrose Park to Astoria a decade earlier. What did the future for hold for him and for TOI? Here are three men who would take on greater significance in the years to come. First, Michael Richardson, a paralegal advocate for the poor and otherwise disadvantaged, had confronted officials of the coterminous townships in Peoria and Galesburg over perceived general assistance malfeasance in the early 1980s. In the 1990s, in Rock Island County and later around the state, Richardson would turn out to be among the most formidable adversaries that township government had yet to face.[149] Second, insurance specialist Jerry Kats had in the mid-1980s convinced Miller that townships needed and TOI ought to sponsor a self-insurance program. On 1 June 1986, the new Township Officials of Illinois Risk Management Association, with Kats as its director, began to accept premiums from member townships. Surplus revenue that Kats and TOIRMA generated would eventually provide a windfall for funding certain TOI operations.[150] Third, Bryan Smith, on 2 April 1985, became one of the youngest township officials ever when he won the clerk's job in Macon County's Long Creek Township. By June of 1987, he was an intern in Astoria as he finished his college degree; and he stayed on to become TOI's first associate director. In 1996, George Miller turned the keys to Fortress Astoria over to Bryan Smith and retired from active duty.[151]

Notes

1. George H. Miller, "From the Executive Director," *Illinois County and Township Official,* December 1993, 65. I cite the magazine hereafter as *ICTO* and as Miller the items that he authored there.
2. "Troy Augustus Kost," *ICTO,* December 1993, 65; "Troy Kost," *Astoria South Fulton Argus,* 10 November 1993, 2.
3. Van Sickle, *Grassroots Democracy,* 44; "Will J. Vieregge Heads Township Association," *ICTO,* November 1941, 9; "Township Officials Annual Convention, September 20–21–22," *ICTO,* 4; Troy A. Kost, "Township Officials, Attention!" *ICTO,* July 1944, 6; "Kost

Urges Members to Send in Dues at Once," *ICTO,* October 1945, 13; Miller, "From the Executive Director," *ICTO,* December 1993, 65. I cite hereafter items that Troy Kost authored as Kost.

4. In 1942, Attorney General George A. Barrett ruled that townships could subsidize subscriptions for their officials by paying dues to the magazine with tax dollars. "Holds Association Dues Payment by Townships Legal," *ICTO,* February 1942, 3.
5. "One Magazine for all Home Governments," *ICTO,* May 1941, 1.
6. "True Bewilderment," *ICTO,* May 1941, 10. Appeals for perpetuation of "home rule" appeared commonly in *ICTO* in the forties, fifties, and sixties. St. Louis received the first home-rule charter in the country in 1876, giving the city greater authority to manage its own affairs. See, Eric Sandweiss, *St. Louis: The Evolution of an American Urban Landscape* (Philadelphia: Temple University Press: 2001), 158–68.
7. Lee Ahlswede never joined Kost at Fortress Astoria. A Chicago public relations executive, Ahlswede had been writing publicity material for TOI prior to 1971, and took on the editorship as a moonlighting opportunity. He soon went out on his own, accepted other accounts, and set up his shop in Northfield (northern Cook County). From Northfield, he edited *ICTO* until 1979. Miller said Ahlswede did a fine job as editor but wanted more compensation than TOI could afford in the inflationary late 1970s. Lee Ahlswede, "Editor's Expressions," *ICTO,* September 1979, 4; George H. Miller, telephone interview by author, 18 December 2001. I cite Ahlswede items hereafter as Ahlswede.
8. "Has National Spirit of Dictatorship Filtered into the Illinois Legislature," *ICTO,* June 1941, 3–4; Kost, "Township Officials, Attention!" *ICTO,* July 1944, 6.
9. "Vieregge Re-elected Head of Township Officials Association," *ICTO,* December 1945, 2; "Officers Re-elected by Township Officials," *ICTO,* October 1946, 3; "Troy Kost Presents Township Officials' Views at Tax Hearing," *ICTO,* April 1948, 10, 14.
10. "Secretary Kost Summarizes the Picture on Legislation," *ICTO,* October 1948, 4–5.
11. "Resolutions at Township Officials Convention," *ICTO,* October 1948, 8–9.
12. Kost, "County Road Units Would Destroy Home Rule and Add to Tax Burden," *ICTO,* March 1951, 7; "Township Road District System Upheld in Saline County Voting," *ICTO,* May 1953, 30.
13. "Macon County, Illinois, May Vote on Township Abandonment," *National Municipal Review* 32 (November 1943): 564; "Township Government Upheld by Voters in Macon County Ballot," *ICTO,* November 1948, 6.
14. "Commission Rule Bid Filed; Supervisors Planning Fight," *Decatur Herald,* 2 September 1948, 3; "Vote on Commission Bid Seen Definite," *Decatur Herald,* 4 September 1948, 3; "Commission Bid Names Sufficient," *Decatur Herald,* 9 September 1948, 3; "Written Opinion Backs Balloting on County Rule," *Decatur Herald and Review,* 12 September 1948, 3. The two dailies combined for a weekend edition, the *Herald and Review.*
15. "It Costs $18 to Pay $36 Bill," *Decatur Herald and Review,* 19 September 1948, 6.
16. "Commission Rule Economics Denied," *Decatur Herald,* 3 September 1948, 3; "Road Commissioner Defends Township Commissioner Government Form," *ICTO,* September 1948, 41; "Priest, County Auditor Debate on Local Rule," *Decatur Review,* 24 September 1948, 20; "Commission Rule Success, Savings Outlined in Talk," *Decatur Review,* 28 October 1948, 32.
17. Ellis D. Roberts, "17-Township Road Program Inefficient and Expensive," *Decatur Herald and Review,* 26 September 1948, 3; "Roads Under Commission Government," *Decatur Herald and Review,* 26 September 1948, 6; "Assessments Under the County Form," *Decatur Herald and Review,* 3 October 1948, 6; "Local Officers: Townships or Pre-

cincts?" *Decatur Herald and Review,* 10 October 1948, 6; "Against the Continuance of Township Government," *Decatur Herald and Review,* 24 October 1948, 6; "Vote 'Against' on the Small Ballot," *Decatur Herald and Review,* 31 October 1948, 6; "Vote Against Continuance of Township Government," *Decatur Herald and Review,* 31 October 1948, 3.

18. "Thomas to Address Decatur Meeting on Township Government," *ICTO,* October 1948, 5; "Hitlerism Seen in Commission Rule for County," *Decatur Herald,* 27 October 1948, 3; "Change in County's Rule Opposed by A. of C. Unit," *Decatur Herald,* 19 October 1948, 3.
19. "Canvass Gives Elbert S. Smith 433-Vote Edge," *Decatur Herald,* 6 November 1948, 3; "Township Government Upheld by Voters in Macon County Ballot," *ICTO,* November 1948, 6.
20. "Clay County Votes to Retain Township Form of Government," *ICTO,* December 1952, 20.
21. "Clay County to Vote on Continuance of Township Organization on Nov. 4," *Flora News-Record,* 24 October 1952, 1; A STANFORD TOWNSHIP TAXPAYER [pseud.], "Further Explanation of Township Versus Commission Form of County of Government," *Flora News-Record,* 30 October 1952, 2; CONTRIBUTED [pseud.], "Letter to the Editor," *Flora News-Record,* 24 October 1952, 1.
22. "County Hospital Maintenance Tax Has 879 Majority," *Flora News-Record,* 5 November 1952, 1.
23. J. M. Sturtevant, "Dr. Sturtevant Favors the Proposed Change," *Jacksonville Journal,* 30 October 1879, 2.
24. "Former Resident Joe T. Boyle Dies," *Jacksonville Journal,* 19 March 1976, 22.
25. "52 Citizens Petition for Vote on Township Government," *Jacksonville Journal,* 7 October 1956, 9, 12. Only two commissioners requested an opinion; the third had died. Morgan County was, at that time, using survey townships (see chapter 1) as precincts from which to elect farmers to serve on farm subsidy program committees. Almost certainly the United States Department of Agriculture would have used civil-political townships if available. "Township Elections Will Be Held Sept. 11.," *Jacksonville Journal,* 2 September 1956, 8.
26. "Petition Filed for Vote on Township Government," *Waverly Journal,* 30 August 1957, 7.
27. "Explain Commission, Township Types of County Government," *Jacksonville Journal,* 26 September 1957, 14. Introducing the debaters that September night for the League was Mrs. Orville Foreman, the same woman who, in 1948, had described and praised the commission form at a League function in Decatur.
28. "2 Kinds of County Government Will Be Explained October 23," *Jacksonville Journal,* 22 October 1957, 5; "A Mighty Poor Time," *Jacksonville Journal,* 6 October 1957, 2/2; "State Senator Speaks on County Government," *Waverly Journal,* 25 October 1957, 1.
29. "A Mighty Poor Time," *Jacksonville Journal,* 6 October 1957, 2/2; "Comparing the Cost of County Government Favors Our Commission System," *Jacksonville Journal,* 20 October 1957, 2/2; "Voters to Decide Township Organization Issue Tuesday," *Jacksonville Journal,* 3 November 1957, 1/12; "Vote Tuesday Against Township Government," *Jacksonville Journal,* 3 November 1957, 2/2.
30. "VOTERS ATTENTION," *Jacksonville Journal,* 3 November 1957, 1/12.
31. "Morgan Voters Turn Thumbs Down on Township Organization," *Jacksonville Journal,* 6 November 1957, 14. Of all the counties that have avoided townships in Illinois, Morgan is the most perplexing. True, Morgan was and still is part of a cluster of counties re-

sistant to the township. On the other hand, Jacksonville attracted a considerable number of well-educated migrants from New England and elsewhere in the Northeast to its liberal arts colleges. In terms of population, it has outpaced most of the commission counties. Proximity to Sangamon County would have helped keep residents of Morgan in touch with developments at the state capital. So isolation cannot be the cause. Maybe residents just liked being different.

32. "County Board Accepts Petitions Seeking Vote on Supervisor Government, Subject to Legal Opinion," *Harrisburg Register,* 14 September 1966, 1; "Commission Form of County Government Explained to Board," *Harrisburg Register,* 5 October 1966, 1.
33. "People of Saline County," *Harrisburg Register,* 1 November 1966, 12; "Township Government is Still the American Way!" *Harrisburg Register,* 4 November 1966, 14.
34. "Vote 'Against' the Township Form of Government," *Harrisburg Register,* 4 November 1966, 16.
35. "Keep the County Board," *Harrisburg Register,* 4 November 1966, 1; "Voters Resoundingly Reject Change in County Government," *Harrisburg Register,* 9 November 1966, 1.
36. Clyde F. Snider, "The Twilight of the Township," *National Municipal Review* 41 (September 1952): 390–96; "Townships Out of Date," *Rock Island Argus,* 29 September 1952, 4; Richard C. Spencer, "Iowa Townships Still Here?" *National Municipal* Review 41 (September 1952): 397–99; James W. Drury, "Townships Lose Ground," *National Municipal Review* 44 (January 1955): 10–13; Paul W. Wager, "Townships on Way Out," *National Municipal Review* 46 (October 1957): 456–60, 475.
37. Illinois, General Assembly, Bills, Petitions, etc., 1957, State Archives, Springfield. H. B. 892 failed to specify a name for this commission. *ICTO* usually referred to it as the Township Government Commission or the Commission on Township Government. Commission minutes and its typeset letterhead stationery refer to it as the Township Government Problems Commission. Perhaps Kost did not appreciate the word "Problems."
38. Irving Howards, "Commission Studies Township Setup," *National Municipal Review* 47 (April 1958): 184–85; Illinois, Township Government Problems Commission, Papers, "Rollo Robbins" (expense account—October 1957–October 1958), State Archives, Springfield; "Will Study Township Government," *ICTO,* September 1957, 43. Hereafter, I cite the commission as TGPC.
39. Illinois, TGPC, Papers, "Minutes of Meeting" (14 October 1957), State Archives, Springfield.
40. Illinois, TGPC, Papers, "Minutes of Meeting" (18 November 1957) and "Testimony Before Commission to Study Township Problems by Maurice W. Scott," State Archives, Springfield. At this meeting, the TGPC decided to employ Howards as its executive secretary. A column by Howards, "Views on Local Government," soon began to appear in *ICTO.*
41. Illinois, TGPC, Papers, "Minutes of the Meeting" (13 January 1958), State Archives, Springfield.
42. Irving Howards, "Views on Local Government," *ICTO,* July–August 1959, 38–39; Irving Howards, "Illinois Township Commission Reports," *National Civic Review* 49 (January 1960): 35–36. Whether the new Commission on Township Government ever met or made recommendations to the general assembly I cannot say. No record of it exists in the State Archives nor did *ICTO* carry reports of its activities.
43. "160 Local Government Studies Under Way," *The Illinois Voter,* July–August 1963, 1, 4; "Know Your Townships," *The Illinois Voter,* February 1964, 3; "Choosing a State Pro-

gram," *The Illinois Voter,* November 1966, 3; "How Do You Study a Constitution?" *The Illinois Voter,* October 1969, 4.

44. Ian D. Burman, *Lobbying at the Illinois Constitutional Convention* (Urbana: University of Illinois, Institute of Government and Public Affairs, 1973), 82; Mrs. Andrew J. Koller, "Statement of Local Government to the Committee on Local Government of the Illinois Constitutional Convention, 18 February 1970, Local Government Committee Papers, Illinois Constitutional Convention Papers, Illinois State Historical Library, Springfield. I cite these papers hereafter as Local Government Papers, and thank a former colleague at Illinois State University, political scientist Thomas Dwight Wilson, for calling my attention to their existence. The official name of the standing committee that considered townships' role in the future of Illinois, however, was Committee on Local Government. For a list of committees, see, Illinois, Sixth Illinois Constitutional Convention, "Journal of the Constitutional Convention, State of Illinois, 1969–1970" (Illinois, Springfield, 16 December 1969, Mimeographed), 15. I cite hereafter as "Journal of the Convention."
45. "One Man, One Vote Rule Expanded to Local Level," *Bloomington Pantagraph,* 1 April 1968, 1. An Internet search for "one man, one vote" yielded just over nine thousand returns while a second for "one woman, one vote" that same day brought in slightly more than nine hundred. *Google Search,* 2001, <www.google.com> (7 October 2001).
46. "Harlan's Dissent Undercuts Tax Bloc," *Bloomington Pantagraph,* 6 April 1968, 3.
47. Harold P. Dunton, "County Board Reapportionment," *ICTO,* November 1969, 11–12; "Changes in County Government," *Illinois Political Reporter,* December 1969, 2–3; "Illinois Settles County Apportionment," *National Civic Review* 59 (February 1970): 89.
48. "Supervisors Initiate Efforts to Preserve Township Government, Block Remap," *ICTO,* March 1970, 40; "The Change Cometh in County Boards," *ICTO,* November 1971, 21–23. See, also, David Kenney, *Basic Illinois Government: A Systematic Explanation* (Carbondale and Edwardsville: Southern Illinois University Press, 1974): 307, 309–10.
49. Cornelius, *A History of Constitution Making in Illinois,* 96, 113; Joseph P. Pisciotte, "How Illinois Did It," *National Civic Review* 58 (July 1969): 291–96. Pisciotte was an Assistant Professor of Political Science at the University of Illinois at Urbana-Champaign.
50. Township Organization really refers to a way of governing a county—based on townships sending supervisors to form the county board. As of 1 May 1972, that situation was changing drastically. Kost had to make sure his townships survived the death of the board of supervisors. Township Organization persists today as the official name for a county divided into townships even though township counties have some other type of "organization."
51. Johanna M. Watson, *Electing a Constitution: The Illinois Citizen and the 1970 Constitution* (Urbana: University of Illinois, Institute of Government and Public Affairs, 1980), chap. V.
52. "Con Con Nominations to Knuppel, Thompson, Killey and Troy Kost," *Lewistown Fulton County News,* 24 September 1969, 1. One of the other candidates from Fulton County was the editor of this newspaper, Dennis Glaser.
53. "Youth, Independence Mark 42nd District Electorate Selection," *Galesburg Register-Mail,* 19 November 1969, 1–2; "Con Con Qualifications," *Lewistown Fulton County News,* 12 November 1969, 2/2; "Vote—November 18, 1969[,] for Troy A. Kost," *Astoria Argus-Searchlight,* 5 November 1969, 9.
54. Burman, *Lobbying at the Illinois Constitutional Convention,* 56, 109; John D. Wenum, telephone interview by author, 21 November 2001; Thomas D. Wilson, "Change in Lo-

cal Government: The Issues Before the 1970 Constitutional Convention, with Emphasis on Issues Affecting County Government," unpublished, 9; cited in James M. Banovetz and Ann M. Elder, *The Local Government Article of the 1970 Illinois Constitution* (n.p.: Illinois Commission on Intergovernmental Cooperation, 1987), 8. Wenum later taught at Illinois Wesleyan University in Bloomington and then retired from academia.

55. "Local Government Committee Meeting," Minutes of 7 January 1970 and of 22 January 1970, Local Government Papers; John C. Parkhurst to Troy Kost, 30 January 1970, Local Government Papers.
56. "Journal of the Convention," (13 January 1970), 10. Here is the official synopsis of the Canfield proposal: "Proposes to abolish township organization of county government and to establish a commission organization of county government with 5 commissioners elected at-large in counties having a population of less than 3,000,000 inhabitants, and 15 commissioners elected at-large in counties having a population of 3,000,000, or more, inhabitants, each to serve in the manner provided by law for county commissioners."
57. Burman, *Lobbying at the Illinois Constitutional Convention,* 55; Troy A. Kost to Township Official, 28 February 1970, Local Government Papers. The Kost letter came into this collection because the clerk of Rockford Township sent his copy to one of the Con Con delegates along with a letter expressing the clerk's dissatisfaction with the "Township System in Illinois." Clerk William W. Peterson felt townships had no place once their population reached 100,000. William W. Peterson to [Betty Ann Southwick] Mrs. Thomas A. Keegan, 10 March 1970, Local Government Papers.
58. "Local Government Committee Meeting," Minutes of 11 March 1970, Local Government Papers.
59. "Local Government Committee Meeting," 11 March 1970; Kelly Smith, "Zeglis Opposes Plea to Eliminate Townships," *Kankakee Journal,* 12 March 1970, 12.
60. Burman, *Lobbying at the Illinois Constitutional Convention,* 101–2.
61. "Petition Against Member Proposal No. 42," ca. 1 April 1970, Local Government Papers; F. H. Englemen to John C. Parkhurst, 8 April 1970, Local Government Papers; "Resolution Presented by Mr. Donald Jones—Annual Town Meeting—Jackson Township—Will County, 14 April 1970, Local Government Papers; Harry Meinert to John C. Parkhurst, 31 March 1970, Local Government Papers.
62. David R. Miller, *1970 Illinois Constitution: Annotated for Legislators,* 4th ed. (Springfield: Illinois Legislative Research Unit, 1996), 61–77.
63. Joan G. Anderson and Ann Lousin, "From Bone Gap to Chicago: A History of the Local Government Article of the 1970 Illinois Constitution," *The John Marshall Journal of Practice and Procedure* 9 (1976): 708–12. Bone Gap was then the smallest incorporated municipality in the state.
64. Thomas Dwight Wilson, "Patterns of Change in Illinois County Government, 1960 to 1974" (Ph.D. diss., University of Illinois, 1975), 79–80; Wenum interview; David R. Miller, *1970 Illinois Constitution: Annotated for Legislators,* 64.
65. "Local Government Committee Meeting," Minutes of 22 January, 28 January, 29 January, 3 February, and 11 February 1970, Local Government Papers; "Statement for Con-Con Hearing, March 6, 1970, at Urbana," Amy Kummerow, 6 March 1970, Local Government Papers; "Township Government: Con-Con Township Statement," Frank J. Kopecky, 6 March 1970, Local Government Papers. Press coverage of the hearings varied, but the 29 January appearance of Richard Friedman caught the attention of at least two reporters. See, "BGA Suggests Hard Look at Townships," *Peoria Journal-Star,* 29 Janu-

ary 1970, C/10; and Jim Williams, "Township Government Inefficient, Wasteful, BGA Official Charges," *Rockford Morning Sun,* 30 January 1970, A/19.

66. Wenum interview; Elmer Gertz and Joseph P. Pisciotte, *Charter for a New Age: An Inside View of the Sixth Illinois Constitutional Convention* (Urbana: University of Illinois, Institute for Government and Public Affairs, 1980), 252; "Local Government Committee Meeting," Minutes of 24 March 1970, Local Government Papers.
67. John C. Parkhurst to Local Government Committee Members, 12 May 1970, Local Government Papers.
68. Charles Hilty, "Con-Con Committee Leaning to Strong County Government" and "Buried," *Bloomington Pantagraph,* 28 April 1970, A/7; Mary Ellen Harkrader to John C. Parkhurst, 28 April 1970, Local Government Papers; John C. Parkhurst to Mrs. Alan Harkrader, 30 April 1970, Local Government Papers; Rosemary Trump to John C. Parkhurst, 19 April 1970, Local Government Papers; John C. Parkhurst to Rosemary Trump, 16 May 1970, Local Government Papers; Alice DeViney to John C. Parkhurst, 8 May 1970, Local Government Papers; John C. Parkhurst to Mrs. Glen DeViney, 16 May 1970, Local Government Papers.
69. "Let New Constitution End Outmoded Townships," *Danville Commercial-News,* 15 February 1970, 6; "Declining Role of Townships," *Moline Dispatch,* 16 March 1970, 4; "Muscles for Townships," *Bloomington Pantagraph,* 17 March 1970, A/4.
70. Illinois, Sixth Illinois Constitutional Convention, *Record of Proceedings: Verbatim Transcripts,* vol. 4 (Springfield: Office of the Secretary of State, 1972), 3392–93; "Journal of the Convention," (30 July 1970), 18–19. The *Record of Proceedings* is a set of seven volumes, of which four contain the word-for-word debates from the floor of the convention. I cite hereafter as *Verbatim Transcripts.*
71. *Verbatim Transcripts,* 3393–94.
72. *Verbatim Transcripts,* 3394.
73. *Verbatim Transcripts,* 3399–3400.
74. *Verbatim Transcripts,* 3394–95.
75. *Verbatim Transcripts,* 3403.
76. *Verbatim Transcripts,* 3400–3401.
77. "Journal of the Convention," (13 August 1970), 12; *Verbatim Transcripts,* vol. 5, 4207.
78. *Verbatim Transcripts,* vol. 5, 4528, 4454–55; John C. Parkhurst, "Local Government," *The Weekly Illinois Constitutional Convention Summary,* 3 September 1970, 13.
79. *Verbatim Transcripts,* vol. 5, 4648–49.
80. *Verbatim Transcripts,* 4649.
81. *Verbatim Transcripts,* 4648–50 and 4652–53. Delegate Kenney later wrote a book about his experience. See, David Kenney, *Making a Modern Constitution: The Illinois Experience* (n.p.: Jackson County [Illinois] Historical Society, 1991).
82. Watson, *Electing a Constitution,* 133–34.
83. "Farm Bureau Files Petitions," *Havana Mason County Democrat,* 7 September 1970, 1.
84. "Changes in County Government," *Illinois Political Reporter,* December 1969, 1–3.
85. Jerry Erwin, "The Referendum to Determine County Government," *Havana Mason County Democrat,* 19 October 1970, 1; Jerry Erwin, "Township Vs. Commission Government—How Do They Compare Economically?" *Havana Mason County Democrat,* 2 November 1970, 3.
86. "Public Information," *Havana Mason County Democrat,* 19 October 1970, 7; "Specimen Ballot," *Havana Mason County Democrat,* 2 November 1970, 3.

87. Kost, "Comments from Secretary Kost's Desk," *ICTO,* December 1970, 12; "Democrats Win County Offices," *Havana Mason County Democrat,* 9 November 1970, 1. By 1 July 1971, boards of supervisors in the 84 counties still having them had to go on record as to the configuration of their replacement county boards. Mason County's 14 supervisors waited until the evening of 30 June and then opted for reducing the board from 14 to 8. Half the new board would come from a district comprising eight townships in southeastern portion of the county and the other half from the five northwestern townships. "Supervisors Reapportion County into Two Districts," *Havana Mason County Democrat,* 3 July 1971, 1.
88. John Crawford Wilson, telephone interview by author, 3 March 1993; Robert Rath, telephone interview by author, 2 March 1993.
89. Wilson interview; Rath interview.
90. Richard Jackson, "Moss Resigns County Board Chairmanship," *Harrisburg Register,* 6 August 1976, 1; Rath interview.
91. "Townships Worth Preserving, Says Vernon Moss," *Eldorado Journal,* 29 October 1976, 1; Vernon Moss, "'Better Keep Township Govt.'," *Eldorado Journal,* 1 November 1976, 2. The 29 October item appeared also in the *Harrisburg Register* on 1 November, but on page 20 instead of the front page.
92. "Attention Voters," *Eldorado Journal,* 29 October 1976, 9; "Are We Misled?" *Eldorado Journal,* 1 November 1976, 6; "Is the So Called Taxpayers Association Misleading Us??" *Harrisburg Register,* 28 October 1976, 2. The 1 November *Journal* item also ran in the *Register* on 30 October.
93. "Saline County Taxpayers It Is Time to Wake Up!" *Eldorado Journal,* 29 October 1976, 5.
94. "Special Propositions: a 'No' and a 'Maybe'," *Harrisburg Register,* 29 October 1976, 1; Saline County. Abstract of Votes for County Officers and Local Referendum. Election of 2 November 1976. "Continuance of Township Organization" carried in 32 of the 34 precincts. In one of the smallest Harrisburg precincts, voters preferred to discontinue; and, in one of the tiny outlying precincts, there was a tie. Saline County. Canvass & Sub-Total by Legislative Districts: Voting Precincts. Election of 2 November 1976. Originals of these last two items are in Office of the County Clerk, Harrisburg.
95. "Townships, Counties Win at Polls," *ICTO,* December 1976, 24–25.
96. "Local Government Committee Meeting," Minutes of 11 March 1970, Local Government Papers.
97. "Secretary Kost's Report on Legislation," *ICTO,* November 1949, 4; "Secretary Kost Summarizes the Picture on Legislation," *ICTO,* October 1948, 5. School consolidation is still attracting research interest, as educators and others question its value for the children and their communities. See, for instance, Emily Talen, "School, Community, and Spatial Equity: An Empirical Investigation of Access to Elementary Schools in West Virginia," *Annals of the Association of American Geographers* 91 (September 2001): 465–86.
98. "Secretary Kost Summarizes the Picture on Legislation," *ICTO,* October 1948, 5; "A Timely Message from Secretary Kost to All Township Officials of Illinois," *ICTO,* December 1950, 4; "Road District Consolidation Continues," *National Municipal Review* 39 (November 1950): 514; "News Briefs from Around the State," *ICTO,* February 1951, 16; "Two Counties Vote Consolidation of Highway Districts," *ICTO,* March 1952, 9; "Merger Defeated," *ICTO,* March 1952, 9. In 1950, Kost battled hard against an Illinois Agricultural Association proposal to eliminate all of the state's township road districts.

See, "Keep Township Road Districts, Troy Kost Tells Macon Officials," *ICTO,* April 1950, 50–51.

99. "A Timely Message from Secretary Kost to All Township Officials of Illinois," *ICTO,* December 1950, 4; "Consolidation of Townships Ripped by Kost," *ICTO,* March 1966, 24; "Local Government Committee Meeting," Minutes of 22 January 1970, Local Government Papers.

100. "Local Government Committee Meeting," Minutes of 11 March 1970, Local Government Papers. Once before, at a hearing of the Township Government Problems Commission, Kost had mentioned the possibility of modifying township lines to improve the lot of tiny townships. See, Illinois, TGPC, Papers, "Minutes of the Meeting" (13 January 1958), State Archives, Springfield.

101. Illinois, General Assembly, *Laws of the State of Illinois Passed by the Seventy-seventh General Assembly and Approved During the Calendar Year 1972* (Springfield, 1973), 1995–97; "Consolidation of Townships," *ICTO,* January 1972, 37.

102. "Keeping Current," *ICTO,* January 1974, 48; "Consolidation of Townships—Where It Stands," *ICTO,* February 1974, 18.

103. "Townships Win Big in Anti-Merger Votes!" *ICTO,* December 1974, 13–14; Mary Gunkel, "Are State Townships on Their Way Out?" *Bloomington Pantagraph,* 29 February 1976, D/17; "Township Topics: 'Township Government Works', Says Alton Telegraph Columnist," *ICTO,* February 1975, 34–35. Since 1974, township consolidations have occurred in two counties. Salisbury, a small township in Sangamon County, northwest of Springfield, went out of existence in the late 1980s, dividing its former territory between Fancy Creek and Gardner townships. Nashua Township, in Ogle County, merged with its more populous neighbor, Oregon Township, in the early 1990s. Todd J. Behme, "Voters to Decide if Nashua Joins Richer Neighbor," *Rockford Register Star,* 13 October 1992, B/1, B/4; Lee Ann Gill, "Downstate Voters Opted for Consolidation," *Crystal Lake Northwest Herald,* 31 October 1994, A/3.

104. Wenum interview.

105. "Township Complaints Filed," *The Illinois Voter,* April 1974, 1–2 and 4.

106. Ahlswede, "Revenue Sharing: Just a Promise?" *ICTO,* September 1971, 7; "Revenue Sharing Comes to Pass," *ICTO,* October 1972, 29; Ahlswede, "Editor's Expressions," *ICTO,* November 1972, 3; "Muscles for Townships," *Bloomington Pantagraph,* 17 March 1970, A/4.

107. James Kloss, "Townships—Who Needs Them?" *Chicago Daily News,* 2 April 1973, 4; Ahlswede, "Opportunities for Townships," *ICTO,* October 1973, 42–43.

108. "Prominent Newspaper Says: Give Townships More Rein," *ICTO,* October 1974, 71–72

109. Illinois, General Assembly, *Laws of the State of Illinois Passed by the Seventy-seventh General Assembly and Approved During the Calendar Year 1971* (Springfield, 1972), 3086–88.

110. Ahlswede, "4 Townships Battle Abolition Referendums," *ICTO,* October 1972, 20–21; Ahlswede, "Editor's Expressions," *ICTO,* October 1972, 3.

111. William Sluis, "Fight Brews Over Effort to Abolish 4 Townships," Chicago *Tribune,* 27 August 1972, 10/3. Sluis did not cover Milton Township.

112. "How Northfield Township Won Its Preservation," *ICTO,* December 1972, 18–19; Kloss, "Townships—Who Needs Them?."

113. "How Northfield Township Won Its Preservation," *ICTO,* December 1972, 18.

114. "Northfield Township," 18–19.

115. "Northfield Township," 19; "Northfield Township Government Okd," *Chicago Tribune,* 8 November 1972, 1/21. It would be a stretch for me to suggest one of the reasons the League chose Northfield Township was the fact that Lee Ahlswede had opened his editorial office for TOI there in May of 1972. So I will not do so. Ahlswede, "Editor's Expressions," *ICTO,* September 1979, 4.
116. "Townships Win Major Victory Over Abolitionists," *ICTO,* May 1973, 6.
117. Kloss, "Townships—Who Needs Them?."
118. "Townships Win Major Victory Over Abolitionists," *ICTO,* May 1973, 6; Daniel Egler, "Most Incumbents Victorious in Area Townships," *Chicago Tribune,* 4 April 1973, 1A/8; Connie Lauerman, "Proposal to Abolish Township Unit Losing," *Chicago Tribune,* 4 April 1973, 1A/8; Daniel Egler, "Incumbents' Victories Bury Bids to Abolish Township Government," *Chicago Tribune,* 5 April 1973, 1A/8.
119. "Townships Win Major Victory Over Abolitionists," *ICTO,* May 1973, 6; Daniel Egler, "Election Set in 131 Townships Today," *Chicago Tribune,* 3 April 1973, 1A/4; Kloss, "Townships—Who Needs Them?."
120. "Townships Win Major Victory Over Abolitionists," *ICTO,* May 1973, 6. Ahlswede commented on the League in his editorial column that month, lumping it with "civil rights groups, economists, ecologists and a variety of social and governmental 'reformers'," which the writer said were constantly stirring up problems. "These gals are obsessed with the idea they can correct certain governmental problems by abolishing townships." He wondered, in closing, if they could also stir up a good batch of "lentil soup." Ahlswede, "Editor's Expressions," *ICTO,* May 1973, 3.
121. "Township Topics," *ICTO,* April 1974, 57; "Township Associations in Counties Getting Stronger," *ICTO,* September 1976, 74–76; "Justify Township Government," *Decatur Review,* 26 August 1975, 4; Ahlswede, "Townships Economical, Work Efficiently," *Decatur Review,* 30 October 1975, 8; "The Case for Township Government," *ICTO,* December 1975, 18–19; "The Case for Township Government (Part II)," *ICTO,* January 1976, 10–13; "Bills Relating to Township Dissolution Fail in Legislature," *ICTO,* July–August 1975, 37; Illinois, Township Government Laws Commission, *Report on the Township Government Laws Commission* (Springfield, 15 March 1978).
122. George H. Miller, telephone interview by author, 18 December 2001. John E. Phillips, the unsuccessful Democrat that started Miller's political career in 1965, later told Miller "no township official from Cook County would ever amount to anything in the Township Officials of Illinois." Miller, "From the Executive Director," *ICTO,* January 1994, 4.
123. "TOCC President Miller Examines 'Township Progress Through Unanimity'," *ICTO,* April 1975, 32–34.
124. "Township Topics," *ICTO,* April 1976, 44; "Township Associations in Counties Getting Stronger," *ICTO,* September 1976, 74–75.
125. Van Sickle, *Grassroots Democracy,* 55. Scheib became president when Miller moved over to be executive director.
126. Miller, "Message from New T.O.I. President George Miller," *ICTO,* December 1977, 35; Tom Smalec, "Exec Boosts Townships," *ICTO,* July–August 1978, 28; Miller, "Townships Must Earn 'Golden Future'," *ICTO,* September 1978, 65; Miller, "A Message from T.O.I. President George Miller," *ICTO,* February 1979, 28. The Smalec item was a reprint from the *Moline Dispatch.*
127. Miller, "Will T.O.I. Survive on Its Deflated Budget?" *ICTO,* April 1979, 40.
128. Miller telephone interview.

129. Miller telephone interview. Stevens Publishing Company of Astoria, now KK Stevens Publishing Company, has printed the magazine for many years.
130. Ahlswede, "Editor's Expressions," *ICTO,* September 1979, 4; Miller, "From the Executive Director," *ICTO,* December 1979, 26.
131. Miller, "From the Executive Director," *ICTO,* January 1980, 9–10. The *Sun-Times* editorial appeared 27 November 1979. Two days later Miller's rebuttal headed to Chicago, as did one from the president of TOCC. The editors ignored the TOCC president's letter, according to Miller's February column, and omitted key parts of his. Miller's sources told him the editor of the *Sun-Times* was married to the president of the Illinois League of Women Voters. Miller, "From the Editor's Desk," *ICTO,* February 1980, 4.
132. "T.O.I. Purchases Office Building," *ICTO,* June 1980, 8–9; "Where the Heck is Astoria?" *ICTO,* July–August 1980, 21. TOI bought its Astoria home from Evelyn Tregellas, one of its employees. They spent $10,000 to buy and that much again to remodel. Miller, looking back, more than two decades later, told me he believed the purchase in Astoria was one of the smartest things he ever did as executive director. Buying there demonstrated to the rank and file TOI's commitment to the rural roots of township government. Membership, that is the percent of Illinois townships paying dues to TOI, jumped immediately after the purchase, from 82 percent to over 90 percent, as small, poor, *rural* townships came aboard. Miller said he was not certain the subsequent move, after he left TOI, to Springfield was the right decision. Even from that office, TOI still has go find the legislators. "They seldom come to you, even if you are a couple blocks away from the capitol." Miller telephone interview.
133. Miller, "Message from the New Magazine Editor," *ICTO,* October 1980, 7.
134. Miller telephone interview; Paul Green, "Spring Training Seminars Taught Thousands of Township Officials, *Township Perspective,* January 2002, 13.
135. "About Our People," *ICTO,* July–August 1982, 3; Van Sickle, *Grassroots Democracy,* 59; Kathy Rosemann, "Lee Moves into New Front Line Position," *ICTO,* December 1981, 28–29; "Good Luck, Grace," *ICTO,* December 1981, 29.
136. "Anti-Township Slate Defeated," *ICTO,* May 1981, 35.
137. "WJBC's Forum Opinions," *ICTO,* June 1981, 32–33.
138. Miller, "Reprint of Letter Sent to All Township Supervisors," *ICTO,* October 1982, 24.
139. "High Price of Government," *ICTO,* September 1982, 26–27.
140. "An Answer to the Sun-Times," *ICTO,* December 1983, 49.
141. "Townships are Useful," *ICTO,* September 1984, 38–39.
142. Miller, "From the Executive Director," *ICTO,* November 1986, 4–6.
143. Thomas V. Ciesielski, telephone interview by author, 2 March 1993.
144. Tony Parker, "Suit Seeks Easier Route to Abolition of Townships," *Bloomington Pantagraph,* 5 August 1988, A/2; *Zemaitis v. DuPage County Board of Election Commissioners,* 1988 U. S. Dist. LEXIS 9469 and 1989 U. S. Dist. LEXIS 7636.
145. Miller, "From the Executive Director," *ICTO,* September 1988, 4–6.
146. Carl Edwards, interview by author, 14 January 1993; Rick Gill, interview by author, 14 January 1993; Horace Brown, interview by author, 14 January 1993.
147. Edwards interview; Mark Mossman, interview by author, 14 January 1993; "Township Issue Goes on November Ballot," *Shawneetown Gallatin Democrat,* 18 August 1988, 1.
148. Edwards interview; "Notice to Gallatin Voters!" *Shawneetown Gallatin Democrat,* 3 November 1988, 19; Bob Walters, "Writer Presents Arguments Against Township Organization," *Shawneetown Gallatin Democrat,* 3 November 1988, 6; "Foster Defeats Dyhrkopp," *Shawneetown Gallatin Democrat,* 10 November 1988, 1.

149. Lori Lau, "Resident Criticizes Township 'Secrecy'," *Galesburg Register-Mail,* 4 January 1983, A/1; Lori Lau, "Township Aid Payments Called 'Cruel Joke'," *Galesburg Register-Mail,* 8 March 1983, A/1; Lori Lau, "Resident Objects to Management of Township Affairs," *Galesburg Register-Mail,* 13 April 1983, A/3; Lori Lau, "Help People or Go By the Book?" *Galesburg Register-Mail,* 30 April 1983, A/1; Michael D. Richardson, interview by author, 10 December 1992.
150. "TOIRMA! What Does It Mean," *ICTO,* May 1986, 6–7; Jerry H. Kats, interview by author, 26 October 2000.
151. "Getting Involved," *ICTO,* July–August 1985, 3; "About Bryan," *ICTO,* July–August 1987, 23.

7 Every Other Summer: 1990–1994

MAYOR OSSIE LANGFELDER issued a proclamation that 13 November 1989 was to be George H. Miller Day in the state's capital city. Miller Day coincided with the annual meeting of TOI, in Springfield, and roughly with the end of a decade of service by Miller as the TOI executive director. Several prominent state political figures, Jim Edgar (secretary of state), Jerry Cosentino (state treasurer), Neil Hartigan (attorney general), and Roland Burris (state comptroller), helped surprise Miller, as did the entire state senate, which had unanimously adopted a resolution to commemorate his tenure in Astoria. Nearly eight hundred of his townships passed congratulatory resolutions and sent letters, which the organizers presented to him on the morning of his special day. George Miller said he greatly appreciated these efforts and profusely thanked everyone, in his first *ICTO* column of the new decade. For his celebration, cooperation among state and local officials had been evident; but he also forecast in his column that "even tougher fights and greater obstacles" lay ahead for the township faithful. "We continue to face onslaught," said the director, "from people who think negatively of township government." He expressed pride in his decade (the 1980s) and pledged to help make "our decade" (the 1990s) similarly successful.[1]

Little did Miller know then how the last decade of the century would test the mettle of him, TOI, and his township brethren. Even as he was expressing gratitude for Miller Day, Illinoisans who thought "negatively of township government" were plotting the first of what became a succession of biennial petition-passing, referendum-requesting confrontations. Every other year, every other summer during the 1990s, in the even-numbered years, the years featuring a November general election, township critics took to the streets, clipboards in hand, determined to find that critical weak spot in the township armor, the first since Williamson had fallen six decades earlier. This chapter chronicles, in two northern Illinois counties, efforts aimed at discontinuance of townships countywide during the period 1990–1994, and pauses here and there to consider special topics pertaining to the matter of township abolition.

Rock Island County

Officials in Rock Island County's 18 townships were peacefully minding the people's business, as they and their predecessors had been doing since the middle of the nineteenth century, when Michael D. Richardson disturbed the peace. Here was a man with the vigor, inclination, time, legal mind, and publicity-driven persona to cast a large shadow locally over township organization and to exert an influence beyond the confines of his adopted county. Not since H. S. Hicks had township forces faced such a literate threat.

Michael D. Richardson

Born in Des Moines, Iowa, in the immediate post-World War II era (13 June 1950), Richardson later lived in North Dakota, finished high school in Nebraska, and started (in Omaha) a college program in law enforcement. He had by 1990 devoted much of his adult life to helping the less fortunate. While in college, he joined VISTA (Volunteers in Service to America) and there got involved in the meals-on-wheels program for seniors. He became a member of the Nebraska Commission on Ageing and later the Illinois Human Rights Authority. He helped create the Bi-State Literacy Council; and, for more than a year in the 1980s, he was an unpaid on-air radio newspaper reader for the blind. After 1980, Richardson earned his living advocating the rights of those unable to do so for themselves. Residing in Galesburg and working out of the Peoria office of Prairie State Legal Services and then residing in Rock Island and working there for Protection and Advocacy (later Equip for Equality), he handled hundreds of charity cases, interacting as a consequence with the Social Security Administration and various state and local offices. It was in this advocacy capacity that he first encountered the sometimes arbitrary, sometimes client-contrary township supervisors who administered temporary relief funds, called general assistance. Prior to his advocacy career, he worked as a U. S. Bureau of the Census field operations agent in a five-county area of western Illinois. There, in addition to coordinating 160 canvassers, he met and briefed the media on a regular basis. On the job, as a volunteer, and in his personal life, he absorbed and put to use a great deal of legal knowledge. Richardson learned how to prepare court briefs and how to deliver them orally. It was Richardson who drafted the bylaws and articles of incorporation for the Bi-State Literacy Council and who argued his own cases when he sued the legal guardian of a juvenile who had allegedly stolen from Richardson and city officials in Knoxville, Illinois, about an incident he said defamed his character. By 1990 he had become an effective writer, a man able to make the case for abolition in a variety of venues.

Even with an altruistic fervor, a public relations bent, considerable legal talent, and a flare for prose, however, Richardson might not have been up to the challenges and frustrations of the nineties had his family life been different. Divorced in the mid-1980s, he moved to Rock Island in 1986, leaving behind in Galesburg his former wife and their three sons. The boys would visit him often at his home in South Rock Island Township; but between visitation weekends, he had a lot of time on his hands. Randy Popp helped him fill that time.[2]

Randy Popp

A native of Rock Island and slightly younger than Richardson, Randy Popp (born 11 January 1953) began to take a serious interest in township-style grassroots government after a friend asked him to help her campaign for the job of supervisor in Rock Island Township. When he asked her what the supervisor did, she said, "Nothing, but someone has to fill the office." Popp had been reading local newspaper accounts of financial problems surrounding area townships but might have gone on working at a local vending company while pursuing his dream of a degree in mass communications at St. Ambrose College if his conversation with the supervisor-aspirant had not reminded him of the butter incident. Several years earlier, Popp's aunt had returned to Rock Island County from Australia needing a place to live. She found lodging with Popp and his family in Coe Township, a rural area northeast of the Quad Cities, but still lacked money for groceries. General assistance seemed the perfect solution, until she could find work and live independently. So she applied for and received a food voucher from the supervisor of Coe Township. With that voucher, at a local market, she purchased, among other things, a pound of butter. A few hours later came a knock at the door of the Popp home. It was the township supervisor wanting to trade a pound of margarine for the butter, because assistance clients, in the supervisor's words, were to buy margarine, not butter.[3]

Popp began talking in 1989 with acquaintances about the roles of township government in their lives, trying to learn how townships functioned and who provided necessary oversight. He learned of Thomas Cieseilski's abortive 1988 petition drive in LaSalle County and spoke with Ciesielski about that effort and the Zemaitis case (chapter 6). Mike Boland, then the social studies teacher of Popp's son and later an Illinois legislator, agreeing that townships might be unnecessary, suggested that Popp contact Michael Richardson, by then a county resident.[4]

Richardson was initially cool to the notion of starting anything like the general assistance controversies he had been part of in Peoria and Knox counties during the 1980s. Popp persisted, calling Richardson "every week" it seemed, until Richardson finally agreed to help Popp with a research effort on local townships. Thus they started to gather and share stories of how township supervisors dealt with their general assistance clients. The duo began to draw in others, who brought more township tales. Out of these conversations came the idea of founding a group to study townships in an organized manner, a group that would use the Freedom of Information Act and any other leverage it could muster to publicize township mischief and malfeasance they were reporting to one another. "I had smelled that stench before," Richardson said later; and, with Popp, in February 1990, they created the Township Study Group (TSG). Other local activists would come and go, but TSG hardly ever consisted of anyone besides Popp and Richardson. "Having a third person present for a strategy session of TSG was special," Richardson admitted.[5]

Township Study Group

TSG is only a memory now, but for several years in the early to middle 1990s it attracted a great deal of media interest, stirred consternation among local township officials, and angered the TOI executive director. Its co-founders, its co-chairs, were masters at keeping the media curious about what TSG was contemplating. Press releases, press conferences, petition drives, lawsuits, public protests, and a barbed and biased newsletter brought TSG local, Chicagoland, Springfield, and Astoria attention. Richardson seemed to relish the opportunities that the TSG era provided him to write for publication, speak to reporters, confront townshippers and other politicians, practice his legal hobby, network with abolitionists around the state, and run for public office. Popp took a much less prominent role but served as an able second chair to Richardson, taking some of the heat, writing letters to various newspapers, editing *TSG news,* and gathering thousands of petition signatures.

Instead of just coming out against township government, TSG took a more cerebral approach, though Richardson told me later that TSG "was really a front from which to organize petition drives in the hope of getting discontinuance on the ballot." Be that as it may, in the spring of 1990, TSG appeared to be a small group of concerned citizens trying to gather the facts about a layer of government that kept a low profile. TSG tried to shine a critical spotlight on Rock Island County's townships so the public could see what Richardson, Popp, and their little band had experienced.

TSG reached out to Illinois's commission counties in a fact-finding endeavor. Randy Popp signed and sent letters to the county clerks of the non-township counties asking for information about the way they conducted county business. What "we needed to do was find out if communities could survive without townships, and surveying those counties seemed to be the most logical first step," Popp told me. In his letter Popp revealed just how little he and TSG knew about township organization in Illinois. For instance, he asked, "[What was] the year your county became a commission county and [what were] the circumstances surrounding the change?" Then he inquired as to "the number of townships eliminated by the change to the commission form of government." The great majority (15 of 17) of the commission counties had never known townships so clerks and others who tried to answer Popp must have been wondering what was going on as they responded. Sue McMaster wrote, "Hardin County...has always been a Commission county…and there has never been any attempts to change." Barbara McDade of Scott County said her county had "always been Commission." Russell E. Morris, from Morgan County, noted his county had "been a commission form of government since its organization in 1823."

Popp's final question drew responses from several as to their "candid opinion of the effectiveness of Commission versus Township government." McMaster, after "talking to other counties that have Township form," had concluded her county's commissioner approach made it easier to solve problems because one had "fewer board members to contend with." Evelyn F. Hogg, Pope County's clerk, cited Pope's "small population area" and said "3 commissioners [were] sufficient." Joan E. Wolfe said Wabash was "very happy" with commissioners "and would not change it." On the other hand, Wabash did have six road districts; and she hoped "some day to unitize them into one." Wabash enjoyed its freedom from "all the various township officials." So, even though TSG had its facts askew, as to just how 17 of the 102 Illinois counties came to claim no township government, fact-finding letters brought relevant intelligence and an encouraging word or two back to Rock Island.[6]

Concurrent with its commission-county correspondence, TSG undertook a local investigative effort by visiting township offices and requesting information, about general assistance, and the chance to examine township financial statements. TSG announced its preliminary findings during April of 1990. In a press release timed to appear just before the annual township meetings, Popp said TSG was that day mailing "notices of violation and demands for compliance" to ten townships in the county. Those townships that failed to heed TSG's written warning would face "possible mandamus actions in the circuit court." TSG took its complaints regarding inadequate general assistance records to the annual meetings of Moline, South Moline, Rock Island, and South Rock Island townships on, asking assembled electors to ask their elected officials to mend their ways and call "special general meetings" at which officials could "explain their positions on general assistance programs." Electors voted down the TSG motions in all four townships, so Popp told the press that TSG would use the petition prerogative to "compel the townships to call the special meetings." For public consumption at least, TSG was, in April 1990, still just trying to amass and assess the facts so its members could spur township officials to do their jobs better.[7]

Nepotism was next. Richardson announced that TSG had been closely scrutinizing the amended annual reports that they had helped force townships to file and that TSG had found officials from one-third of the townships willing to admit that they had relatives on the townships' payrolls. TSG was still, said Richardson, looking for other such marriage and blood relationships in the annual reports. Moline Township Supervisor Larry Foley, according to the release, had won the dubious prize of having the most relatives on his township payroll: two. Richardson promised to review nepotism at the next meeting of TSG, noting such meetings were open to the public and that "township officials [were] welcome to attend."[8]

TSG eventually had to suspend its standing invitation for township officials to attend TSG meetings because, in Richardson's words, township "officials and their attorneys kept crashing the party." Early on, however, TSG welcomed the invigoration and publicity that TSG derived from face-offs with township adversaries. Supervisor Foley came to the TSG meeting on the last day of April to defend the hiring of his son, who was by then "no longer working for the township," and his wife, who was qualified to serve because an insurance company had previously employed her for an extended period. General assistance also came to the floor that evening at the TSG meeting in the Rock Island Public Library. TSG member Jim Lasek said townships were spending more to administer assistance than to help the unfortunate. Frank Youngquist, South Rock Island supervisor, refuted the charge, saying Lasek's numbers were "not even close" to the truth. When Lasek cited the county clerk as his source, Youngquist said he had more detailed information at his office.[9]

A month later George Miller came to join the discussion. State's Attorney Marshall Douglas had scheduled a meeting to review what the laws said township officials ought to be doing. He invited Miller and a representative from TSG, which finally had gotten the attention of the courthouse crowd and Fortress Astoria.

Randy Popp drew the short straw for the honor of being the TSG speaker at the 23 May seminar, but Popp had already addressed the hundred plus in attendance and departed when tempers flared. The flare-up came when TSG member Ed Langdon, Jr., told officials it was their behavior and not that of TSG investigators that was costing townships money to comply with the Freedom of Information Act. Supervisor Bob DeJaegher of Hampton Township told Langdon that he and TSG should have learned all they needed to know about township government in school. Miller defended townshipper reluctance or inability to provide all the information TSG seemed to want but conceded some townships had failed to file annual financial reports, though that failure was "not because they were stealing any money." Langdon walked out, apparently feeling betrayed that Popp had departed so soon, leaving him alone among hostile township officials. Carol Pentuic, managing attorney of Prairie State Legal Services (where Richardson had once worked), reminded officials of their responsibility to the disadvantaged, people who had, she said, the right to sue at the highest courts of the state and nation. Townships were to have on file written, not just commonly understood, guidelines for rejecting general assistance applicants.[10]

TSG put a news release on the street two days later, offering Popp's apology to Douglas for any problems that might have resulted from the activist's early retreat. No questions came up, so Popp had left. He offered to debate township officials and promised such intimidation (as Langdon suffered) would not deter TSG from pursuing its "civic duty."[11]

Popp and Richardson were going to need Langdon and every trooper they could field, because the same day Popp apologized in the news release Richardson announced a petition drive to call a referendum that could abolish all 18 Rock Island townships. The first of this busy decade's five petition-punctuated summers was about to begin. Richardson told the local media that TSG's investigation had raised "serious questions about township governments in the area." He cited incompleteness in filing required documents and failure to meet community welfare needs. TSG had to collect, he estimated, nine thousand signatures to succeed in gaining ballot access and expressed confidence in meeting the 20 August filing deadline.[12] He and his comrades had shifted from investigation to intervention, a transformation that George Miller surely must have expected as he journeyed that week to the Quad Cities.

Flashback to 1973–1974

TSG's 1990 petition drive was not the first citizen effort of the twentieth century to defrock Rock Island townships. Moliner John E. Davis had drawn a bead on those same townships in 1973–1974 when it became obvious to him that the state was going to require that several of his county's townships undergo voter scrutiny in the matter of merging with adjacent townships (chapter 6). To Davis it made sense at the same November 1974 general election to ask voters whether they preferred discontinuance of all townships and a reduction of the county board from 29 members to some smaller number, say 3. Editors at the *Rock Island Argus* agreed.[13] In fact, the *Argus* and *Moline Dispatch* editorial staffs had been nibbling at township heels for years, advocating consolidation of townships and road districts at the least and perhaps abandonment of township government altogether.[14]

Davis, then a member of the county board, cited a number of governmental problems such as Watergate and the resulting public interest in official accountability as good reasons for reform. Rock Island County, he told me in a 1993 interview, had in the 1970s serious financial problems; and he saw township abolition as a way of freeing 10 cents of every tax dollar for county use. As he put it, "Two-thirds of the townships' work went through the county anyway." He had determined that something on the order of nineteen thousand petition signatures would suffice and asked anyone interested in assisting to contact him at home. Davis considered his task monumental but saw no other recourse. The entrenched local politicians, who drew vital support from townships (patronage jobs, for instance), were not about to permit the county board to simply call for such a referendum; and Governor Dan Walker had told the TOI members at their 1973 annual meeting that he "would never sign a bill to abolish township government or severely limit" it. The *Argus* suggested that, since Davis's idea merited "a serious discussion," members of the League of Women Voters "might be interested in pushing" his proposition.[15]

Offers for help did come from individuals and from the League, but most amounted to nothing. Nevertheless, by early July 1974, he had on the order of five thousand signatures, with a 19 August deadline looming. There was still time, with help, of course, to gather the additional thousands that he would need. Then the opposition organized. When they got started, "I was dead overnight," he confided. Threatening phone calls began arriving at his house, and his wife urged him to drop his crusade. For personal protection, over a six-month period in 1974, he never left home without a sidearm. Township supporters, he continued, even smeared him after the abolition campaign ended; and they finally drove him from public office.[16]

He never bothered to file his petitions, but told the *Argus* he hoped to try again in 1976. The *Argus* editors sympathized with his cause and agreed, he "encountered powerful political opposition." In essence, he had been fighting both his own Democrat Party as well as the county's Republicans. All the Rock Island township consolidations failed that fall, and the *Argus* concluded that it would take "a stronger state law" if Illinois wanted "to improve township rule."[17]

In neither his conversation with me nor in press coverage of his petition drive did Davis or anyone else mention the Astoria connection. While TOI was surely aware of the Davis effort, Troy Kost probably felt townships were so invulnerable in Rock Island as to warrant no overt outside intervention.

Petition Signatures

Davis believed, for a 1974 ballot slot, he would need 19,000 signatures while Richardson in 1990 initially said 9.000 would suffice. In 1988, Carl Edwards, in Gallatin County, needed only 250. Why the difference from year to year in one county and from county to county?

The Township Law of 1874, as amended many times, controlled the petition signature requirement against which Davis, Richardson, and Edwards measured their success. From the 1870s until 1933, all who sought to dissolve townships countywide faced the 20-percent rule. They had to secure the signatures on their petitions "of at least one-fifth of the legal voters...to be ascertained by the vote cast at the last preceding presidential election." In 1933, as I discussed in chapter 5, the general assembly reduced the figure from 20 percent to 5 percent in all counties sending fewer than 75,000 to the polls in the preceding presidential election.

That 5 and 20 provision was in effect in 1973–1974 when Davis circulated his petitions. A total of 70,330 voted for Nixon, McGovern, or other presidential candidates in November of 1972. Davis theoretically needed just 5 percent of that number, or 3,517, plus a cushion in case of signature invalidity and other challengeable infractions. With "five to ten thousand" signatures on hand, numbers he quoted for me, Davis may have actually gathered (depending on challenges) enough to have forced a referendum in 1974. Why did he overestimate his target? Somehow he had gotten the idea that he needed 20 percent of the registered voters, which numbered 95,000 or so. If he multiplied 95,000 by .2, he got 19,000.[18]

Tiny Gallatin County, of course, fell in the 5-percent group. Carl Edwards and Mark Mossman, the county clerk and recorder of deeds, would have been working with the Reagan and Mondale election of 1984 to calculate their base vote total. That year Gallatin counted a total of 4,112 who voted in the general election. Edwards told me he was shooting for 250 signatures, so he was aware he needed 200 plus a cushion of, say, 50 signatures.[19]

When TOI's Bryan Smith razzed the abolition movement in his December 1988 *ICTO* column about raising "its ugly face again" and when he said he hoped abolitionists would "receive a lump of coal in their respective stockings" that Christmas, he mentioned no names. Carl Edwards and other out-of-the-way players were of some concern to those at Fortress Astoria, but Patrick Quinn and anyone else that could stir up a multiple-county or statewide storm demanded the full attention of TOI.[20]

Quinn and his Coalition colleagues had taken advantage of a couple aspects of the Township Law, which TOI believed needed attention. First, the rule of 5 and 20 probably was not fair to those in the 20-percent bracket; and more counties would be reaching the 75,000 voter mark as years passed. TOI wanted, according to David Regner, its legislative director, to be sure the legislature and not the courts would be making township-specific laws. Furthermore, TOI leaders still feared Judge Ann Claire Williams might, in her final ruling in *Zemaitis v. DuPage County Board of Election Commissioners,* give Quinn and the Coalition for Political Honesty what they were requesting—a 5-percent signature requirement on dissolution petitions in DuPage and Cook counties. Second, Quinn and the Coalition had apparently "been collecting petitions [signatures] in Cook County for up to *two years*" and still did not have enough to file. What was to stop the Coalition from collecting indefinitely? Unlike Illinois laws, according to Regner, "regarding other types of petitions," the Township Law allowed circulators of township-dissolution petitions to pass them around for years.[21] Davis, of course, had circulated his petitions and accumulated signatures in both 1973 and 1974.

Working in the fall of 1988 with leaders and members of the general assembly, while *Zemaitis* was still pending, TOI helped craft legislation that became H. B. 3355, and finally Public Act 85-1446 on 24 February 1989. Regner and Miller had originally hoped for a narrowing of the petition-signature window from one that had been indefinite in length to a window containing just 120 days. The general assembly presented TOI with a bonus of 30 days, by specifying that petitioners would henceforth have but 90 days to gather names. In addition, the law also standardized the minimum number of signatures for dissolution at 10 percent, of those who voted in the last presidential election, thus bringing this aspect of the Township Law more closely in line with Illinois's Election Code. H. B. 3355, by the way, passed the house, 115-0, and the senate, 59-0. Regner thanked township officials around the state for the help they gave when the time had come to influence legislators.[22]

Public Act 85-1446 dealt TSG a double whammy. This February 1989 modification of the Township Law forced TSG to collect twice as many signatures as would have been the case had TOI not prevailed. Rock Island County, in 1990, would have fallen in the 5-percent category without the new law, because 67,901 had voted there for president in 1988. Public Act 85-1446 also hurt TSG by putting it in a situation that no township abolitionist had ever faced. TSG could not begin collecting signatures until 90 days prior to the petition-submission deadline, a deadline that the Election Code already prescribed.[23] With the passage of H. B. 3355, TOI began a series of legislative maneuvers that greatly changed the playing field on which township supporters and detractors met.

Summer One

TSG was not really ready for the petition season to begin. When Michael Richardson held his news conference to announce the petition drive, the 90-day clock was already into its third day; and TSG did not even know it was ticking. Richardson told me how TSG's printed petitions for the referendum lacked the "necessary language" to confirm for anyone who examined them later that circulation had occurred within the window. Before TSG "hit the streets with the insufficient ballot language," as he put it in our conversation, someone brought the oversight to his attention, so he amended all the petitions with a rubber stamp, reading, "circulated since June 1, 1990." Fortunately for TSG, Richardson kept the receipt for the rubber stamp, because the township side would later challenge the lack of a printed date on the petitions.

Richardson, Popp, and their TSG brethren learned a lot that first summer. For one thing they learned how slow was the going when it came to walking a petition from door to door to door, though they did have some success that way recruiting petition passers who answered their knock. By the end of June, the abolitionists, knowing they needed to move more quickly, began targeting places where people congregated. Many citizens they approached had little or no knowledge of townships, despite the excellent local press coverage, so TSG used precious time to explain the purposes and, no doubt, the downside of township organization in an modern, urban setting.

Having three daily newspapers as well as other media outlets in the Quad Cities, all striving for the latest development in this unfolding controversy, helped the TSG cause; and having a local media generally favorable to a dissolution referendum encouraged TSG immensely. Even before TSG knocked on the first door, the *Argus* editorialized that editors there hoped "the petition drive [would be] successful," not because TSG had convinced them that abolition was necessary but rather that a debate over what could be best for local government efficiency would be in the public interest. A few days into June, WQAD-TV, broadcasting from Moline, also applauded TSG and asked voters if they knew in which township they resided. WQAD offered the notion that, if voters could not get sensible answers to reasonable questions about their townships, they should ask themselves "Why are [we] paying precious tax dollars for something [we] can't justify?"[24]

TSG fed the media's appetite, making the principals available for interviews and by issuing regular news releases over the summer. Popp hit the lack of proper accounting procedures by the townships in his 23 June release. A week later he was complaining about the size of the Moline Township supervisor's salary, as well as payments from town funds to members of that official's family. "The Mayor of the City of Moline receives $15,000.00 per year," he said. "The Supervisor of Moline Township, which represents about half of the city[,] pockets $19,500.00 per year, a sixteen percent increase over his previous year[']s salary." For several weeks, TSG got good press over the issue of official "vandalism" in Hampton Township Cemetery. A 10 July release from TSG found Richardson announcing a slide show at the Rock Island Public Library to highlight damages that he said gravestones had suffered at the hands "of someone responsible for cemetery care taking." Hampton's cemetery found its way into an early August release, as the Study Group reported "mismanagement by township officials" when it came to awarding maintenance contracts.[25]

By mid-summer, even though TSG was only one-third of the way to its announced goal of nine thousand signatures, county, township, and TOI officials were voicing concerns publicly over the mess that they claimed abolition would leave in its wake. George Miller said, "You eliminate townships and you eliminate board members," and that "In an at-large election of three commissioners, county government probably would be represented entirely by commissioners elected by the urban population." Hampton's supervisor, Bob DeJaegher, agreed, saying, "It's my understanding that we would revert to a three-member commission." He feared "taxpayers" and "county government" would see a most "adverse impact" if TSG's plan prevailed. County Clerk Dick Leibovitz and County Board Chair Paul Mulcahey echoed Miller and DeJaegher, while State's Attorney Marshall Douglas told *Dispatch* reporter Lisa Neff, "Townships are at the heart of basic American government" and "Township government serves as a training ground for elected officials."[26]

TSG leaders did what they could to counter such warnings of dire consequences. "The county board is a constitutionally protected office," Richardson told Doug Schorpp of the *Times*. "It will continue whether the townships continue or are discontinued." Neff confirmed, with the Illinois attorney general's office, Richardson's claim about the status of the county board. "County boards aren't connected to townships," an unnamed male source told Neff. "They would have to pick up the duties but the structure [of the board] would not change." Richardson foresaw Rock Island County taking on tax assessment, roads, general assistance, and cemeteries, though he probably erred in the case of roads since road districts were likely to remain in place as separate entities.[27]

The township side not only used the media to fight TSG that summer but also went to court and may have resorted to less appropriate tactics in an attempt to intimidate and otherwise harass the petitioners. The Rock Island County Township Officials Association filed suit in circuit court asking for a ruling as to whether the Association was or was not a "public body" subject to the Freedom of Information Act. TSG, basing its demands on the fact that townships (presumably) used tax funds to pay their membership dues, had been requesting more documents from the Association than it could or wanted to produce. Richardson and Popp told me about several incidents that they believed affected petition circulators that first summer. A shop steward for the United Auto Workers had been carrying TSG petitions until the local union's president reprimanded her for her political actions. She ceased petitioning. Popp lost longstanding accounts at the Moline vending business where he then worked, losses he attributed to townshipper pressures. Richardson's supervisor at the advocacy service where he worked fended off four or five efforts from township officials aimed (he assumed) at jeopardizing his job. The national headquarters of a vending company unsuccessfully ordered the district manager to terminate the firm's Rock Island County manager because the local manager was soliciting petition signatures in his spare time.[28]

Intimidation and fatigue resulted in a slippage of commitment for several circulators in the waning days of the petition window, but not for Richardson and Popp. On 20 August, the two of them celebrated the end of the summer on the streets with a handshake over a stack of petitions resting on the counter of the county clerk's office, petitions that they said contained 7,507 signatures (figure 20). Somehow, in the month since its leaders had reported having just three thousand names on petitions, TSG added another 4,500. Richardson told reporters he thought he had "500 extra signatures in the event there [was] a challenge." County Clerk Leibovitz was still saying nine thousand was the target at which TSG ought to have been aiming, despite the fact that only 67,901 had voted in Rock Island County in the 1988 presidential election. A tenth of that number would be 6,790. TSG's 7,507 should have provided a few more signatures than the cushion of 500 Richardson wanted "in the event" of a challenge.[29]

Challenge and Checkmate

A decade later, in the fall of 2000, when I asked Richardson what one event marked the high point of the 1990s for him, he still pointed back to that

Figure 20 Michael D. Richardson, left, and Randy Popp celebrating submission of their 1990 petitions to the county clerk. (Courtesy of the *Rock Island Argus*)

Monday in August 1990 when he and Popp delivered their petitions: "We were euphoric." Adding to TSG's euphoria was an editorial in the *Quad-City Times* with the title "Get Rid of Townships." There was probably not "a single function performed by any of Rock Island's 18 townships that couldn't be done better and cheaper by the county," said the editorial, which went on to remind Rock Islanders of recent events surrounding an apparently incompetent (at best) highway commissioner in South Moline Township. But, according to the *Times,* the whole township system was faulty, with waste and nepotism seemingly a given. The editors speculated that only those on the township payrolls would bother to fight for the county's 18 townships. "Some of them [on the payroll], in fact, may challenge the petitions seeking to put the referendum on the ballot." Voters, said the *Times,* deserved a chance to get rid of townships, and "Hopefully there will be no such challenge or any challenge will be unsuccessful." Challengers had "until Monday the twenty-seventh to do so."[30]

On Monday a massive challenge did occur. TSG could hardly have known what it would unleash when it chose to threaten the very foundation of partisan politics in a very political county. Ironically, the challenge came not from township officials but rather from the county's Republican and Democrat party chairs. Procedure, not "opposition to the referendum being submitted" was the problem that had prompted Democrat Party Chair John Gianulis to sign on with Bill Smith from the GOP side. Inspectors had found, said the objection, possible forgeries and multiple signatures from the same signers. Over five hundred signatures, objectors said, were those of nonregistered voters—at least not in Rock Island County. Thomas Kilbride, who did the filing for Smith and Gianulis and also represented the township officials, said there had been fewer signatures in the Richardson-Popp pile on deadline day than the statutes required. He claimed that the March 1990 primary election, when 90,600 had voted, was the baseline, not the November 1988 general election (as Public Act 85-1466 specified). TSG really needed 9,060 signatures, said the township attorney. Richardson told Lisa Neff, "'We believe that higher figure is incorrect.'"[31]

What began at the 4 September 1990 hearing of the Rock Island County Board of Elections, as objectors targeted various aspects of the petitions, their circulators, and their signatories, would become a recurring drama in the decade that lay ahead, a drama that both sides took very seriously. At stake for petition circulators was a cause, countless hours of personal time, and expenses. At stake for the objectors and those in the wings was a cause but also political power and people's livelihoods. Rock Island County was mainstream, and a defeat of the township way of life there was not something that any on the township side wanted to contemplate. If Rock Island's 18 townships fell, the general assembly would finally have to live up to its constitutional mandate and pass legislation to accommodate the transition from townships to their successor(s). If Rock Island fell, the courts would surely have a role to play in the ultimate outcome there and elsewhere. In many ways, Rock Island was similar to the state's Collar Counties, where townships were rapidly losing their rural connections. As I said earlier, Michael Richardson and Randy Popp could not have known what they were about to unleash when they started poking around in the business of a local powerhouse.

TSG actually tried a couple of tactics, before the petition review, in the hope of avoiding that fight altogether. On 31 August, Richardson and Popp sent letters to the state chairmen of the two major political parties asking Republican Al Jourdan and Democrat Gary LaPaille to call off Smith and Gianulis, respectively. In the press release announcing the letter campaign, the TSG co-chairs argued that depriving the public of the chance to vote (on an interesting issue) was a sure method of fostering further voter apathy.

They said the discussions their issue would generate would be good for democracy. Next, three days later, a TSG news release previewed an appeal Richardson and Popp were preparing to make to County Board Chairman Mulcahey. In that appeal they hoped to convince Mulcahey to "convene an emergency meeting of the County Board to place the township referendum question on the November ballot," which was a constitutional option every county board possessed. Jourdan and LaPaille apparently did nothing to discourage their county counterparts from objecting to the petitions. Mulcahey was able to dodge TSG's request for a special meeting because the request arrived too late for him to have initiated the process necessary to call an emergency meeting.[32] Euphoria at TSG headquarters (Richardson's modest home on the west side of Rock Island) had by early September given way to dogged determination.

The board of elections met for five hours on Tuesday, 4 September, and reconvened the following day to start the task of going page-by-page and signature-by-signature through the stack of petitions, in an effort to substantiate or repudiate the challenges. Richardson represented TSG; Kilbride spoke for the county party chairs; and County Clerk Leibovitz chaired the three-person board, which included State's Attorney Douglas and Circuit Court Clerk Henrietta Kenney. Douglas's participation in the process drew an immediate protest from Richardson as the Tuesday session began. Richardson wanted Douglas removed because of a potential conflict of interest, claiming Douglas had tried "to obstruct the circulation of petitions." In an attempt to bolster his case for the removal of Douglas, Richardson called Douglas's former campaign manager, Doug House, to the stand. House, however, despite a subpoena from Richardson, did not appear. How could he ignore that subpoena? TSG had neglected, as the law apparently required, "to provide a fee for the witness' cost of appearing." Kilbride, knowing about the fee requirement, had filed a successful motion to quash the subpoena.[33]

Michael Richardson, though familiar with legal matters and a polished orator, was, in the end, not an attorney. The state permitted him to perform in such hearings as an attorney; but lawyering was a sideline for him, an avocation. One has to wonder if TSG did not do itself a disservice by pitting a paralegal against Kilbride and other practicing attorneys, attorneys who belonged to the lawyer fraternity and had law firms to assist them. Perhaps TSG would have folded its tent early on if its principals had needed to go into great personal debt to buy legal services, but the courtroom outcomes in 1990 and subsequent years might have been different if they had chosen to purchase legal assistance. At that first petition hearing in Rock Island, however, Richardson held his ground as Kilbride brought to the forum every objection his clients had made: TSG needed at least 9,600 signatures, some signatures did not match those on file in the voter registration office, quite a few who signed were not registered voters, pages of the petitions failed to carry seals of notarization, pages did not always carry the signature of the circulator on the reverse, some signers had printed where they were to have signed, and at least 50 people had signed more than one petition.[34]

As Leibovitz, Douglas, and Kenney perused the petitions, the Chicago press took notice and linked the Rock Island drive to similar efforts closer to Lake Michigan. *Tribune* writer William Presecky recounted how Patrick Quinn's Oak Park-based Coalition for Political Honesty had tried that same year to place on the ballot, in Worth Township (Cook County, southwest of Chicago), a binding referendum on abolition of that township's government. After summarizing what had transpired as of the first week of September, and allowing Gianulis a chance to proclaim his underlying affinity for townships as the real reason for the petition objections that he and Smith had filed, Presecky gave Richardson his first Chicagoland exposure. Presecky found Richardson willing to say openly for the *Tribune* what he had thus far kept under the surface for the Quad Cities press. The objections revealed to him, said the TSG spokesman, "'the raw political power of the townships and the desire of the political bosses to retain their patronage. They claim it is a bipartisan effort (designed) to keep improper petitions off the ballot....I think it's appalling that the heads of the two political parties would try to take a question away from the voters.'"[35]

Richardson had his public relations persona in place on Wednesday, 12 September, when the board of elections ruled TSG had failed to collect enough valid signatures. The board knocked out 1,246 of TSG's roughly 7,500 signatures, bringing the total of valid signatures down to 6,253, which left TSG 700 or so short of the 7,007 the board finally decided they needed. "'We expected an unfavorable ruling from the board, but we expect a favorable ruling in the court.'" said Richardson to the *Argus* after hearing the decision. He also indicated his dissatisfaction with the board's failure to pull hundreds of voter cards in order to confirm the registration status of petition signers. "'If we can't have it heard in circuit court, we'll take it to the appellate court. The Illinois Supreme Court may get to rule on this,'" he told a *Times* reporter.[36]

An immediate outcome of the board's verdict was an editorial in the *Dispatch* praising Popp and Richardson for "stirring the pot" and all "the volunteers who gathered signatures" that summer. Editors labeled it "a shame" if TSG ultimately were to fail to give voters a chance to abolish townships, even though the *Dispatch* was not yet willing to come out one way or the other on abolition. Editors wished TSG well as it went to court and urged the group to "bring the issue back again next time" if the court system failed to yield a favorable outcome. As a matter of fact, TSG leaders already had plans, if necessary, to make sure there was a next time—and to submit then even more signatures.[37]

Several letters to the editor appeared. Josephine Tady, in a late September letter to the *Argus,* commented at some length about her long service to Rock Island Township as a trustee and (then) clerk, praising the elected nature of those posts and condemning carpetbagger (my term, her characterization) Richardson for moving to her county and almost immediately trying to do away with her job and that of others she admired. Tady also wondered what TSG had to gain and who or what was their source of funding. Richardson's *Argus* reply cited "property tax relief" as the motivation of most TSG volunteers and informed readers that funding came "right out of [TSG's] pockets, the same pockets that township taxes come from." He reminded Tady that his brief residence in the county was no source for disqualification when it came to the political process and suggested she take "a brush-up course on civics" for further elucidation of the residence issue. TSG lieutenant Geoff Michl of Moline also answered Tady, pointing out how townships in Rock Island County were wasting money, he believed, on extravagant salaries for officials, some of which were double dippers or nepotists. Michl bemoaned in his *Argus* letter of early October the role party bosses had played in the abolition scenario and wondered why they seemed so afraid to have the people decide the fate of townships: "This is much too important an issue to be left solely to the politicians." Another Michl letter, which appeared in the *Times* on 17 October, cited salary figures for Moline Township and pointed out that one township supervisor in the county served as a state representative and another as a county board member. He also called into question the job qualifications of many township officials. Teresa Wates, daughter of the Moline Township assessor, answered Michl's second letter in order to defend her mother's qualifications and work ethic. Wates noted also that Michl had paid just $12.58 in 1990 to Moline Township for his share of the property taxes the township collected and wondered if having that back would provide Michl a "windfall."[38]

While letters were bouncing around the Quad Cities, Richardson and Kilbride were busy in the courts. Richardson filed suit contending the board of elections failed to review all records; and he asked the court to enjoin the county clerk from mailing absentee ballots knowing that once these ballots left the courthouse, without TSG's referendum in place, the chance of gaining ballot access later in the fall diminished greatly. Circuit Judge Jeffrey O'Connor denied that TSG request but did agree to move the case up on the docket by a couple weeks in the cause of expediting matters. At that 9 October hearing, Richardson struggled against the problem of having too few signatures. In the process, he "accused Smith and Gianulis of fraud," as Doug Schorpp of the *Times* reported, "because they signed documents stating they had reviewed the petitions but testified they actually had seen very few of them." Kilbride, in response, subtly mocked Richardson's lack of standing in the legal field by saying "even though Richardson is not an attorney," he still needed facts to make such an accusation about these two political figures. Judge O'Connor turned down TSG in his ruling of 11 October, using as rationale the fact that the elections board had kept no transcript of its hearings. Afterward, Popp spoke confidently about an appeal of the judge's ruling;

Kilbride said the chance of reversal was "next to slim" in such matters. Kilbride's only complaint was that TSG remained so stubborn despite such unfavorable rulings. "'They persist in that they are always right no matter what.'"[39]

Sure enough, TSG went immediately appealed O'Connor's decision. Richardson wanted the judges of the Third Appellate Court in Ottawa to demand that a circuit court judge force the county elections board to provide an official record of the meeting wherein the board scrapped twelve hundred signatures. The November election kept getting closer, an advantage the township forces no doubt relished. When the appellate court refused to consider his case before the spring of 1991, Richardson asked the Illinois Supreme Court to stay the distribution of the ballots for the coming election until his appeal made its way through the system. The *Argus* labeled the ploy "purely asinine," despite TSG's "noble fight" to gain ballot access. On the Monday, before the Tuesday on which TSG had hoped to give the people of Rock Island County a chance to vote townships up or down, the supreme court said no to Richardson's final request.[40]

In TSG's last news release of the 1990 campaign, Richardson confirmed defeat in the courts and blamed not the judges but rather the local party bosses for depriving "voters of a ballot question." He pledged not to go away or give up: "The township question will appear on a future ballot…that is a promise." Popp vowed to hit the streets again in 1992 and noted how he resented the way his party's chair, Gianulis, had bullied and shouted him down at a meeting of the Democrat Central Committee, when Popp asked if Gianulis had used party funds to deny voters a chance to voice an opinion on townships. Richardson promised further study of the county's townships and pursuit of a Freedom of Information case against TOI.[41]

Speaking of TOI, just where had George Miller been during the many months of confrontation in Rock Island County? Aside from his appearance at the May seminar and his comments for a Lisa Neff story in July, I can find no mention of township government's man of the 1980s in the local press or the TSG literature. Nor can I find a single contemporaneous reference in *ICTO* to those months of struggle. My guess is that Miller did have a firm hand on the oar as Rock Island's township ship slipped through the potentially devastating currents of 1990. Richardson told me his lawsuit seeking access to TOI records was an outgrowth "of George Miller meddling into Rock Island County's local affairs (my petition drives)." By 1991, the year of the suit against TOI, TSG had managed only one such drive—the one that occurred in 1990. Furthermore, Miller, TOI, and local officials had the luxury that year of laying back from the fray because the county party chiefs and attorney Kilbride were out front objecting to petitions, taking on the press, and going to court against Richardson and TSG.[42]

Pressure Points

Thomas Kilbride was right: TSG did not know when to quit. Not a week had elapsed, following TSG's failure to make the November 1990 ballot, before Richardson was directing the attention of the local press corps toward South Moline Township and its road commissioner, Mel Parker. TSG demanded an "independent investigation" of Parker and his department; and Richardson told the *Argus* he did not think Parker's alleged activities were all that unique: "'I think Mel may be a good example of a bad problem,'" a problem that Richardson thought plagued many townships and their road districts. "'There may be other Mel Parkers out there.'" Nepotism, charged TSG, was yet another South Moline transgression. Perhaps, Richardson offered, one reason that township officials there had "dragged their feet on taking action" on the Parker matter was dirty linen among his South Moline colleagues. Supervisor Larry Burns and Assessor John Hubbs employed their wives to assist with the duties of their respective offices, reported Jackie Chesser for the *Argus*.[43]

Next came TSG's Township Code of Conduct, which forbade hiring of relatives, double dipping, and use of power of public office to influence voters and which, among other things, called for compliance with statutes regarding township business and use of acceptable accounting procedures. Popp, in a December news release, said they were mailing each of Rock Island's townships a copy of the code; and, if officials would follow it, they could help put a stop to what his co-chair called the "seven deadly sins of township government." TSG wanted the board in each township to adopt the code and hold themselves and other officials accountable for ethical behavior. Ralph Erikson told Lisa Neff that his township, South Rock Island, would probably adopt the code. Kilbride said the whole thing was another attempt by TSG "to grab publicity." Carolyn Lund, who was then both a county board member and supervisor of Rock Island Township, said a code of ethics for public officials was a good thing; but she could not support a code that a special interest group had proposed.[44]

The new year (1991) provided no relief for the targets of TSG's scrutiny. January found Richardson planning to file a formal complaint against Lund and one of her township employees for alleged notary public misconduct. For February, TSG released the news that TOI had formed TOIPAC, its political action committee, to influence the election of sympathetic (to TOI) legislators and to collect and distribute campaign contributions for office seekers of the correct mindset. In March, TSG complained about financial reports and the hiring of family members. The group had mailed violation notices and demands for compliance to 10 of the county's townships regarding the required annual financial reports. Nepotism apparently still prevailed despite the light TSG had been shining on the townships. By April, Richardson said TSG was retracting most of the violation notices but took no blame for its previous action. The confusion as to whether townships had complied or not derived from the office of the county clerk, where, Richardson said, employees had misplaced but later found a key file containing the reports.[45]

News releases were by no means the only tactic Richardson and Popp employed to keep their cause in the public eye and townshippers on the defensive. TSG drew Senator Denny Jacobs of East Moline into an exchange of letters to the *Argus*. In his letter, Jacobs accused TSG of spreading misinformation; in theirs, the TSG leadership reminded voters that Jacobs had "a relative, a niece, on the South Moline Township payroll" and stood by their earlier charge that his familial connection to township government might have served to influence his vote on an important township matter before the general assembly. Popp sent a letter to the *Times* in which he poked fun at townships for claiming to be "a true democracy" but having only a handful of citizens participate in the annual opportunity (annual meeting) for a demonstration of the true democracy. Richardson traveled to Springfield in late March, to "sound the alarm over a township effort to get a $20 million chunk of state revenue," wrote Anthony Man for the *Argus*. Joining Richardson at the press conference to oppose a form of state revenue sharing that would have benefited all townships and the road districts in commission counties was Mike Boland of Consumers and Taxpayers Together.[46]

Earlier that month Richardson had launched his effort to have the courts declare TOI "a public body subject to the Freedom of Information Act." He wanted to know TOI's purpose, organizational structure, operating budget, and employee count; and he asked for a "current list of all types and categories of records under control of the Township Officials of Illinois," according to the request he sent to Astoria on 14 March. What he really wanted was TOI's financial records. George Miller, of course, said he could have none of the things he requested, but Miller did send Richardson a TOI brochure and offered to sell him the TOI directory, the Van Sickle history, and various videotapes explaining the township way of life. Once Richardson had sent the same request to TOI President Shirley Green and gotten a similar negative reply, he felt (as he wrote in his "Complaint for Declaratory Relief" for the Fourteenth Judicial Circuit) he had "exhausted administrative remedies under the Freedom of Information Act." The time had come to practice law once more.[47]

Richardson v. Township Officials of Illinois went as far as the Third Appellate Court. TOI's main argument through the circuit and appellate courts was that even though townships often paid their TOI dues from public funds, TOI provided only "limited advice to member townships" and thus did not really qualify as a public body. In short, TOI should not have to give Richardson any information about its private affairs. Motions, responses, and affidavits continued for over a year after the May 1991 filing in circuit court. Eventually Circuit Judge Jay M. Hanson declared, in July 1992, that TOI was "not a 'public body' within the meaning of Illinois law for purposes of the

Freedom of Information Act." On appeal from Richardson, the Third Appellate Court, in what Richardson told me later was "a poorly written and thought-out decision," ruled in favor of TOI. The judges at Ottawa ordered no publication of their decision on Richardson's appeal, which meant (and means) the ruling only applied in this one case and set no precedent for future legal guidance. The Citizen Advocacy Center in Elmhurst, a not-for-profit agency that played a major legal role later in the decade, according to Richardson, "considered filing its own FOI case against TOI in the Second District [Elgin] to try and obtain a more enlightened and legally correct decision." The Center, in Richardson's view, decided not to file because of limited resources and the concern of making bad published case law if TOI were to win again.[48]

Raising the Petition Bar

As Popp and Richardson were conducting their guerrilla actions, TOI had been busy where it really mattered—Springfield. On the day the Eighty-sixth General Assembly adjourned for the final time, H. B. 3310 cleared its second conference committee and won approval of all 59 senators as well as 98 of their house colleagues (98-6-6). Five days later, 13 January 1991, Governor Thompson signed the bill into law, effective the next day, in one of the last acts he performed in his long tenure as chief executive. Public Act 86-1486, "An Act concerning local government, amending named acts," was a catchall into which went a hodgepodge of statutory adjustments for townships and other governmental units. For Popp, Richardson, the rest of TSG, and any abolition-minded citizen who followed, the pertinent section of this lengthy law was rather brief. Legislators simply revised the basis of determining the number of voters under the Township Law of 1874, so that section 2.1 was to read: "upon the petition of at least 10% of the legal voters of any county, as determined on the date the petition is filed." Instead of requiring that interested parties look back two or four years to find the number that had voted in the preceding presidential election, the new method called for taking one-tenth of the number of registered voters on the books on the day the petitioners filed their petitions.[49]

Popp estimated, for the *Times,* that at the stroke of midnight, as 13 January had become 14 January 1991, TSG's burden jumped from around seven thousand signatures to around nine thousand. TSG had tried, via a last-minute letter, unsuccessfully to persuade Thompson to veto the bill, warning that "This type of action only creates voter apathy" by making it more difficult for voters to participate in a democracy. Their letter laid the blame for this raising of the petition bar at the doorstep of TOI, something the co-chairs referred to as "a special interest lobby."[50]

Meanwhile, new TOI Legislative Director Edward Stasiewicz, in his "first major exercise representing TOI interests in Springfield," was tossing kudos for H. B. 3310. He thanked Speaker of the House Edward Madigan; his senate counterpart, Philip Rock; bill sponsors Rep. Terry Steczo and Senator Tom Dunn; and the governor's staff. Stasiewicz especially appreciated the "active participation" of members, like Joan Mruk of Wayne Township in DuPage County, who had talked to legislators about the importance of changing the petition-signature requirement; and he predicted such participation from township officials in the future would allow TOI "to increase its active role in state lawmaking."[51]

Township officials, however, received "The Boo of the Week" from the *Quad-City Times* for their part in "convincing the general assembly to make it more difficult to abolish townships," a layer of local government the editors said was not only unnecessary but also a haven for family-hiring double dippers. Having to gather, for example, at least two thousand more signatures in Rock Island County made getting rid of townships that much harder.[52]

Reflecting, a couple of years later, on this turn of events and other such TOI demonstrations of political resourcefulness, Richardson told the *Joliet Herald-News* he had expected TSG efforts to be "'kind of a boring civics project that would put everyone to sleep.'" What they had done, he continued, "'was wake up a ferocious monster.'"[53]

Transition Legislation

Among the arguments protownship forces used in the 1990 Rock Island campaign to sway voters away from change was the uncertainty about what would happen once townships disappeared. Statutes in force contained vague guidance regarding closure of township offices and disposal of township property; but voters seemed to want more specificity, a firmer notion of what would happen and when. Uncertainty over whether the county board would survive countywide dissolution of townships was an even tougher nut for abolition advocates to crack. Into this debate over transition and transfer of power stepped Rep. Joel Brunsvold, a Democrat from Milan, a Quad Cities suburb south of the Rock River.

During January of 1991, Brunsvold introduced H. B. 68, which was to make it clear that a county could keep its sizable board under a system that substituted "district organization" for township organization. Brunsvold acknowledged his bill came in response to the Rock Island petition drive of the previous summer but insisted he opposed "the idea of dissolving townships." Supporting his bill was Rock Island County Board Chairman Paul Mulcahey, who testified in March before the Counties and Townships Committee of the Illinois House. Mulcahey saw H. B. 68 as only the beginning of a solution to the transition problem that Rock Island or other township counties could be facing: "'We want to let legislators know that the state law here has to be cleaned up.'"[54]

Speaking against the bill and for keeping uncertainty in the mix was George Miller. As long as Miller and township officials could point to the 1874 Township Law as the law of the Land of Lincoln, instead of the 1970 Constitution, when it came to county boards and county commissioners, then they and their cause would continue to reap the fruits of voter fears about having just three officials running a county after abolition of townships. In a conversation we had two years after his testimony on the Brunsvold bill, Miller told me that TOI would remain opposed to transition legislation because TOI wanted the support of voters who "feared going to three commissioners."[55] The Counties and Townships Committee referred the Brunsvold bill for further study, but it never reached second reading. Other transition/transfer propositions had suffered comparable fates in earlier assemblies, and new bills to clean the murky waters would draw fire from TOI as the nineties unfolded.[56]

The Giglio Gig

While Miller and Stasiewicz were putting out the Brunsvold brush fire, a bigger blaze was crackling in Springfield and giving cause for alarm over Cook County's townships. The igniter was Rep. Frank Giglio, a Cook County Democrat from Calumet City and precinct committeeman in Thornton Township. Giglio's intention was to allow Cook's townships the option of deciding unilaterally whether or not to abolish, a course that the 1970 Constitution and subsequent legislation already made possible. Rep. Giglio succeeded in getting the attention of Fortress Astoria and a fair amount of press coverage by pushing his proposition.

On 2 May 1991 Giglio offered an amendment to H. B. 760, a local government measure. The Giglio amendment said that voters living in counties with over one million residents could petition for a vote on abolition of single townships. If abolition were to occur in one township, the county would see to all unfinished business of that dissolved township. The amendment lost, 30-70-7.[57]

A couple weeks later, on 14 May 1991, Rep. Giglio proposed the same amendment to H. B. 798 and came closer, 52-54-0, to getting his way. "Too close for comfort," is the way Miller explained that vote to *ICTO* readers. Rep. Bob DeJaegher of Hampton Township and other township supporters among Giglio's legislative colleagues were hard at work behind the scenes that month. "'Cook County is where you start,'" DeJaegher told reporters, including Diane Ross. "'Once you dismantle townships in one county, you eventually dismantle them in all.'"[58]

Early in June, the Taxpayers' Federation of Illinois came out in support of the Giglio initiative. TOI Legislative Director Stasiewicz expressed surprise at the Federation's position but vowed to continue "working with Republicans and downstate Democrats to defeat" any further Giglio efforts. By that time, Giglio had found an ally, Rep. Jeff Schoenberg, a Democrat from Skokie—on the northern side of Cook County.[59]

The *Chicago Sun-Times* weighed in next with an editorial calling for an end to Cook's "anachronistic" townships and cited Giglio as only the last in a lengthy series of reformers who had tackled the issue. "For 30 years this newspaper has urged abolition of township government in Cook County, finding its functions a duplicative overlay of expensive bureaucracy." *Sun-Times* editors looked back to "the midpoint of the century, when the call for abolition began to be taken seriously." Local politicians, however, had "fought any change and held on to these supposed suzerainties." That mentality, said the Chicago daily, "is the mentality that continues to saddle Illinois with upwards of 6,600 units of local government, more than any state in the nation."[60]

On 18 June, a Giglio-Schoenberg amendment, identical to those that Giglio had attempted with H. B. 760 and H. B. 798, came to the house floor as Amendment No. 1 to S. B. 887, a bill dealing with general assistance that had already passed through the senate. S. B. 887 and its one amendment moved on to third reading in the house. Editorial staff at the *Tribune* had their say as the Giglio-Schoenberg amendment awaited a final decision in the house. Editors liked the fire Giglio had ignited: "Townships are an anachronistic layer of government." In Cook County, where most land had fallen under urban control, the editors felt townships were "doubly redundant." The *Tribune* criticized Giglio's bill for not going "far enough," because one could make the case "for letting voters throughout the state opt to drop township government." Any relief that abolition could provide would "make only a dent in property taxes. But [abolition would] streamline government, eliminate some extraneous operations and recognize that this is, nine decades into it, the 20th Century."[61]

By a vote of 19-77-0, on 25 June, the house failed to agree with the editors of the state's two largest newspapers, defeating both the amendment and the bill—but not before the chamber erupted with fists and shoes pounding on lawmaker desks. Giglio had sensed a bad outcome that day and requested a postponement of the roll call on his amendment. Protests erupted from his opponents (township supporters) who wanted his amendment dead then and there. Rep. Bill Black of Danville, the Republican floor leader, screamed, "'That's an absolute violation of the rules'" and charged down the aisle, trying to reach the Democrat controlling the floor debate. Giglio relented, allowed colleagues to vote on S. B. 887, and accepted defeat.[62]

Waiting for 1992

Although I had then yet to meet Popp or Richardson, I can imagine they must have felt about the free publicity township abolition got in 1991. Some of it was their own doing, but not the abolition-friendly editorials coming out of

Chicago (see above) and Bloomington (see below). Township forces had to expend a lot of effort and political capital to fend off Giglio and his Cook County fire. TOI lawyers had to continue fighting the FOI lawsuit, because Judge Hanson denied their request to dismiss it at the circuit court level. The local Rock Island press corps was still showing an interest in whatever TSG could stir up about township foes. An election year was on the horizon, and the petitions and petitioners were ready to go in Rock Island County.

Township officials could read calendars, too; and, as 1991 gave way to 1992, the alert sounded, in places like Geneseo. Larry Foley, the Moline Township supervisor that TSG had charged with nepotism in 1990, dined with the Henry County Township Officials in December of 1991. He reminded them of what nearly happened in his county, the county next door to Henry, and warned them to get ready for trouble: "'We were doing a lot of things right but no one paid any attention to that.'" Bryan Smith was there, too, for the Christmas buffet in Geneseo, and stressed the need for educating the media and the public before the threat arrived. He discussed the Giglio gig and expressed concern that Cook County might still be vulnerable to such an attack. Robert Hepner, the supervisor of Kewanee Township and a TOI director, said if Cook and Rock Island were targets, so too could Henry. Local Republican lawmaker Todd Sieben of Geneseo said he would be supportive of township government when he got back to the Illinois House of Representatives.[63]

Rock Island Again

George Miller began the second biennium of the twentieth century's last decade with his own warning to the TOI faithful. Rep. Giglio was then distributing petitions requesting a referendum in Thornton Township on discontinuance. Miller predicted Thornton voters, if they got the chance, would defeat such a proposal but cautioned that none in the township camp could "afford the luxury of ignoring his effort." Comparing tax figures and the services that townships, special districts, and Cook County provided for their respective tax bites, Miller claimed Thornton and the other Cook townships accounted for "a savings of millions of dollars in property taxes every year." Consider Giglio or anyone else passing such petitions as "'snake oil peddlers' or worse," said the veteran.[64]

Peddlers of snake oil or purveyors of tax relief (take your pick) were preparing to stir the pot again in Rock Island County. Although Randy Popp, in Michael Richardson's words, was "burned out" from the 1990 campaign and all the personal flack he had suffered as a result of his role therein, Richardson, happy with his success thus far in the lawsuit against TOI, was then "still strong," as he described himself to me several years later. In the hope of reducing the harassment factor, Richardson and Popp purposefully played down their second abolition effort by curtailing seminars and other events that seemed to have led to unnecessary stress among TSG members. TSG, to be sure, had a higher signature hurdle as a result of Public Act 86-1486, so instead of traipsing door-to-door, TSG dispatched fewer (but more determined) circulators to places where the crowds gathered. Taking advantage of better citizen understanding (thanks to all the publicity) of their cause and of the township roles, they were able to move more quickly from one signer to the next. The pedestrian pathway along the Mississippi River turned out to be a great place for TSG to find voters lounging on the benches. They took to jogging or walking alongside potential signers and making their case on the move. Bingo parlors were excellent places to find senior citizens and others with concerns about their tax bills. Petition passers learned church worship schedules and made it their business to slip through the faithful lingering outside after services. TSG again found parade routes a wonderful place to work crowds. One reformer took a side of the street and another the opposite side, and they walked as a team a few blocks ahead of the first parade element asking Rock Islanders to join their movement by signing the papers they carried. At fireworks displays that summer, they moved among the expectant throngs with clipboard in one hand and flashlight in the other. Bored people, they found, made great listeners.[65]

Richardson used his kickoff press conference to vent about the 1991 law making it necessary for TSG to gather nine thousand rather than seven thousand signatures, saying, "'After two local party bosses went to bat for the townships, the legislative trio from Rock Island County went to work for the townships in Springfield'" He admitted the civics classes that kids, like his, were taking in school revealed nothing about "'the rough-and-tumble politics that township officials engage in to prevent a question from reaching the voters.'"[66]

In August, kids and adults got a lesson about how politics can work well; and TSG earned a couple of congratulatory editorials from the local press. As June had turned to July that year, and July to August, TSG's petition drive lacked the spark of 1990. Richardson told the *Argus* they had nine thousand signatures and were planning to get another three thousand for safety sake. Those extra few thousand were looming awfully large when an opportunity to drop out gracefully presented itself. State law permitted no more than three referenda at a single general election. Already set for the Rock Island ballot was a referendum asking voters if they wanted to adopt the executive form of county government. Economic issues were likely to influence the county board to call for a vote, referendum number two, on a quarter-cent sales tax increase to fund renovations aimed at luring thousands of new Department of Defense jobs to Arsenal Island, a part of Rock Island County. Competing with township abolition for the third slot was an idea for a tax increase to assist the county's Health Department. Richardson saw a chance to avoid the fate TSG had suffered in 1990, obtain some favorable press, and do a good turn for the citizens of the county. He announced that

TSG was stepping aside and would be using the remaining days prior to the petition deadline to help the health department pass around its petitions and challenged the local township officials to shift their energies to helping gather enough signatures to put the health tax issue over the top.[67]

Both the *Argus* and the *Dispatch* praised TSG for civic mindedness. "Cheers to the Township Study [G]roup for not only agreeing to delay asking voters whether to abolish townships, but for pledging to help put a Rock Island County Health Department referendum on the ballot instead," said the *Dispatch* editorial. The *Argus* editor reminded readers how some had "roundly criticized" TSG for its abolition stance and noted that this time "congratulations" were more appropriate: "We congratulate the Township Study Group on its noble fight…and congratulate members even more for deciding to postpone the fight."[68]

Not everyone, however, thought TSG deserved a round of applause. Richardson's public challenge to township officials and his subsequent letter to Ralph Erikson, president of the Rock Island County Township Officials Association, asking for assistance in the health-related petition drive, drew a skeptical reaction. Erikson claimed he and fellow officials had already been busy helping "'before Michael Richardson jumped on the band wagon.'" He accused Richardson of manufacturing a "'face-saving thing'" when TSG dropped its own sputtering drive. "'He knew he wasn't going to make the number,'" Erikson said. Filing first and then gracefully withdrawing would have lent much-needed credibility in this gambit. Richardson responded by noting how filing and (thereby) revealing names of his circulators would have exposed TSG members to the same sort of negative pressures from employers and others that they had suffered in 1990.[69]

Giglio Again

Fortress Astoria had its collective eye on Frank Giglio, even as TSG was prepping for, conducting, and abandoning its 1992 signature effort. While his passing of petitions to dissolve Thornton Township and his serving temporarily as the township supervisor were cause for some concern, Giglio really attracted TOI attention because of his position in the general assembly. He was, after all, Rep. Giglio, and as such could cause countywide or statewide mischief.

Senator Harry "Babe" Woodyard, Republican from Chrisman (Edgar County), in a March 1992 guest column for *ICTO,* reminded township officials of what Giglio had almost wrought in 1991. Woodyard recalled how fellow Republicans in the house had "strongly rebuked" Giglio's ideas about township obsolesce. Though the Giglio amendments never came up for a vote in the senate, Woodyard implied that he would have certainly voted no, because once the general assembly and governor approved such a threatening measure for Cook County it would have only been a short time before Giglio or others proposed similar single-township dissolution legislation affecting other counties.[70]

On the second of June, Giglio introduced H. B. 4215, which detailed how a dissolved township's assets and debts were to transfer to a subsequent governmental unit. One of township government's greatest allies, uncertainty over the steps that would follow a vote to dissolve a township, was at risk. Ed Stasiewicz issued a "Legislative Alert!" in his July–August 1992 *ICTO* column. According to Stasiewicz, Giglio wanted a dissolved township's parent county or the affected municipalities to take what was left of township assets after the settlement of township debts. General assistance would devolve upon the county. "[Giglio's legislation could] apply to all counties under township organization," Stasiewicz emphasized. His next column, in September, provided more intelligence about what TOI thought Giglio had in mind. The petition drive in Thornton Township was still alive; and, if Thornton fell, Giglio, said Stasiewicz, wanted a mechanism to allow seamless transfer of assets, debts, and responsibilities whenever any township disappeared.[71]

As it turned out, H. B. 4215 never got beyond a first reading. Thornton Township remained firmly in the township column, because Giglio and allies gathered only about half the number of signatures needed to place abolition on the ballot there in November 1992. Giglio went on to fight other political battles, like losing the November 1994 race to replace the incumbent congressman in Illinois's Eleventh District. Fortress Astoria remained vigilant should another Giglio suggest the general assembly provide a process for transition and transfer of power from the township to whatever came afterward.[72]

Press Problems

Compared to biennia to come, 1992–1993 was not a particularly difficult period for TOI, in terms of petition-passing abolitionists. TSG folded its petition tent early in 1992, and Giglio's Thornton campaign only succeeded in moderately raising the stress level for the Astoria leadership. Several of the state's significant dailies, however, took township government to task in those years.

With an editorial entitled "Abolish Township Government," the opinion leaders at the *Joliet Herald-News* left little doubt about their stance in October 1992. Readers were then telling the editors how big and costly government had grown. Townships seemed a good place to start cleaning up the tax problem: "It is a layer of government that could be abolished and few people would even notice." True, citizen effort would be necessary; but it would be a start toward tax reform. In a companion piece, the editors explained how someone in Will or any township county could launch a movement aimed at abolition, and in other articles reporter Nick Reiher provided background on township duties and explained what TSG had been trying to do since 1990 in Rock Island County.[73]

The *Rockford Register Star* used "Should Township Government Die?" as the title of a pro/con piece by Patty Scully that ran in February of 1993. Featured spokesman for the antitownship position was youthful Ernie Fuhr, a teacher who had recently moved to Rockford from the Quad Cities, where he had met Richardson in 1990 and subsequently worked with TSG in its initial abolition attempt. Fuhr had carried the TSG message up the Rock River to Winnebago County, and the *Register Star* had had no trouble finding the same sorts of township compliance problems as "Richardson said occurred in Rock Island County." Scully contacted TOI for comment on the Fuhr and Richardson criticism, noting in passing that "TOI is widely believed to be one of the most powerful lobbying groups in the state." Ed Stasiewicz told her that petition drives, such as those of TSG, were always unsuccessful because there was "'not earth-shattering public sentiment to abolish townships.'" From Al Zimmer, lawyer for the Illinois State Board of Elections, Scully heard another reason for such failures: the laws "covering petition drives to abolish townships are unusually difficult." Sidebars provided *Register Star* readers with facts and figures about township officials' salaries and duties as well as "What it would take to abolish [a] township." She gave Fuhr's phone number, the address of TSG headquarters, Richardson's home number, and notice that "Fuhr [was] conducting a petition drive to abolish township government in Winnebago County." Part-time columnist Tom Seymour rose to the defense of townships a month later, citing some of the good things (low tax levy, town meeting, knowing your elected officials) about them and admitting, "The big question [was] what happens to township functions and property if townships are eliminated? Nobody knows." Seymour's next column suggested, in jest (I guess), eliminating the United States House of Representatives, saying, "It certainly has more merit than notions of getting rid of our township form of government."[74]

Among the civil township's most determined critics of this era was the *Bloomington Pantagraph,* in no small part because of the ongoing feud between the City of Bloomington Township and the traditional townships that surrounded it. As the City of Bloomington (the municipality) grew in area, so did the coterminous township, at the expense of its township neighbors. Later in the 1990s and into the next decade, the courts would have to get involved in this dispute; but in the early 1990s, the *Pantagraph* editors liked taking shots at the whole township setup. And after each salvo from West Washington Street in Bloomington, letters, from near and far, often in praise of townships would appear.

An August 1992 *Pantagraph* editorial, "It's Time to Review Township System," took a look at the proposed salary of Maxine Schultz, City of Bloomington Township supervisor, and asked if a review as to "whether township government duplicate[d] work that could be accomplished more efficiently by other branches of government" might be in order. A letter from Grace Lee, executive director of the Township Officials of Cook County, appeared a week later. Ed Stasiewicz's rebuttal included a nice analogy about shutting down all newspapers except major metropolitan dailies if being big were the only criterion. A town clerk from Livingston County provided local flavor to the choreographed response sequence.[75]

In May 1993, City of Bloomington Township Trustee Mike Matejka won editorial validation from the *Pantagraph* for having the nerve to suggest his township dissolve itself, and that other townships should follow the leader. Editors cited Michael Richardson extensively and gave TSG a plug. Three days later the first of several adverse reactions to the editorial appeared, from a township road commissioner north of the Twin Cities. Then came one from an "Avoca Township taxpayer" (in Livingston County) and another from Mary Caisley, the supervisor of Normal Township and president of the McLean County Township Elected Officials. Randy Popp joined the parade of letters with one in which he defended Matejka for his courage in view of the "amount of pressure the nearly 12,000 township officials of Illinois will put on him."[76]

Another editorial, "Vote on Townships Would Be Beneficial," appeared while the fallout from the one in early May was still settling. In the second, the *Pantagraph* urged TSG to "continue its push for a statewide referendum on elimination of township government." True, citizens could eliminate townships one at a time or by the county, but "the process could take a long time." Why had a statewide law mandating such a referendum not yet materialized? "Lawmakers are reluctant to make enemies among fellow elected officials—no matter how obscure their positions may be." The editors urged voters to let legislators know voters wanted a chance to express an opinion. Meanwhile TSG "should keep up its campaign." A former managing editor of the paper was one of those writing in defense of townships.[77]

Editors of the *Belleville News-Democrat* likened townships in that area, across the river from St. Louis, to dinosaurs that had outlived their usefulness. They cited a recent efficiency study of Granite City Township, in Madison County, that had revealed "a significant percentage of employees" were relatives of township officials. It was the hope of the *News-Democrat* that voters might somehow manage to get abolition on the ballot, because if they "left it up to the state politicians, township government [would] be around forever."[78]

Winning Friends

Fortress Astoria and its Springfield outpost did not rely solely on rapid responses to damning newspaper copy during those hard times. George Miller was finally able to convince the frugal directors to whom he reported that

TOI needed to raise and spend money if township government was to celebrate its sesquicentennial at decade's end with a full complement of townships intact. On the one hand, TOI would woo the voter through better public relations. On the other, TOI went after the politicians with endorsements and campaign contributions.

Public Relations

TOI hired a Springfield public accounting firm, Kerber, Eck & Braeckel, in the early 1990s, to prepare a strategic plan that could help assure the survival of township government in Illinois and guide TOI in the years to come. After looking at written surveys that selected township officials returned, several meetings at places around the state with officials, interviews with TOI staff and directors, and review and analysis of related materials, the consultants offered a long list of goals, objectives, and actions. Under the first goal, "To assure that TOI provides the leadership necessary to preserve the township form of government," was the objective to "improve the public image of township government." Among the recommendations for action on this item in 1992 and 1993 were better connections with the media and positive publicity about the township story. The emphasis was to be less on reacting to threats and criticism and more on telling the good news about government at the grassroots.[79]

Even before the strategic plan reached his desk, Miller was on record favoring the notion that good public relations were critical. At the 15 February 1992 TOI Continuing Education Seminar in Peoria, for instance, he used some of his time in front of several hundred township officials to share stories of the bad publicity that wayward or ignorant members were bringing down on everyone else in the organization. Hoarding township funds for a rainy day was one flaw he hoped to eradicate, and he gave an example of an Illinois township whose leaders used the interest from a $300,000 certificate of deposit plus a small tax levy to run the township. Spend the money, I heard him say. He noted how another township spent $25,000 to take all of its elected officers to a conference in Washington, D.C. Spend wisely, I also heard him say, because once newspapers and critics get wind of such things, the good news, the kind of news their consultants were about to advocate, had a hard time finding an audience.

Miller's board of directors appointed Joe Browning as TOI's first public relations director. In a 31 March 1992 appearance before the Township Officials of Cook County, Browning, according to Grace Lee, "acknowledged that the major daily newspapers [were] almost immovable in their opposition to township government in Cook County"; but local papers, township newsletters, other township literature, and civic club programs were still good avenues for reaching the voters. He mentioned that day a new public relations manual that TOI would be shipping for $13.00 per copy, a manual "which will give you step by step some things you might implement."[80]

Browning's name is nowhere in that *Media Educational Packet* so I cannot ascertain his role in its preparation. In fact, the advice therein looks generic, like a guide that a city manger or county board chair might just as easily employ. Only on page 20, in a reference to the TOI magazine by name, and on the last page (34), where appears a "Sample Press Release for Snow Removal for a Township," is there any indication of a township connection to this publication. Elsewhere in it are words to the wise about influencing people in a positive way, a sample form on which to keep an updated list of the names and numbers of local reporters, tips on writing advertising copy and taking good promotional photographs, and advice on dealing with an unfavorable story in the media. By the end of the nineties, TOI had published a truly township-focused, media-relations guidebook, entitled *Effective Communication Strategies for Townships,* and made it available to officials, as a single item and through a series of nine excerpts in the magazine, between February and November 1999.[81]

Rep. Giglio should take at least partial credit for one of TOI's biggest public relations efforts. The Giglio threat in 1991 helped Miller, according to Bernard Schoenburg, "obtain from his board $15,000 to spend on public education about townships." Rick Pearson, at the *Chicago Tribune,* said the cost of the big campaign might be closer to $20,000. Half-minute television commercials began airing in August of 1992 in markets like Peoria, Harrisburg, and Springfield; but TOI was not able to penetrate the critical Chicago market because of the high cost of commercial airtime there. The downstate ads featured Miller and township officials telling viewers how dues the townships paid to TOI helped TOI keep townships alive and alert to issues about which people cared. Miller told his directors at their annual retreat in January 1993 that these ads were arousing interest and expanding the consciousness of the public as to the worth of townships and their umbrella organization. "Public education is a must," he emphasized, not only via paid commercials but also through a video that TOI had produced and was in 1993 trying to make available to schools through the Illinois State Board of Education. Service clubs, like the Kiwanis, he said, had borrowed and used the tape for their meeting programs.[82]

TOIPAC

Objective 4 under goal 1 of the 1992 Kerber, Eck & Braeckel strategic plan was "To increase TOI's influence with the legislative and executive branch, and mobilize TOI's 'clout,'" but nowhere in the entire document did the writer(s) mention TOIPAC, which had begun in the fall of 1990. TOIPAC, TOI's political action committee, had come into existence in the days leading up to the narrow escape by the township movement from the potentially devastating effects of an abolition vote in Rock Island County. Bylaws of TOI prohibited official endorsement or condemnation of candidates for public office or the expenditure of funds to accomplish either of those ends. So TOIPAC emerged as a way around TOI's limitations. TOIPAC could solicit donations (not tax dollars, obviously) and pass on funds to aid in the election of township-friendly legislators. Public endorsements would occur, too, even if the funds were not sufficient for a campaign contribution. Several TOI leaders gave $100 or more to get the political action committee going, and the first solicitation of the entire TOI membership appeared in the February 1991 magazine. Ed Stasiewicz reminded readers in his legislative report that the November 1992 general election would be the first to reflect redistricting after the 1990 Census and the first at which TOIPAC would be flexing its muscle. Only one type of candidate would win TOIPAC support of any kind: "one that supports township government." TOIPAC bylaws appeared in December 1991, explaining in detail how endorsements could occur for incumbents and aspirants to the general assembly.[83]

The magazine in the past had carried the names of legislators that the editors felt deserved re-election, but TOIPAC offered much more to township government's friends. TOIPAC would have funds to disburse, and TOIPAC would send a letter to every township official in an endorsee's district informing him or her of the candidate's special status. By January 1992, TOIPAC could list 30 who had given at least $100 and another 32 who had given less. Among the 32 were five state representatives and a state senator. Occasional solicitations appeared in the magazine during 1992, often with the note "This is not a permitted township expense," as a reminder that TOIPAC could accept no tax-derived contributions.[84]

Stasiewicz devoted his entire column in the October 1992 magazine to explaining how TOIPAC was handling its duties. "Our objective is to help retire those [legislators] who don't support township government and help ensure that supporters of township government continue to represent us in the General Assembly." He urged township officials to share the endorsement list with their neighbors and anybody else apt to vote. Because they were new at the game (and perhaps due to limited funds), TOIPAC directors had proceeded "cautiously" this time. For the 177 seats (118 in the house and 59 in the senate), they made 46 endorsements. The paid "political statement" did not indicate which of the 46 received funds that fall from TOIPAC.[85]

Miller said some did find "an additional financial contribution of dollars" in their mail. Incumbents, he acknowledged, had the advantage most of the time; but even former township officials did not get an automatic endorsement. He felt the time spent soliciting responses from candidates, assessing those responses, reviewing voting records, and making endorsements was time they had spent well. TOIPAC is still operating today.[86]

Word War

Michael Richardson and George Miller seldom crossed paths, but their spirited war of words, beginning in 1993 and continuing into 1994, was among the highlights of the decade for me, as an observer. The word war had been brewing before 1993, as both wrote letters for publication and composed items for their almost mutually exclusive audiences, but got started in earnest when an editor provided a prestigious forum. Escalation ensued.

Guest Combatants

On two occasions, in the spring of 1993, the two champions agreed to provide their divergent perspectives as invited spokesmen. Their first face-off appeared in the March 1993 *Illinois Issues*, the second in a pair of daily newspapers, the *Ottawa Times* and the *Rock Island Argus.*

Illinois Issues Editor Caroline Gherandini gave Miller and Richardson less than a thousand words each in which to make their cases. Both began with a brief bit of history, with Miller citing three hundred years of a township tradition in America and Richardson noting how the current township layer did not begin in Illinois until the middle of the previous century. Miller admitted that township supporters had periodically to contend with antitownship energy, usually involving "alleged savings in tax dollars." Richardson dusted off the figure of "6,600" to reference the surfeit of taxing bodies that Illinois then sustained. Both cited court cases to help make their points. Both included code words in their essays, like "anachronistic," "unnecessary," "cost ineffective," "duplicative," "cronyism," and "nepotism," for Richardson, and "progressive," "immediate," "accurate," "lifelines," "state-mandated," and "democracy," for Miller. The TSG co-chair closed by telling *Illinois Issues* readers "The Illinois Constitution provides for voters to abolish townships." He believed the time to act had arrived. TOI's executive director said, "Democracy was never intended to be cheap."[87]

Miller and Richardson, instead of submitting copy, submitted to interviews by Pat Harrison for their second prose confrontation of the spring. Harrison, writing for the *Ottawa Times,* in anticipation of the annual township meetings and the quadrennial election of township officers, posed targeted questions to each man. The full-page spread went out on SNG News Service, and the *Rock Island Argus* picked it up for publication on the second Sunday of April 1993.

Harrison apparently quizzed Richardson first, after having read a TSG flyer entitled "Township Study Group." He asked Richardson how much property tax relief the closing of townships would provide, how township roads would fare without their mother townships, and for examples of patronage at the township level. Richardson got to finish his interview by telling Harrison what he thought would be Miller's "strongest argument." Miller will say, said Richardson, "It is government close to home"; but Richardson felt the signature town meeting had become a joke because of paltry participation, which resulted in part from the fact that citizens did not even know the name of the township in which they resided.[88]

Harrison started Miller with a question about replication of social services between those that townships provided and those that nongovernmental groups using tax dollars had begun to offer. Miller said somebody ought to be watching over such groups more closely (and maybe paying less attention to townships). Harrison quizzed Miller about cronyism and nepotism. Miller said he had seen both but felt critics should focus on running "the rascals out of office" not "condemn[ing] a form of government because of a few bad officials." Harrison had apparently read the March *Illinois Issues* because he asked Miller to explain his statement about democracy not being cheap. Miller said he thought townships were operating about "as efficiently as possible" and cited evidence to support his position. When Harrison asked him how it was that 17 Illinois counties could get by without townships, Miller noted some were among the state's poorest—but did not go so far as to ascribe causality.[89]

Neither guest, facing off for *Illinois Issues* or for Pat Harrison, had a chance to react directly to the words of the other. That would soon change, as Richardson found a forum in a newsletter that he and Popp had started in late 1992 and as Miller employed both his regular *ICTO* column and another he created to help him deal with the TSG menace and Michael Richardson.

TSG news

After TSG dropped its 1992 abolition effort, Richardson and Popp needed a way to fan the Rock Island fires and to expand the influence they were beginning to enjoy beyond their own county. Some townships were adopting newsletters as a simple but effective public relations tool. Why not TSG? Popp, who was working then on his communications degree, and writing for local news outlets, became the editor; Richardson became the chief correspondent. At first the newsletter arrived free for the asking. Later they solicited donations. Later still they sold subscriptions. Their mailing list reached several hundred eventually, and *TSG news* found an audience across the state.

TSG published 15 issues of its newsletter, beginning in the winter of 1992 and ending in the summer of 1994. All issues featured black text on white paper, in a pleasing, (usually) three-column format with headings in larger type. For the most part, *TSG news* remained devoid of illustrations. Copy filled both sides of the 8½ by 11 pages, which a single staple secured. A majority of the 15 issues (11) contained 3 sheets of paper (6 sides), while 3 had 2 sheets, and 1 had 4. Popp almost always featured a couple of stories on the first page, several of which continued to a subsequent page. Some front-page material focused on Rock Island County, especially in the first months of publication. Non-local stories began in the first issue but became more prevalent on the first page as TSG connected with abolition groups around the state. Inside *TSG news* Popp consistently ran a segment he called "Township Tid-Bits," wherein appeared untitled items that tended to make township officials look bad and abolition look good. Longer stories also appeared inside, along with an occasional "correction." Corrections were really a device that Richardson and Popp employed to bring back to the reader an item that had previously appeared on the newsletter's pages. One such correction revolved around an apology from TSG for not labeling Al Jourdan as the "former" Illinois Republican Party Chairman, in a story about the use of Jourdan's McHenry County auditor office for a meeting with "Dorr Township Road Commissioner/Car-parts Thief Leroy Eddy" (April–May 1994). The back page of most issues carried a reprint of an editorial or other commentary critical of townships. Some of these items were current when TSG used them. Others dated back to earlier days of the 1990s, when the media had been especially critical of townships. Exceptions to the media reprints included a letter to Popp from Rep. Giglio explaining how Giglio's views on townships had changed since he tried to dismantle them in Cook County (January 1994); Moliner Marla R. Redmond's letter to TSG about her problem getting service from her Moline Township supervisor (March 1994); and a letter from John Gianulis, chairman of the Rock Island County Democrat Central Committee, asking townshippers to be alert in 1994 for another anti-township run by TSG (April–May 1994).

Blow-by-Blow

What follows is an abbreviated account of the verbal sparring that took place between Miller and Richardson as *TSG news* grew in significance. Richardson, the champion of the underdog, needed all the public exposure he could get for his cause. Miller, the entrenched and celebrated champion of a seemingly invulnerable institution, avoided *ICTO* acknowledgement of Richardson as long as he could. Eventually, however, Richardson went too far in his quest for press; and Miller felt he had to answer.

When Michael Richardson sent his half of the March 1993 *Illinois Issues* pairing to several Illinois newspapers, and some editors published just his side of the story, George Miller called Richardson's use of the copyrighted column improper, at the least, and suggested that township officials ought, in return, take seriously Popp's invitation for free copies of *TSG news*. Miller provided address and phone number for getting on the mailing list, saying, "Anyone who is interested in what TSG is up to should request the free subscription. TOI's Executive Director has, why don't you."[90]

Miller devoted his entire magazine column in June 1993, running nearly two pages, to the TSG threat. He included a reply he sent to the *Sterling Gazette* after the editor there had contacted him for his side of the story, which Richardson was telling in his *Illinois Issues* mailing. Miller repeated his invitation for any and all to read *TSG news* "at no cost" and promised his readers that "the next few issues of this publication" would counter the "false claims and allegations" Richardson, Popp, and others were then making. He challenged Richardson and Popp to send him their "commentaries for unedited publication" in *ICTO.*[91]

Miller filled his column in the next issue with TSG material, noting how he had not previously realized "just how much garbage Mr. Richardson and his cohorts would try to sell as fact." He said he had failed at first to recognize the TSG leaders' ability to "twist" the facts and turn them "into out and out lies." *ICTO* readers, however, should not dismiss Richardson as a "nut," because he, in Miller's words, was "intelligent, articulate and astute." If Richardson could get by calling township officials "townshippers," then Miller was going to start calling TSG followers "groupies." Again he urged his faithful to call or write Rock Island and request a free newsletter."[92]

By September 1993, TSG had become more than just a local nuisance for township officials in Rock Island County. George Miller not only fired at TSG in his regular director's column but that month began a second column that he would devote over the next year just to TSG and its abolition-minded ilk. TOI had been trying unsuccessfully to gather information on Richardson and Popp. When "several members of the Illinois General Assembly and at least one member of Congress" joined TOI in an attempt "to learn more about the agency [where Richardson worked] and Mr. Richardson's income," Richardson's defenders told the "petitioners" to "back off" or face litigation. Randy Popp, "current editor of the 'Tall Story Group' newsletter," apparently had two addresses—one from which to vote and one for use in registering for school. On page 12 of the September *ICTO,* Miller introduced the "TSG Bulletin Board," whose purpose was to pair, item-by-item, "TSG's Story" with "The Truth." A not-so-flattering image of Richardson would appear at the top of each edition of the new feature.[93]

Richardson finally reacted to Miller's incessant plea for TOI members to apply for a free subscription. Yes, TSG had made the offer through "several daily newspapers and *Illinois Issues* magazine," and they were mailing "courtesy subscriptions to legitimate members of the news media," but Miller had gone too far by extending the offer "to all township politicians in the state." TSG had received a flood "of demands for free subscriptions" and had to take action or fail to honor their own offers. "Unless you have donated money to the Township Study Group, taken the pledge to circulate abolition petitions in 1994, responded to one of our offers, or are a member of the news media, this will be your last free issue of *TSG news."* For $10 per year, anyone else was welcome to remain on the mailing list.[94]

In October 1993, Miller again focused his regular column on TSG but said that in the future he would only do so "under unusual or extreme circumstances." Yes, Richardson and Popp were irritants to Miller "personally and to the association and its members generally," but they were really like "a flea to an elephant or a fly to a horse."[95]

This spirited give and take spilled over into 1994. Richardson continued to point out the worst in township officials and the worst aspects of this form of grassroots government. He also did his best to irritate Miller and thus keep the exchange going. Miller made the mistake in January 1994 of sharing with his *ICTO* readers the fact that three decades earlier he and "a group of amateur campaigners" in Leyden Township had called themselves "The Pooches." By February, Richardson was referring to the TOI leader as "Pooch Miller." By April, Miller was chiding Richardson for using a term that "just a handful of close friends" had once used among themselves. "Mike Richardson is no friend of mine," said Miller. As 1994 unfolded, Richardson became more and more involved in the third Rock Island campaign and in a campaign that was quickly developing to the east, in one of the Collar Counties. *TSG news* numbers became less frequent and finally ceased with the June–July number. Miller kept the "TSG Bulletin Board" going through the summer of 1994 and instituted "TSG Commentary By George," the September 1994 version of which suggested "law enforcement officers" might be able to provide *ICTO* readers with "background on Richardson." The last "TSG Commentary" appeared in November 1994. By then, Miller was devoting space to township government's next nemesis.[96]

Gearing up for Another Summer

Michael Richardson had business in Peoria on 26 January 1994 and offered to take the time to meet there with me as a means of helping me catch up with all that had been happening on the abolition front lines as the new year, a general election year, began. I accepted his invitation, and we sat down in a quiet corner of the public library. My notes from that meeting indicate my impression of how proud he was of *TSG news,* whose circulation had exceeded 750, including 75 going to reporters and newspaper editors, one for each member of the general assembly (except Rep. Bob DeJaegher), 50 to township officials who had paid for or somehow gotten a gratis subscription, 200 to supporters of abolition in Rock Island County, and the other 300 or so to folks around the state with a reformation mentality. The two TSG co-chairs were "having a lot of fun" putting together their periodic township-condemnation piece, and were not, as Popp once feared, in danger of "running out of news." Richardson had high hopes for an impending meeting of the Ross Perot/United We Stand Party, at which Richardson was going to make his case for having that third party adopt township abolition as another tax-saving plank of its statewide platform.

Then we got to his forecast for the abolition efforts he anticipated for 1994. Of course, Rock Island was at the top of the list. Madison County was there, too, mainly because of Collinsville maverick Mel Jones, who had a long history of adversarial action against governmental tax-and-spend practices. McLean County was there, largely due to the longstanding stance of the *Bloomington Pantagraph* against the township layer; but Richardson could name no person in the county who might lead a petition drive. Richardson had been in contact with Gallatin County and was hopeful that Carl Edwards might try again what had failed at the polls in 1988. Ernie Fuhr had gotten some antitownship press in Winnebago County and seemed to have retained the reformational zeal he had exhibited in Rock Island in 1990. A township trustee from St. Clair County, after losing faith in townships, had requested a sample petition from TSG and was already at work in Belleville and other communities. A farmer in Jo Daviess County, had, along with a neighbor, taken West Galena Township to court for destruction of private fencing along a township road, won the case, bought ad space in the local newspaper to criticize township government, and seemed poised to lead a taxpayer revolt in northwestern Illinois. And then there was Bob Anderson in McHenry County, the county that in 1849 had led the Yankee invasion against county commissioners by casting 1,923 votes for township organization and none against. "Anderson has several people on his side," said Richardson, "including some environmental groups and organized political groups. He may take it on."[97]

McHenry County

Born Robert Gordon Anderson on 28 August 1937 in Waupaca, Wisconsin, Bob Anderson, a long-time resident of Wonder Lake, in the northeastern part of McHenry County, did take on townships in 1994. In fact, long after Richardson, Miller, and Popp laid down their arms, Anderson still maintained the reformer's zeal with which he first put grassroots government on notice in his county, notice that revealed his determination to make a difference for the taxpayer. He had lived near Wonder Lake (the reservoir) since 1947, when his father, then a Kroger employee, came to the county as a corporate transferee with his family from central Wisconsin. At Wonder Lake, in a lake frontage home, Bob Anderson and wife Judy raised five children on the proceeds from his barber shop, his work as a bus driver for nearby Harrison School, and her work as a purchasing clerk. Though not as comfortable in front of audiences or as facile with the written word as Richardson, Bob Anderson comes across as a sincere man, a community fixture, on a long-term mission.[98]

The Barber from Wonder Lake

Anderson did not suddenly emerge in the spring of 1994 from his modest shop and decide to declare a vendetta against McHenry's 17 townships. By 1994 he had gained a local reputation as a taxpayer advocate. While his children attended public schools, he put in nine years as a member of the McHenry High School Board of Education (District 156). Later he (as he put it to me) worked hard "to knock off school bond issues" and raised money to buy advertising to counteract the "excessive" salaries of school administrators. Then, minding his barber business one day in 1993, he received a telephone call from a reporter who was looking for Anderson's reaction to a legal notice the reporter had in hand. The notice dealt with a warming shelter for snowmobile riders that the board of Anderson's township was proposing to build, at a park the township operated, for around $130,000, partly from grant funds. That shelter was to have, in addition to a fireplace for warming rider extremities, a grill, toilets, and a parking lot for wheeled vehicles. At first it was a warming shelter; then it became a pavilion. Anderson got busy and formed Concerned Taxpayers for Integrity in Government (CTIG), along with two of his like-minded neighbors. The trio recruited others, and CTIG was able to present McHenry Township Supervisor Albert Adams with over three hundred signatures on an advisory petition, suggesting he scrap the shelter. Adams reacted by telling Anderson, in Anderson's words, "Even if you bring in ten thousand signatures, we are going to build our shelter."[99]

They built it so Anderson began regularly attending township board meetings. At one stretch, he was present for 13 straight monthly meetings of the McHenry Township Board of Trustees and became conversant in what Adams, Anderson said, often referred to as the "Howdy form of government." "Just stop in," said Adams; and Anderson did, while gradually the notion of doing something more about McHenry and the other 16 townships began to take shape.

How did one go about doing something about township government? A call to the state election board proved less than helpful. In fact, he "sensed a bit of ill will" over the telephone when he called Springfield. Then he got lucky. Rep. Cal Skinner, a McHenry County legislator with a tax-watchdog label who had read newspaper accounts about Anderson and the warming shelter, passed to him an item that Skinner had gotten in the mail—*TSG news.* Reading TSG's newsletter, Anderson learned something about the abolition process; but, more significantly, he discovered that he and his group were not alone in their distaste for and distrust of township government. Most important, however, from that hand-me-down newsletter, he found in late 1993 the name and number of Michael Richardson.[100]

Richardson once told me that getting to know Bob Anderson and working with him on a common cause "was the high point of a long decade." To Richardson, Anderson was "an inspiration, honest, and determined." For Anderson, the linkup in anticipation of the 1994 petitioning period provided not only the chance to learn from someone who had been there, but also gave him someone with whom to compare notes and with whom to compete at the serious and draining task of collecting thousands of signatures. He "needed Mike" in 1994, he told me in 2000, to keep him pushing on to the next house or the next public parking lot. From Richardson, he also received in 1994 something else—a sample petition into which Anderson would plug McHenry County wherever Richardson had plugged Rock Island County.

Unfortunately for Anderson, Richardson had used wording from TSG's 1990 and 1992 drives, failing to realize that state law had changed with respect to petition wording as of the result of Public Act 88-62, effective 1 January 1994. "I owe Bob an apology" for leading him astray on his petition and for costing him personally for legal fees to overcome the error, said Richardson to me.[101]

George Miller may have learned that TOI had a problem in McHenry County when he read this tidbit in his January 1994 *TSG news:* "Bob Anderson, head of Concerned Taxpayers of McHenry County, has informed TSG news of his intention to spearhead a county-wide township abolition drive this summer." More likely than not, however, the TOI network would have already advised Fortress Astoria that an ill wind was blowing across Wonder Lake from a tiny barber shop near its eastern flank. Miller's first acknowledgement in the magazine of the McHenry matter came in June of 1994 when he attributed the petition drive there to TSG, noting that "a spokesman for the petitioners," undoubtedly Anderson, had appeared before a committee of the McHenry County Board to inform board members of their role should the petition drive succeed. The next *ICTO* named names, and correctly credited CTIG, not TSG, with the McHenry drive. CTIG's leader, "local businessman Bob Anderson" had set out to collect twelve thousand signatures, said Miller. He poked fun at Anderson for reporting that California and Texas did not have townships, noting "Illinois doesn't have earthquakes and hasn't had any since townships were organized here in 1848." Of course, Illinois had had many minor earthquakes since the first 24 Illinois counties organized township in 1850. Miller did allocate part of his July–August *ICTO* column to an admonishment for complacent township officials "from Jackson to Jo Daviess" about the need to help defeat the Rock Island and McHenry efforts and "perform our [township] duties according to Statute."[102]

Spring Training

The kickoff press conference for CTIG's 1994 abolition petition drive took place on Monday, 23 May. In the weeks preceding, however, Anderson left little doubt that he was going to attempt to collect more signatures than had any other township abolitionist in Illinois—ever. He clashed publicly with Supervisor Adams at McHenry Township's February board meeting, a meeting at which Adams allegedly called Anderson "'a jerk'" and asked the man, who had attended 11 straight such meetings "not to attend any more meetings." Adams, at the March board meeting, denied calling Anderson nasty names; but Town Clerk Thomas Birmingham recollected the utterance of "some unpleasant words," words that the clerk omitted from the official minutes. Anderson pressed Adams at that March session for answers to Anderson's questions about a proposed addition to the township's community center, which Anderson said he had not heard discussed in previous meetings. Anderson kept after Adams, about this and that, until Assessor Theodore Pitzen, probably verbalizing the feelings of Adams and colleagues, said "'You're very obnoxious, Mr. Anderson.'"[103]

When the McHenry Township electors convened for their annual get together on the second Tuesday of April that year, Anderson was present to express his opposition to an official resolution in support of township government. More than a hundred citizens voted to support the resolution, which was similar to resolutions that electors in at least two of McHenry's other townships also approved that night. Pressed to explain to electors his distaste for this most proximate of local governments, Anderson said, "'I'm not here to answer questions.'" After the meeting, he told the press that the audience "was not an accurate representation of the residents."[104] At least that is what he must have sincerely hoped.

Later that month, Anderson drew concerted fire from local township officials and county board members as he trotted his radical ideas from one county board committee to another. At the Finance Committee's meeting on 13 April, members admonished him "for campaigning without statistics." Board member Mike Brown asked Anderson to bring answers instead of ignorance to their attention. Member Don Doherty agreed with Brown, telling Anderson that he, Brown, and the rest of the county board would need to know much more before they could make decisions in the best interest of "'the public.'" Anderson told Brown and Doherty that getting rid of 168 township officials ought to be "'a savings'" to start. Committeeman Richard Meyers, also clerk of Nunda Township, admitted he had "'known Bob for a long time'" and that previously they had "'agreed on many things'"; but this time he and Anderson were on opposing sides. "'I'm prepared to go out and campaign against you from one end of this county to the other,'" Meyers told his long-time acquaintance. Comments such as those of Meyers, a man serving a county and a township, caused some observers to question "whether those county board members who have links to township government may be in a conflict of interest." At the Transportation Committee hearing the following Monday, Anderson tangled with several township officials as they confronted him over what Algonquin Township Supervisor Tom Schober called a "'half-baked idea.'" For Schober and the other officials at his side, Anderson, in Schober's words, lacked "'any credibility.'"[105] As for the barber from Wonder Lake, it would take more than mere words to deter him from his civic cause.

By the middle of April, the *Chicago Tribune* had picked up the scent of a story brewing out in the northwestern suburbs. *Tribune* reporter Martha Russis produced a balanced article featuring petitioners Anderson and Richardson and defenders Ersel Schuster (supervisor, Seneca Township, McHenry County) and Donna Mayberry (McHenry County supervisor of assessments). The paper also ran a companion item focused on good deeds of Cook County's townships. Anderson predicted a victory for CTIG, even before he collected the first signature. A photograph of him beside a sign declaring that Albert A. Adams was the supervisor of McHenry Township allowed *Tribune* readers to put a face with a name they would be seeing again and again in that paper and in others around Chicagoland.[106] CTIG had made it into the big arena; and township officials across the metro area must have then known that if McHenry County were to shuck its township layer and survive so might Cook, Lake, DuPage, Kane, Will, Kankakee, and counties on down the map.

Except perhaps for Figaro, the barber of Seville, no barber received as much press as did Bob Anderson during the 1990s. Quite a bit of it, like the *Tribune* items of 17 April, showed balance; some was clearly, editorially, favorable to his point of view. On the other hand, where he really could have used the sort of antitownship editorials that had been appearing around the state was in the principal daily paper of McHenry County. "They are still killing me," Anderson confided much later, after the *Crystal Lake Northwest Herald* had taken yet another shot at an idea of his about limiting government. Opposition from the *Northwest Herald* began during the spring of 1994, with the editorial "Simple Only on Surface." His "argument has a neat logic," the editorial conceded; but "someone will have to pick up—and pay for—the services." Citing a recently released DuPage County study, which revealed the likelihood of a net negative impact of abolition on that more populous county, the *Northwest Herald* urged a similar study to see if McHenry County would suffer the same or an even worse fate. "Taxpayers deserve sound information," said the editors; but the careful reader (or agitator/barber) could read the direction the *Northwest Herald* was that spring leaning.[107]

Around the first of May, County Board Chair Dianne Klemm released a memorandum from the office of her county's state's attorney, dated 13 April and entitled "Dissolution of Township Government." Using primarily Illinois statutes then in force, the author, Assistant State's Attorney Mary Kay Walter, struck another blow to the Anderson case for abolition by noting how the existing county board (with 24 members) would give way, following a second election, to a three-member county commission. Furthermore, according to Walter's interpretation, that subsequent election to choose the trio of commissioners could only occur "at the next general election in an even-numbered year." So, to make the case even worse for CTIG, Walter was predicting two years of a lame-duck regime for the county board should Anderson's notion prevail. She went on to list 34 township functions, duties, and authorities from chapter 60, the township portion of the Illinois Compiled Statutes. In most cases, but not all, Walter found a comparable (county-specific) section of other statute chapters. Walter took care to point out that this response to the Klemm request did not constitute a position from the county's legal office as to the desirability of township dissolution.[108]

Mladen Rudman covered Walter's report for the *Wonder Lake Star,* and led with the Walter prediction that the 24-member board would revert to a tiny commission of only three members. He followed with commentary from Randy Popp, "the editor of a Rock Island-based newsletter dedicated to writing the final chapter on townships statewide." Popp, and Richardson, had, of course, heard that opinion before; and Popp noted, as would Richardson and as did Anderson, that supervisor-populated county boards had disappeared around 1970, that there no longer was an automatic "'relationship between the county and townships'," and that McHenry should be able to elect a new board of up to 24 members. The problem was that the general assembly had neglected to establish, by law, a clear path from one board to its townshipless successor. Nor had the courts gotten the chance to rule on such legislation or succession—because no abolition movement had been successful since one man, one vote (chapter 6). So critics of Anderson and other reformers could continue (effectively) to threaten the public with the consequences of having just three commissioners "run a rapidly growing McHenry County" or other county in the state, knowing that, if abolition occurred, the outcome was at best uncertain and perhaps likely to set a precedent by producing a board similar in size to that which governed McHenry County.[109] Uncertainty tended to favor the status quo.

His First Season in the Sun

Spring training came to an end on a sunny Monday, the twenty-third of May, when the barber from Wonder Lake, on the lawn outside his shop, with Michael Richardson in the audience for support, announced his intention to collect twelve thousand signatures. "Excited," is the word Anderson later used to characterize his mood that day; but had he known then "the strength of TOI," he guessed would not have been so thrilled. He told the audience of more than fifty reporters, moms, dads, and babes in arms that CTIG expected "'in the spirit of democracy'" that all elected officials and registered voters would be eager to sign the petition sheets that he and his 180 or so volunteers were about to circulate. "'Without a doubt, it will be on the ballot," he predicted. Cynical Seneca Supervisor Schuster, presumably not in the audience for Anderson's first press conference, told Rich Carter of the *Northwest Herald* she too advocated less government but that the Anderson plan "'would create bigger government and more costs to the taxpayer.'" Schuster said she had "'to question [Anderson's] morals'" because he was kicking off a signature drive "'without knowing the consequences'" of his actions.[110]

So began the signature solicitation season in McHenry County. Bob Anderson found he had a flair for gathering signatures, as he and Richardson kept one another posted on the progress of their respective drives and motivated each other by telephone. Mid-summer brought the impact report from County Administrator Bill Barron, which the legislative and public affairs committee had requested of him back in April. Later in the summer, McHenry County's Democrats offered to help Anderson with his campaign.

Barron had solicited predictions from his departmental managers, "elected and appointed," as to whether township dissolution would have any impact on their operations. All 28 responded, and 12 predicted "no discernable impact." For the other 16, expected impacts varied from the thousand dollars per year the sheriff would lose in an agreement with Nunda Township for management of abandoned vehicles to the more than one million dollars the supervisor of assessments forecast takeover of township assessing duties would cost that department. Clear even then to Barron and the county's leadership, however, was the fact that there was not much clarity on the matter of township roads and bridges. Would dissolution of the 17 townships mean concurrent dissolution of their road districts? State's Attorney Gary Pack had asked the Illinois Attorney General for a legal opinion on the matter; but had heard nothing, and uncertainty tended to favor the way things were. Pending an opinion from Springfield, the director of the county highway department put the cost in the first year after county takeover of roads and bridges at more than six million dollars and at about six million annually in the years to follow. If the county could sell the township road equipment, those receipts would help somewhat to offset the startup costs. All in all, mainly due to the additional personnel the county would need to hire, the ongoing extra cost for the county would likely be on the order of nine million plus a one-time transition cost of six million. Barron did note that the expected reduction in the size of the county board might lead, in his words, to "an initial reduction in annual County Board expenses…of $245,350.00."[111]

Democrats played the role of the loyal opposition to McHenry County's ruling Republicans in the 1990s. For the party out of power in 1994 to ally with abolition leader Anderson late in his petition drive was neither surprising nor foolhardy. Little else seemed to be working for Democrats in the conservative northwestern suburbs, so why should they not try to help Anderson put his project over the top? By early August, CTIG was reporting it had 10,650 signatures and expecting challenges to the validity of some. Democrat Chairman Frank McClatchey and his organization permitted CTIG to collect signatures at the Democrats' booth at the county fair the weekend of 6–7 August because, as he put it, "'this thing is going to be challenged like nothing has been challenged before. It's not for sure on the ballot yet.'" Anderson obviously welcomed the support no matter what the real reason for the offer might have been. "'In the spirit of democracy,'" he told Rich Carter, "'I would like to see the Republican Party [his party] do the same.'" Ersel Schuster, however, "blasted the Democrats for getting involved." Seneca Township's outspoken supervisor said she thought Anderson had duped the minority party, saying, "'They are playing the public for fools when they do these things.'" *TSG news* ran with the late-inning intervention of the Democrats as its top story in the June–July number and condemned Rock Island

County Democrats for failing to assist TSG that summer, particularly "Democrat overlord John Gianulis."[112]

Township supporters were not idly standing by, waiting to offer the next reporter a suitable quotation, as Anderson and crew canvassed the more densely populated portions of the county in search of signatures. Instead, the township side "counterattacked," as Philip Franchine of the *Chicago Sun-Times* wrote, "with mailings, bumper stickers and a political action committee," which would later become Citizens for Township Government, or CTG. Over the course of the summer, Algonquin Township's Tom Schober took to shadowing Anderson, trying "to talk people out of signing petitions." Up near the Wisconsin line, at Hebron, an anti-Anderson activist posted a four-foot-by-eight-foot message for passersby that read: "Abolish Bob Anderson." Fortunately, no one took him or her seriously. When I asked the Andersons about personal threats, as of early September that year, neither had any to report.[113]

Challenge Phase

No one accused Democrat Frank McClatchey of being psychic in early August when he predicted a stiff challenge to the CTIG petitions. Michael Richardson had already warned Wonder Lake's barber to expect a formal challenge and to overshoot on signature acquisition. CTIG attorney, Michael Walkup, told a gathering of abolition activists in May of 1995 that the attorneys for the township officials in the 1994 McHenry challenge had been meeting for some time before the objection hearings began and had developed or purchased a computer program to simplify the signature checking, expecting to find about one of every three signature lines flawed. Walkup said he had heard that CTG leaders hoped the challenge process would occupy several weeks, between filing day and election day, and force CTIG to expend most of its war chest on attorney fees, leaving little time or money for campaigning.

In effect as of 1 January 1994, Public Act 88-62 (H. B. 2120) revised the Township Law of 1874 to produce the Township Code, and filled nearly two hundred pages of the laws emanating from the Eighty-eighth General Assembly. When lawmakers got to section 25-5, dealing with discontinuance of townships, they used the wording of Public Act 85-1446, from 1989, regarding the base figure for calculating the minimum number of petition signatures needed for access to the ballot (those that voted in the preceding presidential election) rather than the (almost certainly) higher 1991 hurdle of Public Act 86-1486 (those registered to vote at the time of petition filing). The discrepancy, however, never became an issue in the 1994 campaigns. Richardson said later that their discovery of the discrepancy in the middle of the summer gave TSG "a psychological boost." The general assembly fixed the problem in 1995 with Public Act 89-235 (H. B. 1733), effective 4 August

1995, by reverting to a signature minimum based on the number of voters registered as of filing day.[114]

Monday, 22 August 1994, provided Anderson one of the high points of the decade. Late that afternoon he and fellow CTIG members, wearing big yellow buttons that proclaimed their allegiance to the cause of tax relief, placed on the counter of County Clerk Katherine Schultz a stack of petitions measuring 7.5 inches high and "containing 13,347 signatures," according to Kurt Begalka, covering the top local story that day for the *Northwest Herald.* "It took them over an hour just to stamp the whole stack," Anderson recollected for me six years later. He had delivered on part of his promise; and, like Richardson and Popp four years earlier, he hoped there would be no challenge to affect delivery of the other promise—a ballot slot. "'Other than a voter or taxpayer off the street,'" he said, no one should challenge those petitions. If township officials challenged, "'it would be an outrage.'" Even before Shultz and crew had finished the stamping marathon, CTG had requested copies of the entire CTIG petition pile.[115]

Mark Guerra of rural Cary in Algonquin Township, in the southeasternmost corner of the county, served as president and a principal spokesperson for CTG. Raised in Chicago, Guerra had moved his family from Joliet to their Cary address in 1985. Compared to what Guerra had experienced in Chicago and Joliet, the services he found in Algonquin Township and its coterminous road district were exceptional and by 1994 costing the family only eighty dollars annually on their property tax bill. He said he could not believe anyone would want to curtail such superior public services, so he responded to Anderson's petition campaign of 1994 with township-friendly letters to the *Northwest Herald.* Sometime later his highway commissioner, Bob Miller, and supervisor, Tom Schober, approached him about leading a group they were forming as an antidote to CTIG. Although he had not previously known Schober and only barely knew Miller, Guerra, a machinist who earned a living by repairing large pumps for Seaquist Dispensing of Cary, accepted their offer. Guerra had not been politically active beforehand, but he had an interest in public service and a history of volunteerism locally in such organizations as the high school's booster club. Subsequent to 1994, however, he became a successful petition circulator in his own right and ran for the county board, the fire protection district board, and GOP committeeman.[116]

Previously strangers, Anderson and Guerra met several times in the weeks that followed, as they traveled the circuit promoting their points of view. From Anderson, Guerra said he learned opponents could agree to disagree but still remain on good terms. Guerra told me he held Anderson in considerable regard, noting that what Anderson attempted in 1994 (and subsequent years) forced rapidly changing McHenry County to reflect on how its governmental institutions were performing and to make adjustments to accommodate new settlers. Townships are not perfect, Guerra said; but he felt their services were in 1994 "a steal at the price." Bob Miller and Tom

Schober had chosen well when they found Mark Guerra, who comes across not as a politician but, like Anderson, as a citizen stepping forward.[117]

Although I did not ask him, I doubt that Guerra was among the some thirteen thousand who signed the CTIG petitions in the summer of 1994. A solicitor did approach him, he told Kurt Begalka; but he felt he "'was wrongly approached.'" Given what he considered deceptive practices by CTIG and his belief in township services at bargain prices, he had "'no qualms challenging [Anderson's] petitions.'" George Miller said he had confidence in Guerra and the rest of CTG, feeling certain his "'petition watchdogs'" would make a formal objection. On the other hand, he expressed less confidence in the voters for whom emotion "'and not the facts'" could be the deciding factor.[118]

On the same day as the Begalka piece citing Mark Guerra and George Miller appeared, the *Northwest Herald* published the first of over a hundred letters the paper would carry on the topic of McHenry townships between the stamping of Anderson's petitions and the opening of the polls on 8 November. One of the first two letters took Anderson to task for asking township officials to refrain from opposition to his petitions. The other letter took the newspaper to task for failing to give the township side front-page coverage on 3 August, as it had the Anderson side. I counted the first of this pair as protownship and the latter as ambivalent on the principal issue. A quick tally of all the letters on the topic between the first pair and election day found 27 on the side of abolition and 100 expressing a positive position on township government, or a ratio of slightly more than three to one favoring retention.[119]

While CTG members pored over Anderson's petitions, the county board took the proactive stance of authorizing the state's attorney to gather a list of constitutional law specialists from whom the board could choose one or more to "make recommendations on the best way to dismantle township government." Chairwoman Dianne Klemm continued to caution members that Anderson's referendum had yet to pass muster, but members like Mike Brown thought it none too soon to begin preparations. Brown used the occasion to go public with his perspective on abolition: "'In 77 days, my grandchildren will get on a bus traveling township roads that will need salt.'" As he saw it, voters could on 8 November bring an immediate end to township services—or at least that was what he seemed to want uninformed voters to believe.[120]

Lori Martinek echoed Brown's gloomy forecast in her 26 August column from rural Marengo, out to the west, beyond the booming portion of McHenry County. She had been talking to Riley Township Highway Commissioner Bob Moehling, who told her, "'The next morning [after 8 November] you'll wake up and it'll be gone.…It's going to be bad news. There's just not going to be any service out there.'" He went on to lament the fact that the county had "'never plowed gravel roads'" nor "'country subdivisions'" (like the one in which Guerra lived). "'What's going to happen out in the country where the wind blows and you have to keep trucks out longer in order to keep the roads clear?'" School buses, Moehling told Martinek, were not "'going to go anywhere [that] winter'" because county highway plows were unlikely to lead buses through snowdrifts, as did his plow crews. Martinek recounted dire details from the Marengo mayor and the Marengo township supervisor, and closed with telephone numbers of CTG so readers could volunteer to "check petitions and pass out literature."[121]

Objectors had until close of business on Monday, 29 August, to object to the Anderson petitions. CTG muscled up with multiple attorneys, to find fault regarding compliance by CTIG with state law, and with hundreds of "concerned citizens," hoping to find errors in the signatures or any other aspect of the petitions. County Clerk Schultz spent part of Friday, the twenty-sixth, researching procedures that her electoral board would have to follow at any challenge hearing. Anderson, observing all the turmoil his thirteen thousand signatures had precipitated, told the *Northwest Herald* that CTG was nothing more than "'a front for township officials fighting to keep their jobs.'"[122]

On challenge day, the *Arlington Heights Daily Herald* ran an in-depth story about what was to happen that Monday in Woodstock, the seat of McHenry County, under the headline "All Eyes on McHenry County Anti-township Referendum." George Miller and the TOI leadership around the state were watching and speculating that, if Anderson came even close to closing down McHenry's townships, others would make "'a run for its someplace else.'" Township officials from Cook, DuPage, Lake, and other heavily urbanized counties were watching and wondering if they had done enough in their own domains to make townships relevant to taxpayers. Lawyers were watching because the murkiness of state laws, about what would happen if a county lost townships, promised lots of work for lawyers. And, of course, the people of the affected county itself were waiting to see if they would have a chance to make modern history.[123]

As most expected and the barber from Wonder Lake had feared, CTG came down hard on the product of CTIG's long summer in the sun. Speaking for the 16 other objectors (one per township?), Mark Guerra expressed his outrage at how Anderson and comrades had "'illegally'" obtained many of the signatures. He (they) felt most McHenry voters did not want the issue on their November ballot. Lawyers "volunteering time to work with township supporters" numbered three that Monday, according to Jayna Legg. CTG attorney Mike Poper said only 8,727 of the 13,347 signatures "at the most" were valid—and Anderson needed more than ten thousand. CTG attorney Patrick Coen, noting how Anderson and CTIG had not collected any signatures from some townships, told Legg "at least 10 percent of the voters in each township should have signed the petitions." CTG attorney David McArdle told the *Chicago Tribune* Anderson had used outdated wording in his petition, wording that had ceased to be appropriate after a revision of the Township Law went into effect the first of January that year. There were other flaws, said the lawyers; but the message was clear that Bob Anderson needed his own attorney and probably should have had one on the job from the outset.[124]

By Thursday, 1 September, the day Clerk Katherine Schultz and the rest of her electoral board (State's Attorney Gary Pack and Circuit Clerk Bill Kays) convened their hearing of petition challenges in the handsome new courthouse on the north side of Woodstock, CTG had five attorneys ready for the fray. Barber Bob Anderson, as of the day before, had one, Michael Walkup (also a candidate on the Democrat ticket for the Thirty-second Senate District). The five worked for free, though their firms performed remunerated legal services for various townships. Anderson's attorney, although sympathetic, was not free.

Township attorneys filed a total of one dozen objections. The first eight focused on petition wording and how CTIG had circulated the petitions. Walkup and McArdle spent considerable time arguing the matter of wording. McArdle, of course, wanted strict adherence of the wording for petitions as suggested in Public Act 88-62 and, therefore, disqualification of all Anderson's petitions. Walkup argued that the new wording was merely advisory to the county clerks of Illinois, as they went about the preparation of ballots containing such a dissolution referendum and that no voter, according to Walkup, could possibly misconstrue the intent expressed in the McHenry petitions.

After lunch, township attorney Coen presented constitutional objections, even though the electoral board had the authority to base its ruling only on one or more of the eight objections that McArdle had delivered in the morning. Coen's main point, which would have near-future implications as TOI pushed for legislation that could derail abolition activists, was that townships in urbanized areas should not have the power to dominate rural townships, perhaps thereby denying rural residents a governmental form vital to their interests. Coen said the 1970 Constitution required signatures from at least 10 percent of registered voters in every township and not merely 10 percent of all the county's registered voters. When reporter Mark R. Madler questioned him at hearing's end, Anderson admitted most signatures on the petitions were from urbanized areas of McHenry County: "'Where are you going to go, farm to farm?...You have to go where the people are.'" By 1998, anyone seeking dissolution of all townships in an Illinois county would have to have the signatures of at least 10 percent of the registered voters in every township on the petitions calling for a referendum (chapter 8).[125]

Schultz, Pack, and Kays based their decision to rule CTIG off the ballot on the fact that that Anderson had employed outdated wording on the petitions. Guerra said, as far as he knew, the board's ruling ended the matter. The county board, by a 17-2 vote, confirmed the ruling. Anderson said, if he could raise enough money, he would appeal.[126]

CTIG found the money. With the Andersons putting up at least half of the $15,000 and promising to secure the balance, they hired Michael Lavelle, a noted election law attorney. On Monday, 12 September, Anderson announced plans for the appeal of the electoral board's decision to deny him ballot access. Lavelle, who had served as chairman of the Illinois State Board of Elections and the Chicago Board of Election Commissioners, told the press that day he had faith in legal precedence favoring flexibility in the wording of the petitions for referenda. In Lavelle's words, for the *Chicago Tribune,* both the electoral board and the county board were "'putting form over substance.'" Anderson expressed, to the *Tribune,* his amazement at the "'lengths to which those sloughing at the public trough would go to protect their wallets and privileges.'" McArdle maintained that Anderson should have followed the new law in preparing his petitions. Guerra said CTG welcomed the fight but would rather not waste taxpayer dollars when CTG was in the right. He told Dennis Mathes of the *Woodstock Independent* that "'Bob Anderson cannot win in court.'"[127]

By week's end multiple lawsuit proceedings were underway. CTIG sued for a reversal of the electoral board decision and a mandate for Clerk Schultz "'to publish the current form of the question upon the ballot after certification of the referendum by the County Board.'" CTG sued to have the electoral board revisit its decision to ignore two of CTG's original eight procedural objections, one of which concerned where in the courthouse Anderson should have filed his petitions. Judge James C. Franz of the Nineteenth Circuit agreed to consolidate the two suits and to accelerate his calendar, as critical deadlines approached vis-à-vis the 8 November election. CTIG tried widening its list of targets the next week, by filing an amendment to its original suit, asking the judge to include the 24 county board members as defendants. A couple of days later, county board members individually avoided being a target for CTIG legal action by agreeing "not to pursue further court action" if a judge were to order the referendum to proceed.[128]

As the judicial system ground on toward resolution of the conflict, many people had opinions to share. Henry Hemmingsen said folks out in Coral Township needed grassroots government more than they needed Bob Anderson. Ersel Schuster, Seneca Township supervisor and chair of the McHenry County Association of Township Officials, asked whether voters wanted "distant and bigger government" or "local and smaller government." "It is a well-known fact," said Schuster, "bigger government has never resulted in better or less expensive government." Robert A. Korslin shamed the election commission for discounting the wishes of thirteen thousand voters, those who had signed the petitions, and for refusing to use the microphones in the 1 September hearing "so interested people could hear what they were saying!"

R. W. "Bill" Baruth Jr. wondered why the "pro-township group" did not stand behind its "convictions" and let the people decide the system's fate. "Or," Baruth asked, "are you afraid township government is so bureaucratic and outdated that we, the people, will throw the fat-cats out?" Carole Ann Kaiser, Woodstock, spouse of Don, the supervisor of Greenwood Township, in a long letter to the *Daily Herald,* reminded readers how hard her husband worked in performing his township duties. Judy Anderson, Wonder Lake, wife of Bob, the activist, told how the "five township lawyers working at no charge," on the day of the objection challenges to the petitions, had shaken any faith she had left in the "government closest to the people." Raymond Wells, of Algonquin, had a few choice words for Ersel Schuster, recalling her stance, as a member of the county board, against "too much government" while she argued now for keeping an extra layer of government intact. Wells also suggested getting rid of the townships' "latest flunky, Mark Guerra." Carole Ann Kaiser, in a second letter, suggested that Bob Anderson suffered from such an obsession for "changing laws to suit his self-serving agenda" that he should "move to another country." Gerry Wish, president of Taxpayers of McHenry County, likened "the form of government closest to the people" to pickpockets, who also try to get close to the people. Anita Sherwood, supervisor of Nunda Township, wrote about the fairy-tale emperor and his imaginary suit of new clothes. Bob Anderson, with "his slanted, vindictive effort to destroy the most complete, economical and best government this country has," represented, of course, the lazy tailor. Earl R. Wilcox took Carole Ann Kaiser to task for her efforts "to try and pound Bob Anderson and his reputation into the ground" and asked her to start on Wilcox, who had collected over two thousand of the thirteen thousand signatures.[129]

Judge Franz heard arguments and rendered his decision the same day, 14 October. In his view, the electoral board had erred when it ruled against CTIG. While the court reversal obviously pleased Anderson greatly, he knew he had the daunting task of trying "'to reach as many people as possible'" with almost nothing in the war chest. Algonquin Township Supervisor Tom Schober told the *Northwest Herald* that CTG was ready to kick its "education campaign into high gear to convince voters that townships serve a purpose." That campaign would include "mass mailings, radio advertising and debates." An appeal, of course, was still a possibility; but for some reason (overconfidence?) the township side had not (unlike Michael Lavelle) prepared an appeal in advance. Furthermore, because CTG had agreed to let stand the some four thousand challenged signatures, in order to obtain a quick decision from the electoral board back in September, CTG could not in mid-October return to the electoral board to try and head off the referendum on the basis of improper signatures, bad addresses, unregistered signatories, etc. Every sign, the evening of 14 October, seem pointed toward an 8 November ballot box battle.[130]

Compressed Campaign

A flurry of activity followed. On Monday, 17 October, CTG's attorneys filed the appeal to the Franz ruling that they had failed (apparently) to prepare in advance. They filed, however, without the backing of the electoral board, which decided at the last moment not to join the suit, according to David Duschene of the *Daily Herald.* Tuesday of that week the county board certified the question of township organization for the November election, with one member abstaining, one missing, and Mike Brown voting not to certify.[131]

George Miller dined in Crystal Lake the night of Monday, 24 October, and spoke at the annual dinner of the county's township officials. Rallying the troops and disparaging Bob Anderson were, of course, on his public plate. Behind the scenes, no doubt he had some pertinent suggestions for winning a fight that could lead to a slippery slope for township government all over the state. Anderson called the Miller plea for money and manpower as well as his appearance in Crystal Lake evidence of an effort to "'come in and squash us.'"[132]

Neither side had much time left to spread its message, once the Second Appellate Court finally agreed with Judge Franz's order mandating a ballot slot for the referendum. Wonder Lake's locally famous barber continued to express faith in a system that, although awkward, was going to give people a choice. Vibrant yard signs, demanding that voters "Support Townships," sprouted like magic across the rolling glacial terrain of McHenry County. T-shirts admonishing observers to "Abolish Township Government," an institution that the artist depicted as a pig in a top hat astride a pork barrel, became commonplace as CTIG spent some of its modest cash reserves. CTG paid for radio spots to caution voters that "abolishing townships will raise your taxes," "cut vital services," and "ensure your voice will no longer be heard." Letters to editors continued to entertain and enlighten, like the one from Sherry Johnson of Crystal Lake beginning with "I love my township," or one from Caryn Corp of Cary, warning that abolition would bring Chicago-style politics to her suburban county. Township officials tried luring Anderson into spending his time helping a new task force study the ramifications of abolition. Instead, he vowed to devote the remaining few days to getting out the vote by knocking on doors. Algonquin Township Supervisor Schober warned that in a month there might be no general assistance for the clients he was then helping and no one to plow the snowbound roads.[133] Surely Schober must have known that, in a county so progressive and prosperous as McHenry, welfare clients and users of the township roads would not suffer long (or at all) if voters axed Algonquin and its 16 sister townships. Uncertainty, however, favored the status quo.

Editors at the *Northwest Herald* were having a hard time, toward the end of October, concealing their distaste for dissolution as their reporters covered the campaign's last days. Halloween finally brought the official editorial-staff position, which was to "Continue Townships: Yes." They gave Anderson credit for "dogged, if misguided, determination," but court battles had provided the public a scant 13 days to find answers "to all the questions that surround dissolution of townships in McHenry County." To dissolve them would, said the editors, have "voters flirting with disaster." Nobody really knew, they continued, "what would happen"; and they had been waiting patiently since the spring for Anderson to explain how the transition would work and who would "pick up the pieces." Anderson's strategy, ran the editorial, was to "shoot first, ask questions later." It went on to belabor the issue of roads and the apparent fact that maintenance would suffer or maybe an ambulance would not be able to "access one of those roads this winter." But that was "not his problem. That's the county board's problem. That's *your* problem, if you happen to live on that road. Figure it out yourself." Consolidation, said the *Northwest Herald,* might not be such a bad idea and deserved serious study; but Anderson's way was wrong. "In this case, people have good reason to fear the unknown."[134]

Early in November the *Chicago Tribune* and the *Woodstock Star* took their turns. *Tribune* editors opted for retention of McHenry townships. In Cook County, said the state's largest newspaper, one could probably make a case that townships had become outdated, but "in substantially rural counties like McHenry" townships were still vital. Someone would have to provide the services townships offered, and taxpayers would still have to foot the bill. Besides, residents seemed there to prefer the "intensely personal service from township officials," and, if townships went, so would the county board. Out in Woodstock, at the center of much of the abolition action, editors at the *Star* could not make up their editorial minds, a common tactic when a publisher wants to avoid offending either side. "We're not going to tell you how to vote. We're not going to recite the facts or say who's right. We can't because we don't know. And take our word for it: nobody else knows either." The *Star* made one good point, however, when the editor(s) noted the real question on the ballot should be "Are townships indispensable or simply too expensive?"[135] Maybe a divorce from townships and the big county board, left over from the board of supervisors (circa 1970), would be best. Yet divorces are fraught with uncertainties, and uncertainty tends to favor doing nothing.

Letter writers peppered the *Northwest Herald* with last-minute opinions, mainly supportive of the status quo. A total of 67 appeared between the first and sixth of November, with 21 going to press on the sixth alone. Only 6 of the 67 letter writers painted abolition in a favorable light.

Besides the late letters, CTG had hundreds of volunteers going door to door and bought several pages in the *Northwest Herald* for advertising the township point of view. Among the canvassers in rainy Algonquin on Saturday, 5 November, were Senator Steve Rauschenberger of the Thirty-third District and officials from Hanover, New Trier, and Maine townships in

Cook County. Identical full-page ads began on Thursday, 3 November, and ran again on the Sunday and Monday before the election. Under the title "Township Government: Worth Saving Because It Works," the CTG ad informed readers how county employees earned twice as much and had better (more costly) benefits than comparable township workers and that elected township officials worked "24 hours a day" instead of closing on weekdays at 4:30 p.m. and not opening at all on weekends. Township proponents used about half the ad to personalize the question for voters by listing the names of the hundreds of township roads that might suffer "a manpower shortage if township government" failed to prevail Tuesday. Barnard Mill Road, the site of Bob's Barber Shop, was, by the way, on the hit list.[136]

CTIG could not match CTG dollar for dollar or its cast of hundreds knocking on doors around the county, but Anderson did receive free publicity as a participant in a Crystal Lake debate on Thursday, the third of November.[137] Cyndi Wyss reported that "More than 250 people packed Lakeside Center" for the League of Women Voters forum that evening. With Anderson appeared CTIG Vice-president Michael Rakestraw, also of Wonder Lake, to oppose the likes of Bob Miller for CTG and George Miller for TOI. Diane Klemm and Bill Barron provided the county's perspective. Michael Richardson had planned to participate, but "cancelled at the last moment," according to *ICTO*. In his prepared remarks, Anderson continued to hammer at what he considered the excessive costs of the township model. Both Millers warned of the dire possibilities should abolition occur. "'Nothing would preclude [three county commissioners] from coming from one largely populated area,'" said Road Commissioner Miller. Executive Director Miller compared such a county commission to a "'politburo,'" which people "'all over the world'" were trying hard to avoid. As for Barron and Klemm, they said they feared the repercussions of township dissolution; but, for the public record, they told the audience McHenry County would "adapt if the chips [fell] to the non-township side." To her credit, Klemm, whose ex-husband was state Senator Dick Klemm, an outspoken friend of township government and former supervisor of Nunda Township, said it was "'silly to think everything would grind to a halt.... A court ruling and cooperation between governments would take care of the necessities of life.'" I doubt, however, that Diane Klemm's words swayed any of the hundreds who heard her that night, or read the Wyss report the next day, especially the unnamed debate attendee that asked Anderson and Rakestraw who would pick "'up the road kill until every thing [got] straightened out.'"[138]

Decision Day

Yes votes (to continue township government) outpaced no votes by roughly three to one: 42,438 to 13,540. CTIG's proposal failed to win in any of the 17 McHenry townships, with voters wishing to continue (what their predecessors had begun in 1849–1850) ranging from 72 percent of those voting in Algonquin Township to 92 percent in Riley. In general, Anderson fared better on the urbanizing eastern side of the county that he did toward the west—as observers expected. Mark Guerra praised the public for listening and weighing the facts. Diane Klemm praised Bob Anderson for the courage to bring a little-known entity to the public's eye and promised the county board would continue its study of possible intergovernmental cooperation. Editors at the *Northwest Herald* liked the vote of confidence that its McHenry readers had provided townships, a vote that ought to "put many questions and allegations about township government to rest." Anderson said he had only begun to fight in his own county and had plans to share his abolition message with activists in other Collar Counties.[139]

Rock Island Blues

Bob Anderson's petition-passing competitor, out where the Rock joins the Mississippi, had not lost his own knack for drawing a crowd of reporters or for pushing township buttons. Michael Richardson put the pressure on South Moline Township's board for its failure to confront the alleged criminal activity of a former highway commissioner. Calling the board members "'accomplices'" in the apparent misdeeds of Mel Parker, Richardson told Jim Dyer of the *Quad-City Times* that it was just another reason why TSG wanted South Moline and the other Rock Island townships obliterated. He and Randy Popp fanned the flames of a controversy that summer in Rock Island Township over the official residence of Supervisor Carolyn Lund, who took to spending weekends at her new home elsewhere and rooming during the week with a friend in Rock Island. Richardson ran as an independent candidate for the new post of county executive to provide another point of access to the local press corps, so that he could tout his real agenda—abolition.[140]

Right on schedule, the week of 23 May 1994, TSG announced its third petition drive, "Operation Tax Relief," and established a signature target of 9,414. "'It's not only doable,'" Richardson told reporters, including Cheri Bustos of the *Times,* "'It's as good as done.'" County Democrats had already sent letters to every Democrat precinct committeeman asking for information "about any petition activity" so the party leadership could prepare again to challenge the TSG petitions. Richardson interpreted the letters as the first step in putting party pressure on petition circulators, as had been the case in the 1990 campaign, and had already warned circulators to expect to receive hate mail, have their relatives contacted, and find their personal vehicles "egged." Gianulis told Bustos "he would never think of interfering with the [d]emocratic process." Rock Island County Republicans came out against the idea of such a referendum a couple days later.[141]

TSG needed to average about one hundred signatures each day, and by the end of the first 14 days of collecting had collected only a thousand. Hoping to accelerate the process, Richardson took his clipboard inside The Mark of the Quad-Cities (a civic center) and tried to intercept some of the eleven thousand attending a concert one summer evening. Mark authorities told him to leave. By the second Tuesday of August, two weeks before the Monday deadline, TSG was at just over five thousand signatures. They were, however, "'working like madmen,'" Richardson told a reporter, who calculated TSG then needed 340 signatures per day to reach the minimum figure. Ten days later, TSG had barely passed the six thousand mark, and ended with only 6,111. "We just came up short of signatures," he told me later, "but we did get thousands to sign and felt Bob's victory was a plus" for all of us.[142]

Not satisfied with the consolation of having McHenry County vote on abolition, Richardson went to court to argue that the signature requirement for someone seeking an abolition referendum was excessively high. To petition for a vote to establish townships in a commission county still took only 50 signatures, and the signatures needed to amend article IV of the Illinois Constitution was just 8 percent of those voting in the previous gubernatorial election. Judge Thomas C. Berglund of the Fourteenth Judicial Circuit denied Richardson's request and dismissed the suit on 14 October that fall. Once again, TSG had failed to give the voters of Rock Island County a chance to eliminate what to Richardson and Popp was such an obvious waste of taxpayer dollars and a source for governmental abuse of the people's trust.[143]

Unfinished Business

Even if the Rock Island judge did not buy Richardson's argument, TOI foresaw a problem because of the issue that Richardson raised—the huge difference between the petition signatures needed to bring about a referendum to establish townships versus those necessary to call for a vote on disestablishment of townships. As part of its next legislative agenda, therefore, TOI lobbied for dropping the requirement of just 50 signatures, which had been on the books since 1851, in favor of the same 10 percent figure that abolitionists had to attain. Besides, no county had voted on adopting townships since Morgan in 1957; and the likelihood of any of the 17 commission counties voting on that issue again seemed remote. Why not take credit for offering to even the playing field? The TOI signature suggestion became H. B. 212, under the sponsorship of Rep. Ann Hughes of McHenry County, and eventually part of H. B. 438, about which I will have much more to say in chapter 8.

Notes

1. Miller, "From the Executive Director," *ICTO,* January 1990, 4–5.
2. My first meeting with Michael Richardson occurred 10 December 1992, after we had spoken several times that autumn on the telephone. Over the rest of the decade, we talked by phone and met numerous times; and we have kept in contact since via letters and

email. Much of my background on Richardson has come from these personal interactions. He and his cause generated a tremendous amount of local press coverage, some of which contained biographical information. See, for instance, Lisa Neff, "Co-chairs Dig in for Long Haul," *Rock Island Argus,* 16 September 1990, A/2, and John Beydler, "Richardson Exudes Tenacity," *Moline Dispatch,* 3 November 1996, A/9. The court cases that Richardson argued *pro se* and lost at both the circuit and appeals court levels were these: *Michael D. Richardson v. James R. Grundel,* 85 Ill. App. 3d 46, 406 N.E.2d 575 (3d. Ill. App. Ct. 1980); *Michael D. Richardson v. Lola Dunbar et. al.,* 95 Ill. App. 3d 254, 419 N.E.2d 1205 (3d. Ill. App. Ct. 1981). I cite these newspapers hereafter, respectively, as *Argus* and *Dispatch.*

3. I first heard the butter story on 10 December 1992, when I met with Popp and Richardson in Rock Island. Further clarification of the incident occurred via email from Popp on 9 April 2002. See also, Lisa Neff, "Co-chairs Dig in for Long Haul," *Argus,* 16 September 1990, A/2.
4. Randy Popp, email correspondence with author, 9 April 2002.
5. Michael D. Richardson, interview by author, 10 October 2000.
6. Popp and Richardson provided me with a sample of the letters Popp sent to county clerks of commission counties on 25 February 1990 and with copies of the responses they received. These items remain in my possession.
7. Randy Popp, interview by author, 10 December 1992; Township Study Group of Rock Island County, "News Release," 3 April 1990; Randy Popp, email correspondence with author, 9 April 2002; Nicholas Thompson, "Citizens' Group Says Townships Violate State Laws," *Argus,* 3 April 1990, A/1; Nicholas Thompson, "Citizens' Group Questions Handling of Some R. I. Township's Relief Funds," *Dispatch,* 27 April 1990, A/6.
8. Township Study Group of Rock Island County, "News Release," 26 April 1990. These releases strike me as readable and humorous, something the working press would have found attractive. TSG was not afraid of naming names. Richardson told me that he and Popp loved to sharpen their pens and go after township officials, but tried to stop short of libel.
9. Nicholas Thompson, "Township Study Group, Supervisors Clash at Meeting," *Argus,* 1 May 1990, A/3.
10. Nicholas Thompson, "Tempers Flare During Seminar," *Argus,* 24 May 1990, A/1–2.
11. Township Study Group of Rock Island County, "News Release," 25 May 1990.
12. Nicholas Thompson, "Group Seeks Vote to Abolish Townships," *Dispatch,* 26 May 1990, A/1.
13. "Move to Abolish Townships," *Argus,* 29 October 1973, 4.
14. "Why Not County Commissioners," *Dispatch,* 11 November 1920, 20; "Time to Consolidate," *Argus,* 8 August 1951, 4; "Townships Out of Date," *Argus,* 29 September 1952, 4; "County Road Districts?" *Argus,* 4 March 1957, 9 March 1957, 4; "County Commission—Are People Ready?" *Dispatch,* 12 November 1963, 6; "Township Rule Fades," *Argus,* 4 April 1969, 4.
15. John E. Davis, telephone interview by author, 21 March 1993; "Move to Abolish Townships," *Argus,* 29 October 1973, 4; "Hostility to Township Move," *Argus,* 24 August 1974, 4.
16. Davis telephone interview.
17. "Hostility to Township Move," *Argus,* 24 August 1974, 4; "Township Mergers Fail," *Argus,* 8 November 1974, 4.

18. Howard W. Allen and Vincent A. Lacey, eds., *Illinois Elections, 1818–1990: Candidates and County Returns for President, Governor, Senate, and House of Representatives* (Carbondale and Edwardsville: Southern Illinois University Press, 1992), 497; "Will Circulate Petitions to Abolish Township Government," *Argus,* 26 October 1973, 2.
19. Allen and Lacey, *Illinois Elections,* 526.
20. Bryan Smith, "From Bryan's Desk," *ICTO,* December 1988, 13.
21. David Regner, "Township and Road Districts Legislative Reporting," *ICTO,* November 1988, 12. Regner was from Cook County, where he had served as a township auditor in Elk Grove Township before winning a seat in the house of representatives in 1966. He moved from the house to the senate and then became the TOI legislative director in 1982, a post he held until 1989. In 1993, he became the supervisor of Palatine Township (Cook County). "Regner Rejoins the Ranks," *ICTO,* May 1993, 19.
22. Illinois, General Assembly, *Laws of the State of Illinois, Eighty-fifth General Assembly, 1988* (Springfield, 1989), 4,058. For comparison of the 1989 legislation affecting townships with the relevant section of the Election Code then in effect, see Illinois, General Assembly, *Laws of the State of Illinois, Eighty-second General Assembly, 1982* (Springfield, 1983), 67. David Regner, "Township and Road Districts Legislative Reporting," *ICTO,* February 1989, 8.
23. Allen and Lacey, *Illinois Elections,* 527 and 538. The 90-day window for signatures did not just apply to township dissolution petition drives. Several other drives had already succeeded by 1990 in Rock Island County and statewide under that constraint on circulators. See, Lisa Neff, "Petition Process: Views Differ on Effectiveness," *Argus,* 16 September 1990, A/1–2.
24. "Editorial," WQAD-TV, 6 June 1990.
25. Township Study Group of Rock Island County, "News Release," 23 June, 10 July, and 7 August 1990; Doug Schorpp, "Township Group Presses Police for Investigation," *Davenport Quad-City Times,* 2 June 1990, C/2. While Hampton Township was probably no more guilty of mismanagement, and was in some ways more progressive, than other Rock Island townships, Richardson and Popp seemed to take special pleasure in handing Supervisor DeJaegher a hard time. DeJaegher, after all, was a double dipper, serving simultaneously in his township role and as a member of the general assembly, a legal but cozy connection. Hereafter, I cite the *Quad-City Times* as *Times.*
26. Lisa Neff, "Abolishing County's Townships Would be a State First," *Dispatch,* 21 July 1990, A/1–2.
27. Doug Schorpp, "Group: Eliminate Townships," *Times,* 19 July 1990, C/4; Neff, "Abolishing County's Townships Would be a State First."
28. Patrick Seltz, "Township Association Files Suit Against Study Group," *Argus,* 21 July 1990, A/2; Randy Popp and Michael D. Richardson, interview by author, 10 December 1992.
29. Lisa Neff, "Group Files Petitions to Abolish Townships," *Argus,* 20 August 1990, A/3; Allen and Lacey, *Illinois Elections,* 538.
30. Michael D. Richardson, interview by author, 16 October 2000; "Get Rid of Townships," *Times,* 23 August 1990, A/6. By that Thursday, 23 August, TSG knew township officials were combing the petitions in search of reasons to discredit the summer's catch. In a news release, Popp pleaded with townshippers to "Let the people vote." He cited possible legal issues because Thomas L. Kilbride, the paid attorney of the Rock Island County Township Officials Association, was doing the combing, at the apparent expense of the taxpayers. "Thousands of county voters want this question on the ballot," said TSG's co-

founder. Township Study Group of Rock Island County, "News Release," 23 August 1990.

31. Doug Schorpp, "Parties Join Forces to Save Townships," *Times,* 28 August 1990, C/2; Lisa Neff, "County Party Leaders Challenge Township Referendum Petitions," *Argus,* 28 August 1990, A/1–2.
32. Township Study Group of Rock Island County, "News Release," 31 August and 3 September 1990; Lisa Neff, "Each Petition Challenge Under Review," *Dispatch,* 5 September 1990, C/10.
33. Lisa Neff, "Township Ballot Battle Begins," *Argus,* 4 September 1990, A/1–2.
34. Neff, "Township Ballot Battle Begins"; Neff, "Each Petition Challenge Under Review."
35. William Presecky, "Pressure Builds on Townships as Rock Island Battle Brews," *Chicago Tribune,* 1/7.
36. Lisa Neff, "Board Rules Township Issue Won't Be on Ballot," *Argus,* 12 September 1990, A/1; Doug Schorpp, "Electoral Board Rejects RI Township Petition," *Times,* 13 September 1990, B/2.
37. "Vote Brings Debate," *Dispatch,* 14 September 1990, A/4; Lisa Neff, "Co-chairs Dig in for Long Haul," *Argus,* 16 September 1990, A/2.
38. Josephine Tady, "Township Government is Best Alternative," *Argus,* 30 September 1990, A/4; Michael D. Richardson, "Township Study Group Works to Give Voters Choice," *Argus,* 5 October 1990, A4; Geoff Michl, "Double Dippers Are Wasting Our Tax Dollars," *Times,* 17 October 1990, A/6; Teresa Wates, "Letter Writer Doesn't Know Facts," *Times,* 24 October 1990, A/6.
39. Doug Schorpp, "Judge Will Rule on Push to Ban RI Townships," *Times,* 4 October 1990, B/2; Doug Schorpp, "Judge Should Rule Today on Township Referendum," *Times,* 10 October 1990, C/2; Lisa Neff, "Circuit Court Rejects Township Group's Request for a Review," *Dispatch,* 12 October 1990, A/6.
40. Ed Tibbetts, "RI Study Group Keeps Alive Fight to Get Township Issue on Ballot," *Times,* 17 October 1990, C/2; "Township Study Group Appeals R.I. County Circuit Court Ruling," *Dispatch,* 16 October 1990, C/4; "Voting Delay Tactic Wrong," *Argus,* 30 October 1990, A/4; Ed Tibbetts, "Township Group Ends its Fight, for Now," *Times,* 6 November 1990, C/2.
41. Township Study Group of Rock Island County, "News Release," 5 November 1990.
42. Michael D. Richardson, undated note to author, circa January 1993.
43. Jackie Chesser, "Township Group Wants Investigation," *Argus,* 12 November 1990, A/7; Jackie Chesser, "Group: Nepotism Common Practice in South Moline," *Argus,* 13 November 1990, B/8.
44. Township Study Group of Rock Island County, "News Release," 5 December 1990; Lisa Neff, "Townships Balk at Proposed Rules," *Dispatch,* 6 December 1990, C/6.
45. Lisa Neff, " Township Study Group Co-chairman Charges Supervisor with Misconduct," *Argus,* 5 January 1991, C/1; Township Study Group of Rock Island County, "News Release," 6 February, 25 March, and 18 April 1991.
46. Denny Jacobs, "Be Aware of Information from Township Group," *Argus,* 17 February 1991, E/5; Michael Richardson and Randy Popp, "Jacobs Off the Mark with Township Group Criticism," *Argus,* 7 March 1991, A/4; Randy Popp, "Dump Townships," *Times,* 18 February 1991, A/6; Anthony Man, " RI Anti-township Activist Takes Message to Capitol," *Times,* 2 April 1991, A/12; Diane Ross, "Groups Oppose Plan to Include Townships in Revenue Sharing," *Dispatch,* 30 March 1991, D/1.

47. Diane Ross, "Township Officials Fail to Halt Lawsuit," *Dispatch,* 13 November 1991, A/12; *Michael D. Richardson v. Township Officials of Illinois,* 91 MR 86 (Rock Island County), "Complaint for Declaratory Relief."
48. *Michael D. Richardson v. Township Officials of Illinois, 91 MR 86 (Rock Island County),* "Opinion"; Michael D. Richardson, email correspondence with author, 19 April 2002.
49. Illinois, General Assembly, *Laws of the State of Illinois, Eighty-sixth General Assembly, 1990* (Springfield, 1991), 3,998.
50. Harvey Berkman, "Dissolving Townships Gets Tougher," *Times,* 11 January 1991, C/2; Lisa Neff, "Group Wants Governor to Veto Township Bill," *Argus,* 13 January 1991, A/7.
51. Ed Stasiewicz, "Township and Road Districts Legislative Reporting," *ICTO,* February 1991, 6. Stasiewicz joined TOI on 17 September 1990 and set up TOI's legislative offices at the organization's "recently acquired office building in Springfield, Illinois." He came to TOI after earning his degree in public policy administration at Governor's State University and working as a Republican staff analyst for committees in the house dealing with local government. "TOI Team Announces New Acquisition," *ICTO,* November 1990, 28.
52. "Political Job Security," *Times,* 20 January 1991, B/4.
53. Nick Reiher, "Rock Island Area Has an Ongoing Township Battle," *Joliet Herald-News,* 4 October 1992, 1/8.
54. Harvey Berkman, "Law Hampers Township Vote in RI County," *Times,* 25 January 1991, C/3; Diane Ross, "Officials Worry About Possible End of Townships," *Dispatch,* 21 March 1991, C/6.
55. Ross, "Officials Worry About Possible End of Townships"; George H. Miller, interview by author, 21 January 1993.
56. Illinois, General Assembly, House, Bills, 1991, State Archives, Springfield. Examples of earlier legislation that met a similar fate include several from the Seventy-ninth General Assembly. Rep. Joseph G. Sevcik introduced H. B. 93 to provide "for transfer of duties, functions, assets and liabilities of township coterminous with a municipality to that municipality in the event of dissolution of that township or abandonment of township government in county." It failed in committee. Rep. Harry Yourell co-sponsored H. B. 1484, which would have simplified the shutting down of a single township. It died on roll call in the house. Senator James C. Soper's S. B. 801 provided for proper disposition of "duties, functions, assets and liabilities" when voters in a conterminous township in a county of more than 100,000 population did away with that township or voters countywide voted to abolish all townships. The Counties and Townships Committee of the house defeated the Soper bill. Illinois, General Assembly, House and Senate, Bills, 1975, State Archives, Springfield.
57. *House Journal,* 87th G. A., 1991 sess., 1992, 1597–98; Diane Ross, "Legislator Wants End of Cook Townships," *Argus,* 9 May 1991, A/7.
58. *House Journal,* 87th G. A., 1991 sess., 1992, 2229–30; Miller, "From the Executive Director," *ICTO,* June 1991, 4; Diane Ross, "House Vote Fails to End Cook Townships," *Argus,* 16 May 1991, B/16.
59. Diane Ross, "Township Value Questioned Again," *Argus,* 14 June 1991, B/10.
60. "Time to End County's Township Government," *Chicago Sun-Times,* 14 June 1991, 31. Giglio explained his position, and the opposition he was receiving, in a letter to the *Sun-Times* a week later. He wrote: "Legislators who are responsible to the taxpaying public are heavily influenced by their township officials, who badger, threaten and cajole them

into maintaining the status quo by defeating any measure that would move toward eliminating township government." Frank Giglio, "Township Government is Taxpayer Burden," *Chicago Sun-Times,* 20 June 1991, 44.

61. *House Journal,* 87th G. A., 1991 sess., 1992, 4262–63; "Rural Government in an Urban Age," *Chicago Tribune,* 23 June 1991, 4/2.
62. *House Journal,* 87th G. A., 1991 sess., 1992, 5637; Hugh Dellios, "Legislative Outburst Dashes Bill to Allow Killing Townships," *Chicago Tribune,* 26 June 1991, 2/9.
63. "Annual Meeting of Henry County Township Officials," *ICTO,* January 1992, 25.
64. Miller, "From the Executive Director," *ICTO,* January 1992, 4.
65. Michael D. Richardson, interview by author, 16 October 2000; Michael D. Richardson and Randy Popp, interview by author, 10 December 1992.
66. Lisa Neff, "Petitions Seek Change in Township, Board Vote," *Argus,* 21 May 1992, B/2.
67. Dennis Moran, "Group Pushes for R.I. County Health Referendum," *Argus,* 5 August 1992, A/8; Doug Schorpp, "Group Ends Bid for Township Vote," *Times,* 5 August 1992, C/2.
68. "Township Study Group Makes the Right Move," *Dispatch,* 6 August 1992, A/4; "Township Study Group Decision is Right One," *Argus,* 6 August 1992, A/4.
69. Dennis Moran, " Referendum Petition Support Questioned," *Argus,* 13 August 1992, B/5. Erikson probably had a good point here, in that TSG was not at the nine thousand mark, much less heading for a big cushion of signatures. "We had an out," Richardson told me, "and we gladly agreed." Michael D. Richardson, interview by author, 16 October 2000.
70. Harry "Babe" Woodyard, "Township Governments Face Many Challenges," *ICTO,* March 1992, 6.
71. *House Journal,* 87th G. A., 1992 sess., 1993, 2,972; Ed Stasiewicz, "Township and Road Districts Legislative Reporting," *ICTO,* July/August 1992, 7; Ed Stasiewicz, "Township and Road Districts Reporting," *ICTO,* September 1992, 21.
72. Bernard Schoenburg, "Townships are Bargain in Government, Says Official," *Springfield Journal-Register,* 13 December 1992, 15; *D.C.'s Political Report,* n.d., <www.dcpoliticalreport.com/ 1994/IL94.htm> (20 April 2002).
73. "Abolish Township Government," *Joliet Herald-News,* 4 October 1992, 1/10; Nick Reiher, "If Township Government Were Erased, Who Could or Would Do Its Work?" "Rock Island Area Has Ongoing Township Battle," *Joliet Herald-News,* 4 October 1992, 1/1.
74. Patty Scully, "Should Township Government Die?" *Rockford Register Star,* 22 February 1993, A/1-2; Tom Seymour, "Why Township Government is Worth Keeping Around," *Rockford Register Star,* 25 March 1993, C/8. Tom Seymour, "Cutting Deficit Means Eliminating the House," *Rockford Register Star,* 1 April 1993, B/5.
75. "It's Time to Review Township System," *Bloomington Pantagraph,* 7 August 1992, A/6; Grace S. Lee, "Township Government Fills Voids Left by Others," *Pantagraph,* 14 August 1992, A/6; Edward Stasiewicz, "Township Government is Best for the People," *Pantagraph,* 22 August 1992, A/12; Anne Schultz, "Township Government is By and For the People," *Pantagraph,* 1 September 1992, A/8. I cite hereafter as *Pantagraph.*
76. "Township System Outlives Usefulness," *Pantagraph,* 3 May 1993, A/6; Loren Freed, "Taxpayers Should Keep Their Well-run Townships," *Pantagraph,* 6 May 1993, A/8; Helen Hish, "Township Government is Alive, Vital, Efficient," *Pantagraph,* 10 May 1993, A/6; Mary Caisley, "Townships Important to Citizens' Well-being," *Pantagraph,* 23 May 1993, A/8; Randy Popp, "Matejka Right to Suggest Dissolution of Townships," *Pantagraph,* 23 May 1993, A/8.

77. "Vote on Townships Would Be Beneficial," *Pantagraph,* 21 May 1993, A/6; Gene Smedley, "Pantagraph Fails to Look at True Role of Townships," *Pantagraph,* 25 May 1993, A/8. Editorial reference to an effort by TSG to influence the general assembly to pass a law mandating a vote on dissolution in every township county derived from a news conference Richardson held in Springfield that month to push the issue. No legislator would step forward, however, to sponsor such a measure. George Miller attended the news conference and said he favored the existing petition process over the plan Richardson was suggesting. Tom Prendergast, "Townships Referendum Bill Sought," *Pantagraph,* 18 May 1993, A/2; "End to Township Layer Debated at State Capitol," *TSG news,* June 1993, 1. Only once in Illinois history had a mandate come from the general assembly for all counties to vote on whether or not to have township government—1849. Even then, just 59 of 99 counties voted (chapter 2).
78. "The Township Dinosaur," *Belleville News-Democrat,* 13 September 1993, A/4.
79. Kerber, Eck & Braeckel, "Proposed Strategic Plan for Township Officials of Illinois (TOI)," (Springfield, 11 March 1992), 16–18.
80. Grace Lee, "News from Cook County," *ICTO,* June 1992, 44–45. The first of several identical full-page advertisements for the *Media Educational Packet* ran in the May 1992 issue of *ICTO* on page 53. Browning was then the supervisor of Benton Township, Franklin County.
81. *Media Educational Packet* (n.p.: Township Officials of Illinois, 1992); "Why Develop a Communications Strategy?" *ICTO,* February 1999, 10–12; "Components of a Communications Strategy: Correcting Errors in the Media," *Township Perspective,* November 1999, 28–30; Bryan E. Smith, email correspondence with author, 22 April 2002.
82. Bernard Schoenburg, "Townships are Bargain in Government, Says Official," *Springfield Journal-Register,* 13 December 1992, 15; Rick Pearson, "And Now a Word from Townships," *Chicago Tribune,* 11 December 1992, 3/10; "Township Government...It Doesn't Cost, It Saves!" *ICTO,* September 1992, 42.
83. Kerber, Eck & Braeckel, "Proposed Strategic Plan for Township Officials of Illinois (TOI)," 19; "TOIPAC!! What Is It and Why?" *ICTO,* February 1991, 8; Ed Stasiewicz, "Township and Road Districts Legislative Reporting," *ICTO,* November 1991, 6; "TOIPAC By-Laws," *ICTO,* December 1991, 12–13.
84. James R. Cannell, "Your Legislators—How They Voted for You," *ICTO,* March 1950, 2–3; "TOIPAC—What Is It?" *ICTO,* January 1992, 65; "Your Money Cheerfully Accepted!" *ICTO,* May 1992, 59.
85. Ed Stasiewicz, "Township and Road Districts Legislative Reporting," *ICTO,* October 1992, 6; "TOIPAC and Township Government," *ICTO,* October 1992, 14–15.
86. Miller, "From the Legislative Director," *ICTO,* November 1992, 4–5.
87. Michael Richardson, "Abolish Townships: Anachronistic and Unnecessary Today in Illinois' Massive Political Pyramid," *Illinois Issues,* March 1993, 8; George H. Miller, "Townships Defended: 'Last Bastion of Pure Democracy in an Otherwise Bureaucratized Republic,'" *Illinois Issues,* March 1993, 9.
88. Pat Harrison, "Township Government: Bargain or Bogus?—Against: Michael Richardson," *Argus,* 11 April 1993, F/5.
89. Pat Harrison, "Township Government: Bargain or Bogus?—For: George Miller," *Argus,* 11 April 1993, F/5.
90. "A Win for TOI," *ICTO,* May 1993, 19. I believe this item contains the first mention of Michael Richardson by name in *ICTO.*
91. Miller, "From the Executive Director," *ICTO,* June 1993, 4–5.

92. Miller, "From the Executive Director," *ICTO,* July–August 1993, 4.
93. Miller, "From the Executive Director," *ICTO,* September 1993, 3; [Miller] "TSG Bulletin Board," *ICTO,* September 1993, 12.
94. Michael Richardson, "A Message from the Chair," *TSG news,* September 1993, 5.
95. Miller, "From the Executive Director," *ICTO,* October 1993, 4.
96. Miller, "From the Executive Director," *ICTO,* January 1994, 4; "Township Activist Disputes TOI's 'Truth,' Again," *TSG news,* February 1994, 4; [Miller] "TSG Bulletin Board," *ICTO,* April 1994, 6; [Miller.] "TSG Commentary By George, "*ICTO,* September 1994, 5.
97. Michael D. Richardson, interview by author, 26 January 1994.
98. My first meeting with Bob and Judy Anderson took place in Woodstock and Wonder Lake on 1 September 1994. Bob and I have corresponded, talked on the telephone, and met several times since then.
99. John Kelly and Christopher Wills, "Township Provides Hometown Service to Needy," *Pantagraph,* 21 December 2000, A/4; Robert G. Anderson, interviews by author, various dates.
100. Robert G. Anderson, interview by author, 8 December 2000.
101. Anderson, interview by author; Michael D. Richardson, interview by author, 16 October 2000.
102. "Township Tid-Bits," *TSG news,* January 1994, 3; [Miller.] "Commentary, By George," *ICTO,* June 1994, 15; Miller, "Township Abolition Petition Drives Begin!" *ICTO,* July–August 1994, 9.
103. Mladen Rudman, "Sparks Fly Between Tax Reform Group, Township Leader," *Elgin Courier-News,* 17 March 1994, A/3.
104. Jayna Legg, "Area Township Meetings Draw Support from Voters," *Crystal Lake Northwest Herald,* 13 April 1994, C/1–2. I cite hereafter as *Northwest Herald.*
105. Mladen Rudman, "Group Continues Push to Abolish Township Government," *Wonder Lake Star,* 20 April 1994, 7; Mladen Rudman, "CL's Harper, McHenry's Meyers Strongly Oppose Group," *Wonder Lake Star,* 20 April 1994, 7; Rich Carter, "Township Officials Blast Consolidation Plan," *Northwest Herald,* 19 April 1994, C/1.
106. Martha Russis, "Government Glut?" *Chicago Tribune,* 17 April 1994, 18/3; "Serving the Community's Special Needs," *Chicago Tribune,* 17 April 1994, 18/3.
107. "Simple Only on Surface," *Northwest Herald,* 1 May 1994, 8.
108. Mary Kay Walter to Dianne Klemm and the McHenry County Board, 13 April 1994, personal copy.
109. Mladen Rudman, "Townships Abolishment Movement Contested," *Wonder Lake Star,* 4 May 1994, 4.
110. Robert G. Anderson, interview by author, 8 December 2000; "Township Abolition Drive Gears Up," *TSG news,* April–May 1994, 1; Rich Carter, "Anti-township Petition Drive Begins," *Northwest Herald,* 24 May 1994, D/1. I encountered a lot of criticism of abolitionists and their schemes, but that by Schuster had to be the first instance in which a critic called the morals of an Anderson or a Richardson into question.
111. Office of the McHenry County Administrator, "The Possible Statutory Dissolution of Township Government: An Impact Analysis on McHenry County Government," (Woodstock: 11 July 1994).
112. Rich Carter, "Dems Lend Support to Campaign to Abolish Township Governments," *Northwest Herald,* 4 August 1994, 6; "Democrats Lend Hand in Ballot Drive," *TSG news,* June–July 1994, 1. Leona Nelson, Crystal Lake, noted later that month that the

McHenry Democrats had been critical of township government in the past. Nelson quoted extensively in her letter to the editor from a February 1967 article (presumably from the *Northwest Herald*) in which the county's Democratic Central Committee had called townships "'archaic, cumbersome and entirely unnecessary'" and urged its Republican counterpart to join in "'demonstrating that service to the people of McHenry County, rather than the preservation of useless political jobs, is the main concern of both political parties.'" Needless to say, Nelson's letter went into the pile of those supporting Bob Anderson's cause. Leona Nelson, "Still No Use for It," *Northwest Herald,* 28 August 1994, A/8.

113. Philip Franchine, "Townships Under Attack by Wonder Lake Barber," *Chicago Sun-Times,* 14 August 1994, 8; Robert G. Anderson, interview by author, 8 December 2000; Robert G. and Judy Anderson, interview by author, 1 September 1994.
114. Illinois, General Assembly, *Laws of the State of Illinois, Eighty-eighth General Assembly,* 1993 (Springfield, 1994), 888; *Laws Eighty-fifth General Assembly, 1988,* 4,058; *Laws Eighty-sixth General Assembly, 1990,* 3,998; Michael D. Richardson, email correspondence with author, 17 June 2002; Illinois, General Assembly, *Laws of the State of Illinois, Eighty-ninth General Assembly, 1995* (Springfield, 1996), 2,950.
115. Robert G. Anderson, interview by author, 8 December 2000; Kurt Begalka, "Anderson Delivers Petitions," *Northwest Herald,* 23 August 1994, A/1–2.
116. Mark Guerra, telephone interview by author, 1 June 2002.
117. Guerra, interview by author.
118. Kurt Begalka, "Township Fans Check Signatures," *Northwest Herald,* 24 August 1994, A/1–2
119. Dorothy Miller, "Civics Lesson," *Northwest Herald,* 23 August 1994, A/6; Michael P. Greenberg, "Anti-township Bias," *Northwest Herald,* 23 August 1994, A/6.
120. Sean Bret Thomas, "County Takes Action in Case Townships Go," *Northwest Herald,* 25 August 1994, A/1–2.
121. Lori Martinek, "Townships Play a Vital Role in Rural Areas," *Northwest Herald,* 26 August 1994, C/3.
122. Jana Legg, "Township Supporters Ready Objection," *Northwest Herald,* 27 August 1994, A/1–2.
123. Jill Janov, "All Eyes on McHenry County Anti-township Referendum," *Arlington Heights Daily Herald,* 24 August 1994, 1/9. I cite hereafter as *Daily Herald.*
124. Jayna Legg, "Township Fans Object to Petitions," *Northwest Herald,* 30 August 1994, D/1–2; Charles Mount, "McHenry Township Vote is Challenged," *Chicago Tribune,* 30 August 1994, 1/1 and 7. Public Act 88-62, as McArdle noted, had caught Anderson and Richardson by surprise, altering the suggested wording of the petition from "For the continuance of township organization" or "Against the continuance of township organization" to the yes/no question: "Shall township organization be continued in (name of county)?" This *Tribune* piece appeared in the Metro Northwest edition. I have not designated editions in subsequent notes, but the reader should seek such a localized edition if the main edition does not carry a cited item.
125. Mark R. Madler, "Hearings Begin on Petition Challenge," *Northwest Herald,* 2 September 1994, C/1–2.
126. Mark R. Madler, "Panel Knocks Township Question off Ballot," *Northwest Herald,* 7 September 1994, A/1.
127. Mark R. Madler, "Township Foes Plan Appeal," *Northwest Herald,* 13 September 1994, A/1–2; Charles Mount, "Township Foes Enlist Expert to Lead Fight," *Chicago Tribune,*

13 September 1994, 2/1 and 8; Dennis Mathes, "Anderson to Appeal Referendum Decision," *Woodstock Independent,* 14 September 1994, 1.

128. Mark R. Madler, "Township Foes Go to Court to Get Referendum on Ballot," *Northwest Herald,* 16 September 1994, A/1–2; Mark R. Madler, "Township Backers File New Suit," *Northwest Herald,* 17 September 1994, A/1–2; Kurt Begalka, "Township Battle Soon May Widen," *Northwest Herald,* 23 September 1994, A/1–2; Mark R. Madler, "County Board Won't Battle Township Ruling," *Northwest Herald,* 27 September 1994, D/1.

129. Henry Hemmingsen, "Abolish County Next?" *Northwest Herald,* 15 September 1994, A/10; Ersel C. Schuster, "Bigger Never Better in Any Government," *Northwest Herald,* 18 September 1994, A/11; Robert A. Korslin, "Civics Lesson," *Northwest Herald,* 18 September 1994, A/11; R. W. "Bill" Baruth Jr., "Arrogance," *Northwest Herald,* 21 September 1994, A/6; Carole Ann Kaiser, "Townships Justified," *Daily Herald,* 24 September 1994, 1/9; Judy Anderson, "Targeting Townships," *Daily Herald,* 24 September 1994, 1/9; Raymond L. Wells, "Records Tell It All," *Northwest Herald,* 26 September 1994, A/6; Carole Ann Kaiser, "Anderson Cries 'Foul' after Striking Out," *Woodstock Independent,* 28 September 1994, 4; Gerry Wish, "Townships Outdated," *Daily Herald,* 9 October 1994, 1/13; Anita Sherwood, "The Naked Truth," *Northwest Herald,* 11 October 1994, A/6; Earl R. Wilcox, "Anderson Not to Blame for Costing Taxpayers," *Woodstock Independent,* 12 October 1994, 4.

130. Mark R. Madler, "Judge Oks Township Vote," *Northwest Herald,* 15 October 1994, A/1–2; Robert G. Anderson, telephone interview by author, 14 October 1994.

131. David Duschene, "Township Supporters Appeal Referendum Ruling," *Daily Herald,* 18 October 1994, 1/8; Kurt Begalka, "Township Question Gains Nod," *Northwest Herald,* 19 October 1994, C/1.

132. Amy R. Mack, "Township Backers Get Boost from State Group," *Northwest Herald,* 25 October 1994, D/1.

133. Charles Mount, "Township Vote Rises from Dead," *Chicago Tribune,* 27 October 1994, 1/3; Bryan E. Smith, letter to the author, 4 October 2000; Sherry Johnson, "Hooray for Townships," *Northwest Herald,* 27 October 1994, A/10; Caryn Corp, "Machine Politics," *Northwest Herald,* 30 October 1994, A/9; Rich Carter, "Township Leaders Seek Task Force," *Northwest Herald,* 28 October 1994, C/1.

134. "Continue Townships: Yes," *Northwest Herald,* 31 October 1994, A/6. CTG made good use of this favorable editorial by printing copies of it on colorful paper. On the reverse, CTG asked questions about what would happen should the voters turn their backs on townships on 8 November. CTG warned that residents would see their "access to elected officials…virtually eliminated" if a county commission of "three" replaced the 24 then serving on the county board. Citizens for Township Government, "Too Many Questions??? Not Enough Answers???" campaign literature, fall 1994.

135. "Preserve Townships in McHenry," *Chicago Tribune,* 2 November 1994, 1/12; "Voters Must Roll Township Dice," *Woodstock Star,* 2 November 1994, 8.

136. Eric Peterson, "Pro-township Forces Striking Back on Street," *Northwest Herald,* 6 November 2994, B/1; Grace Lee, "News from Cook County," *ICTO,* December 1994, 22–23; "Senator Rauschenberger Walks the Streets for Townships," *ICTO,* February 1995, 5; "Township Government: Worth Saving Because it Works," *Northwest Herald,* 3 November 1994, C/3.

137. CTIG had collected donations totaling about ten thousand dollars by the beginning of November, almost all of which went to pay the lawyers that argued against the CTG challenge of petitions. Robert G. Anderson, telephone interview by author, 4 June 2002.
138. Cyndi Wyss, "Township Issues Scrutinized," *Northwest Herald,* 4 November 1994, A/1–2; Miller, "From the Executive Director," *ICTO,* December 1994, 4. In Anderson's words, despite Klemm's affection for the townships, she refused to join the chorus and proclaim, "The sky was going to fall." Robert G. Anderson, telephone interview by author, 4 June 2002.
139. Kurt Begalka, "Anti-township Forces Undeterred," *Northwest Herald,* 10 November 1994, C/1–2; Cyndi Wyss, "Local Abolition Drive Fails Miserably at County Polls," *Northwest Herald,* 9 November 1994, A/3; "A Resounding 'Yes'," *Northwest Herald,* 10 November 1994, A/10; Chris Fusco, "Township Critic Spreads Message, Targets April Vote," *Northwest Herald,* 9 November 1994, A/3.
140. Jim Dyer, "Board's Inactivity Riles Activist," *Times,* 5 January 1994, M/1; Cheri Bustos, "GOP, Reformers Want Township Supervisor Out," *Times,* 15 July 1994, M/4; "Lund Moves but Refuses to Quit," *TSG news,* June–July 1994, 1–2; Sonja West, "County Exec Hopefuls Look Toward Fall Vote," *Argus,* 16 March 1994, A/1 and 5.
141. Cheri Bustos, "Township Group Begins a Push for Referendum," *Times,* 25 May 1994, M/3; Shayla Thiel, "Township Opponents Push for Ballot Spot," *Argus,* 26 May 1994, B/1; "GOP Opposes End to Townships," *Argus,* 26 May 1994, A/1. Richardson and Anderson agreed to use the expression "Operation Tax Relief" in their otherwise separate 1994 campaigns, a portent of cooperation that was to occur in the next biennium.
142. "TSG Seeks Referendum," *Times,* 8 June 1994, M/4; "Activist Can't Lobby Inside The Mark," *Times,* 2 July 1994, M/2; Cheri Bustos, "Township Opponents Gather Signatures," *Times,* 10 August 1994, M/3; Cathy Hensley, "RI County Ballot Could Fill Up Quickly," *Times,* 17 August 1994, M/1; Michael D. Richardson, interview by author, 16 October 2000.
143. *Michael D. Richardson v. Richard "Dick" Leibovitz, et. al.*, 94 MR 144 (Rock Island County), "Order."

8 STOP (Or Not): 1995–1999

NEITHER SIDE TOOK the 1994 outcomes in McHenry and Rock Island counties as the end of the struggle for signatures and votes. George Miller, Bryan Smith, Ed Stasiewicz, and others in TOI believed what Bob Anderson had accomplished meant the rules of engagement needed changing. Anderson, Michael Richardson, Randy Popp, and a few new faces on the reformer scene believed the thirteen thousand voters that said no to townships in McHenry's 8 November referendum and the rumblings of taxpayer discontent at the grassroots boded well if only they could approach the 1996 campaign in a more unified posture.

A Friend Indeed

TOI needed a little help, again, from its friends in the general assembly. If a barber with meager resources, living in an up-scale county where many residents still held rural institutions and traditions in high regard, could bring abolition to the ballot, what might happen in a more urbanized county with little affection for farmers and township officials or in a rural county down on its luck and out to trim taxes or in a county facing a scandal involving township officials? Once one township fell, whatever the circumstances, even if township officials had been doing their jobs well, the repercussions could be disastrous. TOI had, in the early 1990s, encouraged legislators to raise the petition-signature bar from 10 percent of those voting at the last presidential election to 10 percent of a county's registered voters on a particular day, but either of these standards meant that heavily populated townships in a county could control the fate of the less populated rural townships in the same county.

When it came to general assembly matters, TOI could claim experience in abundance. Legislative Director Stasiewicz had by then been manning TOI's Springfield firebase for four years. George Miller had two decades of general assembly lobbying behind him and often made the hour plus drive from Astoria to the state capital on township business. For sponsors of TOI's referendum-related legislation, somewhat ironically, they turned to Ann Hughes, Bob Anderson's representative in the house, and Dick Klemm, Anderson's senator. Hughes, a member of the Illinois House of Representatives since 1993, lived near Woodstock, had an agricultural-business background, served as the chair of the McHenry County Board from 1988 to 1992, and chaired the Committee on Counties and Townships for the Eighty-ninth General Assembly (1995–1996). Dick Klemm, from Crystal Lake, had in 1992 (after 12 years) given up his house seat, which Hughes then occupied, in order to run for and capture the senate seat in the Thirty-second District. In 1995–1996, Klemm chaired the Local Government and Elections Committee on the senate side of the capitol. Before moving to Springfield in 1981, Klemm chaired the McHenry County Board. In the Eighty-ninth General Assembly, Dick Klemm was one of seven senators who had served as a township official; but Ann Hughes was not among the seventeen representatives with a township office on their resume.[1]

Even before Rep. Hughes introduced her township-friendly legislation on 24 January 1995, which became H. B. 438 (amending the Township Code and thereby making abolition much more difficult), George Miller revealed that TOI felt the existing setup, section 25-10 of the Code, was unfavorable to the township cause. In an interview with Kurt Begalka of the *Northwest Herald,* just after Bob Anderson piled his 1994 petitions on the county clerk's counter in Woodstock, Miller lamented the fact that so many people in McHenry's townships were new to the area and did not "'know that much about township government.'" What he recognized was that, like Mark Guerra, former residents of counties such as Cook, with little township connection, could be voting out something that long-time residents treasured. "'If it had to be approved in every township in the county, then I would have no concern at all on having it put on the ballot,'" he told Begalka.[2]

The Hughes-Klemm bill breezed through Springfield and into the statute book. As Hughes introduced it, and Stasiewicz paraphrased it for *ICTO* readers in March of 1995, H. B. 438 sought to amend the Township Code by "requiring that every township in the county with an abolition referenda [*sic*] on the ballot shall approve the discontinuation of township government for township government to cease in the entire county." In that mode, requiring that a majority of voters in each township approve, H. B. 438 came out of Hughes's committee on 16 March with a recommendation to pass. In April, the house added, first, an amendment (essentially H. B. 212) calling for 10 percent of the registered voters rather than just 50 to sign petitions to call for a referendum on establishment of township government in a county, and, second, an amendment reducing the number of townships in which an abolition referendum would have to pass from "each" to "at least three-quarters." Hughes and her committee agreed to the amendments; and on 24 April, the house voted 100-0-8 to approve H. B. 438 and sent it to the senate. Klemm's Committee on Local Government and Elections took the house bill, modified it by adding a stipulation that the townships making up the "three-quarters" had to contain at least a majority of the population of the county, and secured full senate approval (54-0) on 15 May. Once the house had concurred (on 24 May), the senate amendment made sure that lightly populated townships, containing a small portion of a county's population, could not gang up on a township or two with most of the people in the county. Nevertheless, TOI got what its leaders wanted: substantial protection of rural interests and an almost insurmountable target for Anderson, Richardson, and anybody else seeking to dethrone this Illinois institution. Governor Jim Edgar signed the bill into law 18 August 1995.[3]

Writing his September *ICTO* column, even before Edgar affixed his signature, George Miller said that township supporters could "be fairly certain that [their] form of government [would not] be wiped out on a whim or a local issue affecting very few highly populated townships." I hope, he continued, "it is safe to say, for all practical purposes, the township abolition agony is also behind us."[4] A month later, Bob Anderson, Michael Richardson, and two others sued in DuPage County, challenging the constitutionality of Public Act 89-365 (H. B. 438).

Starting STOP

The barber from Wonder Lake was doing more than cutting hair and driving a school bus as his state representative and state senator shepherded the TOI proposal to require most townships approve a referendum to dissolve townships countywide. Letters needed writing, such as the one he mailed to Senator Klemm requesting legislation to provide for "an orderly transition to county government" should an abolition action be successful and the one to the *Daily Herald* intended to help spread his message to other Collar Counties about how the Republican Party and TOI, "using Gestapo-like tactics," had "saved their self-serving government jobs and political turf, for the time being." He and spouse Judy had fundraisers to attend, in the hope of paying off 1994 bills, such as the one for just over twenty thousand dollars that the firm of Lavelle, Juneau & McCollom, Ltd., had presented CTIG to compensate Michael Lavelle and other attorneys earning $150 to $200 per hour, against township attorneys working pro bono. Anderson had a newsletter to write and distribute to the thousand or so on the CTIG mailing list, warning them in the March 1995 issue that Rep. Hughes, after promising "to resolve the conflicting laws concerning township abolition," was in reality working hard for her H. B. 438, which would make overwhelming voter support necessary in order to pass an abolition referendum. He and Judy had to take a day off from work on 1 March to make the four-hour drive to Springfield in anticipation of testifying against the Hughes-Klemm bill, only to learn when they arrived that the Hughes committee would be hearing testimony on 2 March. Township officials, Anderson claimed in a letter he sent to all members of the general assembly, did learn of the change in advance and were able to testify on the substitute day. The Andersons, back at work in McHenry County, were not present.[5] Of all the business to which the barber/bus driver attended in the early months of 1995, the one that promised to pay off most handsomely was the decision to join forces with TSG, and others, in a multi-front war against TOI and its allies. Out of that alliance grew STOP, Send Township Officials Packing.

Beginning with its Seminar on Township Government Abolishment, in Crystal Lake on 21 May 1995, and continuing with strategy sessions through 1996, many of which the Andersons and Richardson invited me to observe, abolitionists in the northern part of Illinois formed a coherent and effective alliance and acquired their own free legal counsel from the Citizen Advocacy Center (CAC). In turn, that warm May Sunday afternoon in Crystal Lake, Michael Richardson, Randy Popp, Bob Anderson, and Michael Walkup provided personal perspectives on what had transpired in Rock Island and McHenry counties since each had entered upon the abolition crusade. Anderson offered a number of tips for anyone in the audience thinking they might want to start down the petition-passing path: hire a knowledgeable campaign manager, treat the press honestly, hold fundraising events and otherwise ask openly for donations, get a lawyer to review the petition wording in advance, wear a campaign button when circulating, put only 10 signature slots on each petition (as a morale booster), and go where the people congregate (not house to house). From the audience came good suggestions such as development of a white paper with answers to questions voters invariably ask. More than thirty, almost all from the Collar Counties, signed the roster that day, and the consensus was to meet again in midsummer.

The cadre of this new alliance gathered on 23 July at the Elmhurst offices of CAC. Michael Richardson distributed his white paper, "More is Really Less." In it he evaluated the various duties of Illinois townships, both mandated and permitted, and editorialized on issues he felt showed the unnecessary nature of township government. Walkup told listeners how there was nothing to report about the transition legislation that they so fervently wished would emerge from Springfield. CAC legal intern Myrha Guzman discussed the importance, for any such group, of choosing a name that would endure and serve as a meaningful umbrella for all seeking its shelter. Several name suggestions had come to the attention of Guzman: Send Township Officials Packing (STOP), Citizens Alliance to End Township Taxes (CATETT), Citizens Opposed to Township Government (COTG), and Stop Wasting Our Township Taxes (SWOTT). STOP won unanimous approval, joining TSG and CTIG in the TOI hall of shame. In fact, at the 1995 annual meeting of TOI, some clever member or members brought in a casket and posted on it a sign that read, "HELP! HELP! Put TSG, CTIG & STOP to Rest Once and for All! Make Your Contribution to TOIPAC. Bury the Bums![6]

By far the most serious topic of STOP's Elmhurst meeting in July of 1995, however, was H. B. 438. Intern Joni Jacobsen, a second-year law student at The University of Chicago, had taken the lead at CAC in researching the Hughes-Klemm bill, which was then awaiting Edgar's signature. In a one-page handout, Jacobsen summarized the bill and predicted that the section boosting the signature requirement to call a referendum to establish townships would survive any constitutional test because "The constitution…grants the General Assembly the power to decide the manner in which [a] referendum may be initiated on the question of township formation." She might have added that the change would also bring the signature standard for an establishment referendum even with one calling for abolition, as had not been the case since the 1870s. On the matter of requiring that "a majority of the voters in at least three-quarters of the townships" would henceforth be necessary for cessation of township organization, Jacobsen reported what was likely a serious flaw. The 1970 Constitution did not permit such a "supermajority requirement" for that sort of question.[7] Discussion followed among Jacobsen, Walkup, Richardson, and CAC Director Theresa Amato, the upshot of which was to authorize Jacobsen to write the governor a letter.

On 3 August, two weeks before Jim Edgar signed H. B. 438 into law, the Jacobsen letter departed Elmhurst. She wrote on behalf of CAC, requesting a veto of the bill; citing her "review of the relevant constitutional and statutory provisions, the legislative history, and existing case law"; and using the term "supermajority" several times. Jacobsen quoted from the transcript of the house debate of 20 April, during which (she said) Rep. Hughes said, "'I believe a super-majority is a reasonable approach to providing the ability of the voters to determine whether they will continue to have township government or not.'" In her letter, the intern told the governor that she believed neither Hughes nor anyone else among general assembly members was aware that the constitution's framers had not authorized a supermajority in such a situation. She warned the governor that his signature would precipitate a prompt legal challenge.[8] He signed; STOP sued.

STOP held a strategy session, on 17 September at CAC's Elmhurst headquarters, approximately three weeks prior to the filing of its suit challenging Public Act 89-365. Here are some of the items I jotted in my notebook that Sunday. Gail Dunham, a disgruntled, former township supervisor (Naperville Township) and then employee of CAC, spoke effectively against townships and seemed destined to be a problem for TOI. Rosanna (Rosie) Fitzpatrick, an aspirant to political office in DuPage County (from Bolingbrook Township), first came to my attention at this meeting. She and her cochair for STOP in DuPage County, Dennis Hamm, had written to the DuPage County Board of Election Commissioners indicating their intention to pass petitions 1996 and asking if the board planned to enforce the supermajority rule of Public Act 89-365. Bob Anderson and Michael Richardson had written to the Illinois State Board of Elections asking that body to settle the issue of the supermajority rule in advance of 1996. Which would the state board use: the constitution or Public Act 89-365? Theresa Amato said she considered the STOP press conference, which was to follow the filing of the lawsuit on 5 October, to be the more important of the two events they were planning for the DuPage County governmental complex in Wheaton. Bob Anderson announced he would be targeting McHenry Township for abolition in 1996 because state law forbade a countywide referendum any sooner that 1998 (four years following 1994). Membership in STOP had doubled since the May seminar. CAC had become involved in a legal wrangle over the rights of township electors to challenge spending plans in Maine Township (Cook County). More than half those present that day in September were from Maine. STOP leaders were still pushing for transition legislation but could not find a legislator brave enough to sponsor it, given the township-friendly climate in Springfield. STOP planned as much public education as possible prior to the 1996 petition-passing season and had set up a speakers bureau.

D (DuPage) Day

Theresa Amato chose DuPage County as the locale for the STOP suit challenging the supermajority law partly because CAC had its headquarters there and partly because DuPage County appeared to be a prime target for success of a dissolution drive, due to its almost entirely suburban character. Few farmers remained, and many residents had little history with township government. In her prepared statement for the press, on 5 October, she said Public Act 89-365 was "an act of desperation to disenfranchise the public" that grew directly from the McHenry County abolition movement of the previous year. "There is no doubt that the General Assembly passed this bill to chill citizens from undertaking petitions to place the question of township government on the ballot." CAC, for those who did not already know, was a nonprofit, nonpartisan organization committed to facilitating citizen involvement in government.[9]

Bob Anderson, Rosie Fitzpatrick, and Michael Richardson (all among the plaintiffs in the suit) also issued press releases that October morning. Anderson said STOP counted four hundred members and had by then spread to seven counties (figure 21). Fitzpatrick presented figures that showed cash on hand in Bolingbrook Township at over seven million dollars, far more that she felt prudent. Richardson said, "STOP will help taxpayers rid themselves of township tax parasites" and "stop this unconstitutional law."[10]

Press coverage for D (DuPage) Day was extensive. Dave McKinney, state government writer for the *Arlington Heights Daily Herald,* took advantage of prior notice of the filing and press conference, and had a story running about the events that morning. McKinney quoted Hughes as saying her bill would "'stand up to a court's test if for no other reason than the principle involved'" of not allowing a "'heavily populated township...to determine alone whether township government across the entire county should cease to exist.'" Charles Mount had covered the 1994 McHenry battle and was on the job again for the *Chicago Tribune.* Hughes, who lived in (rural) Hartland Township, told Mount that constitutionality issues "'never came up'" when

Figure 21 Bob Anderson outside the DuPage County courtroom where he and others sued over the supermajority law, Public Act 89-365.

she was working the bill. "'It would be like California starting a move to abolish state government and teaming up with the four most populous states.'" She added that Amato was wrong in blaming McHenry County officials for egging "'her into pushing the bill.'" McHenry County's *Northwest Herald* had Kurt Begalka on the township beat once more. Rep. Hughes told Begalka that her bill was like the law that "'allows school districts to unite. Voters in each must agree.'" Editorially, a few days later, Begalka's employer accused Anderson of "Still Tilting at Windmills," and called the Hughes-Klemm law, "sensible legislation" that would "eliminate the chance for a couple of urban townships from ganging up and eliminating township government from an entire county." Editors reminded readers how they had "discredited Anderson's movement last year," and said, "The courts ought to do it this time." Papers in DuPage County were, of course, interested in the local furor over grassroots government, too. Stories ran within the week in papers out of Elmhurst, Hinsdale, and Naperville, with Dunham and

Fitzpatrick receiving the bulk of the ink. Downstate, the *Bloomington Pantagraph,* and no doubt other papers, carried the story within a day or so, and probably, like the *Pantagraph,* played a local angle.[11]

Still a Year to Go

The next general election would occur on 5 November 1996. As 1995 wound to a close, both sides prepared once more for battle—both sides operating from positions their leaders considered stronger than in 1994. TOI had the Hughes-Klemm law and a resounding victory in McHenry County on the books. STOP had returning veterans and a group of rookies ready to test themselves against TOI and local officialdom. STOP also had the complimentary services of a tough young attorney, then watching their lawsuit make its way along the judicial path.[12]

George Miller said nothing in *ICTO* that fall about the STOP suit, though his telephone must have rung numerous times on the fifth of October as reporters from around the state sought his take on the story unfolding in Wheaton. TOI presented Rep. Ann Hughes with its Special Recognition Award at the annual meeting in November 1995. Miller described her afterward as "diligent and very effective in guiding our legislation through the House Committee and to total approval of the entire House.[13]

STOP stayed busy in a low-key way that fall of 1995, getting ready for the May 1996 announcements of abolition campaigns. In late November, Richardson told me via letter that the STOP steering group was meeting periodically to prepare a joint brochure targeting township excesses, to ascertain where abolitionists were serious about petitioning, and to prepare a letter imploring Rep. Hughes to "request introduction of a transition plan." STOP was anticipating a large gathering of potential activists in February for a "how-to" seminar. At the end of 1995, Richardson announced he was switching his party; joining the Republicans; and running for Congress against the veteran incumbent, Democrat Lane Evans of Rock Island, partly because the local Democrats had been so nasty to him during his abolition campaigns. In announcing his party switch and congressional aspirations, he revealed he had run for a seat in the unicameral Nebraska legislature, finishing last in an eight-candidate field, behind even one aspirant who died during the race. Congressional candidates, no matter how unlikely their chances of winning, get press attention, which to Richardson was a good thing.[14]

Optimistic, even confident, STOP leaders entered the general election year of 1996 believing their side would prevail in the DuPage County lawsuit and expecting to see several abolition referenda on the ballot in the counties where STOP membership was most active. Had Richardson, Anderson, and associates known then of two totally independent abolition efforts, far downstate, they might have been even more sanguine as they held their winter and spring preparatory sessions.

Wary as 1996 began, but determined to reach the impending sesquicentennial of township government with all 1,433 townships intact, TOI leaders asked to become a co-defendant in the DuPage suit, continued preparing local officials for what could be a momentous year, and started thinking of a simple change in Township Code that could render countywide petitioning efforts on the issue of abolition virtually impractical in most counties. George Miller also made plans to retire before the end of the coming summer. Thus began the fourth biennium of the last decade of the second millennium.

The Third of June

A pair of township-significant decisions became public on 3 June 1996. A DuPage County judge ruled on the STOP lawsuit, and TOI's directors announced the name of the new executive director. First, I turn to Wheaton, then to Astoria.

DuPage Decision

The original defendants in STOP's suit, challenging the constitutionality of the Hughes-Klemm law requiring a supermajority to approve dissolution of all townships in a county, had no evident stake in the outcome, other than protection of themselves against damages; but TOI and the state's civil-political townships did. In November 1995, therefore, Wayne Township (one of DuPage County's nine townships) and TOI filed a petition to intervene and become defendants. Wayne Township, said the petition, was "directly interested in the continuing existence of townships"; and the lawsuit presented "a challenge to this existence." TOI, as the representative of over 98 percent of the townships in Illinois, felt it had to have official standing in order to "promote and support" its constituent townships. Furthermore, Wayne Township needed financial assistance if it was to speak for all Illinois townships. Judge Rodney Equi of the DuPage (Eighteenth) Circuit Court granted the intervention petition in early January 1996, despite Amato's memorandum in which she argued neither nondefendant had a right to intervene nor could they show how a decision of the court would directly affect them.[15]

Once TOI and Wayne Township entered the fray, the pace of legal maneuvering increased. The newly approved intervening defendants moved to "strike and dismiss" the original complaint from STOP. TOI and Wayne Township asked for a favorable summary judgment, basically because plaintiffs had yet to suffer any damage to their cause. Defendants' attorney Richard A. Cowen offered an alternative to complete dismissal, which was to leave the case open and see, after the November election, "whether there [was] any issue, constitutional or otherwise, that [was] ripe for judicial decision." In other words, Cowen was saying this: gather your forty thousand or so signatures in DuPage County, get the issue on the ballot, let the voters have their say, and then check to see if the referendum lost because of the supermajority law. Theresa Amato's motion for summary judgment on behalf of her clients revolved around the problem that abolition-minded citizens would face in DuPage: having "to wage nine individual township campaigns because plaintiffs [could] not afford to lose in more than two townships." No longer was the issue really countywide; it was township by township.[16]

Last word in the DuPage County Circuit Court matter fell to Judge Robert E. Byrne. On Monday, 3 June, he issued a five-page finding in which he declared a portion of Public Act 89-365 not in compliance with the 1970 Constitution. "It is clear," he wrote, "that the drafters [of the new constitution] intended to protect the majority of people living under township government from a minority that might try to interfere with the majority's attempt to abolish township government." In Byrne's view, drafters did not expect that a majority of voters in a majority of townships should approve abolition.[17]

The Illinois press covered the Byrne ruling as zealously as it had the filing back in October 1995. Mike Burke of the *Daily Herald* tracked down Wayne Township Supervisor Joan Mruk, who expressed surprise at the decision. Bob Anderson told the *Chicago Daily Law Bulletin* that the ruling marked a pleasant respite in "'a long battle without many victories.'" A crew of *Chicago Tribune* reporters talked to Anderson; George Miller; and Herb Wehling, supervisor of Milton Township in DuPage County. Wehling, not a party in the case but interested in the outcome, warned that a county-focused bureaucracy would make matters worse if STOP eventually prevailed with its abolition agenda. Michael Richardson told Kurt Begalka of the *Northwest Herald* he had not worried too much over the outcome but had thought, "'the power of the townships was going to work'" to the disadvantage of the plaintiffs. Tom Schober predicted an appeal when Begalka called him for the Algonquin Township take on the matter. Cheri Bustos connected her story to the Rock Island County petition drive then in progress, quoting Richardson as being hopeful about gathering fifteen thousand signatures—just to be comfortable. Bustos noted that Rep. Donald Moffitt of Galesburg had helped Hughes with H. B. 438 and that every Quad Cities lawmaker, who voted, had voted for the measure. Steve Zalusky, writing for Press Publications, caught up with Gail Dunham, who told him the ruling would allow the DuPage branch of STOP to "'start collecting signatures,'" something they had been delaying because of the pending case. Richardson told John Kanthak of the *Rock Island Argus* that the decision of Judge Byrne was a boost for his signature drive, while Miller said TOI and its attorneys were considering an appeal.[18]

Director Decision

Losers in the race to replace George Miller did not have the luxury of appealing the other 3 June decision. TOI's board of directors that day announced the appointment, effective 1 August, of Bryan E. Smith to become the new executive director.[19]

Though still feisty and a force with which to reckon, at age 68, Miller had been expressing a desire to retire and end his weekly commute from Leyden Township in Cook County to Astoria Township in Fulton County. To assist in this replacement process, TOI hired the PAR Group out of Lake Bluff, Illinois, to conduct a nationwide search to find Miller's successor. Nearly 150 candidates stepped forward to present credentials, with 18 of them emerging from the first cut and 7 from the next. Board members interviewed those on the short list, among which were Smith and Ed Stasiewicz, the legislative director. Not among those making the list with Smith and Stasiewicz, however, was Michael Richardson, who applied, I presume, in good faith, offering in his letter of application to put "aside rhetoric of the past" and work "with township officials for the improvement of township government."[20]

When he took the reins two months later, Bryan Smith, born 21 July 1965, was just barely 31 years of age (figure 22), two decades younger than George Miller had been in 1979 when he replaced Troy Kost. In his valedictory column, Miller praised Smith for his nine years of "on the job training" as associate director, noting how almost four dozen township officials had written letters to the search committee on Smith's behalf. Miller reminded *ICTO* readers how effective a team player his protégé had been and that Smith had "campaigned for and won elective office in three elections." Only one of the other finalists for the position could claim to have been an elected township official.[21]

A native of Macon County and a graduate of Milliken University, Smith came from a family that could claim a long history of public service in Long Creek Township, which stretches eastward from the eastern margin of Decatur to the Piatt County line and encompasses much of the watershed of Long Creek. His grandmother had served as a township cemetery trustee, and his father was an assistant township supervisor before assistant supervisors went away in the early 1970s (see chapter 6). Father R. C. Smith remained an elected member of the Macon County Board and eventually chaired that body. In 1985, when Bryan Smith was a sophomore at Milliken (in Decatur), the Democrat Party leaders of Long Creek Township were meeting at the home of his parents and looking for likely candidates. A lady suggested that a young man like Bryan might make a good addition to the ticket and later asked him personally to run for town clerk. He agreed, secured the nomination at his party's caucus, and won a four-year term at the township election that spring. He stayed in that township job while commuting a hundred miles from Long Creek to Astoria for his TOI internship and afterward as Miller picked him to be the first TOI associate director. Smith ran again for the clerkship in 1989, won a second term, and agreed to fill a vacancy and become the Long Creek supervisor in 1992. The next year voters elected him to a full term as supervisor, and he served until 1996 when he resigned to take over for Miller. Thus the tradition continued of TOI always having as its leader a man with experience as an elected township official: Kost as highway commissioner, Miller as auditor (trustee), and Smith as clerk and supervisor.[22]

Figure 22 Bryan E. Smith, left, and George H. Miller soon after Smith replaced Miller as the executive director of TOI. (Courtesy of the Township Officials of Illinois)

As TOI's associate director, Smith generally avoided the abolition battles that plagued TOI after his tenure there began. Instead, he used his regular column in the magazine to inspire officials by sharing tales about his travels and making observations that assisted him in getting across his point. The first "From Bryan's Desk" appeared in the September 1987 magazine and dealt with his chancing upon two township road district workers doing their jobs on a hot summer day. His October column that year chronicled impressions of his first National Association of Towns and Townships conference.

In November he reported about the annual TOI education conference and in December sent seasons greetings to the membership. Beginning in January of 1988, "From Bryan's Desk" joined "From the Executive Director" and other reports that appeared regularly, month after month. Voter turnout, annual township meetings, and educational opportunities for township officials were among his rather generic subjects in those early years at Astoria. Beginning with the February 1991 column, however, Smith focused more on the nuts and bolts of being a township official. Reminders and technical talk became commonplace for Smith, while his boss worked hard to fend off the likes of Richardson, Popp, and Anderson. As the time to name a successor for Miller neared, Smith did take a more aggressive stance toward the abolition movement. In a speech to the Whiteside County Township Officials, for instance, he labeled "those opposing township government 'a bunch of losers.'" As evidence he cited the fact that Richardson had finished last in his run for the Nebraska legislature and that others active in STOP had "lost township elections."[23]

Road trips, such as the one to Whiteside County, allowed Smith to show the TOI flag and to let the membership meet him. He delivered lots of after-breakfast, -lunch, and, -dinner addresses to inform and inspire local officials all over the state. In the process, he demonstrated his connection to the day-by-day chores of the town clerk and township supervisor. By being out on the circuit, he allowed the rank and file to connect with an affable, respectful, and able young man. When the time came for George H. Miller to retire, Bryan E. Smith was ready for higher office.

Anticipation

For STOP, 1996 looked like a year in which to make history, a year for taking down the township apparatus in one or more of the populous counties of northern Illinois. Theresa Amato had instilled a sense of confidence regarding the likely fate of the supermajority law. Availability of new talent among the abolitionists seemed to bode well for multiple-petition drives—thus dividing the attention and resources of Fortress Astoria. Turnover at the top of TOI might just help someone to secure that first, critical victory and in the aftermath demonstrate that transition to county control could succeed without adverse consequences. Furthermore, 1996 was a presidential election year, so STOP could expect higher voter turnout in November—usually a good circumstance when trying to secure votes in favor of abolition.

The months leading up to 22 May, the day that the petition season opened for 1996, were busy ones for the activist leadership. February found the STOP steering committee hosting a kickoff seminar, on petitioning, raising money, and other pertinent topics, at the Holiday Inn in Elgin. STOP provided potential petition circulators with (this time) the correctly worded model petition and "Instructions on How to Collect Properly the Signatures to Put the Abolish/Continue Townships Question on the Ballot." Its author (likely Michael Richardson) told reformers they should "aim to collect 20% of the relevant number" rather than the 10 percent minimum the law required. February also found Bob Anderson carrying a sign that read "Hughes Wanted for Tax Crimes," as he led a group of about forty in what the *Chicago Tribune* called "a fiery protest" outside a fundraiser for the chief sponsor of H. B. 438. Governor Jim Edgar ignored Anderson and his comrades as the governor made his way inside the McHenry Country Club to join about three hundred Hughes supporters who had gathered to help bankroll her campaign for re-election. In March, Anderson ran for and won a seat at the local GOP table as a precinct committeeman in McHenry Township, a seat that he told me he hoped would provide him "better access to insider information." Also in March, he addressed the Elgin chapter of the League of Women Voters, seeking recruits for STOP and to explain why he felt CTIG lost so badly in 1994. At the annual meeting of McHenry Township electors on 9 April, he saw his resolution to call an advisory referendum, on the issue of dissolving just that one township, go down to defeat by a three to one margin. A few weeks later, Dundee Township (Kane County) GOP committeemen asked him to deliver his dissolution dissertation at their regular meeting, in part because some of the Republican leaders in that township were unhappy with how township officials had conducted the recent annual meeting. Richardson called a news conference that Tuesday to thumb his nose at such (in his opinion) meaningless events, calling them "a joke," and announce his fourth campaign to rid Rock Island County of annual meetings and other vestiges of the township phenomenon.[24]

Newspapers, cognizant of the DuPage County lawsuit challenging the constitutionality of the law that resulted from H. B. 438, and of the likelihood of petition drives in several counties, kept townships in the public eye with coverage of single events such as the Hughes fundraiser in McHenry, periodic pieces, in-depth studies, and timely editorials. The *Elmhurst Press,* for instance, ran a mid-February series of a half dozen articles with titles like "Is STOP Making Sense?" Editors at the *Press* then took townships, Republicans, and Governor Edgar to task, calling the township "outdated, inefficient, and in its worst cases, wasteful," blaming Republicans "for creating legislation that [made] it more difficult for a county to abolish township government via a referendum," and declaring the governor guilty of "political hypocrisy unbecoming a Republican." The *Edwardsville Intelligencer* reported the local League of Women Voters was getting set to study townships before deciding whether or not to take a stand one way or the other on their usefulness. Richardson's most consistently supportive metropolitan daily, the *Quad-City Times,* recounted his previous failed petition drives but quoted him as believing 1996 would be different because that year the local effort would enjoy the validation of drives in other counties.[25]

Solicitation and Litigation

The signature solicitation season ran until 19 August. Then the litigation ordeal began. Campaigns occurred in five Illinois counties that year. STOP had a hand in three: DuPage, McHenry, and Rock Island. Other abolition-minded Illinoisans were busy in Madison and Saline.

DuPage County

Only after Judge Byrne ruled a portion of the Hughes-Klemm law (H. B. 438) unconstitutional did the DuPage County arm of STOP set about the heroic task of gathering between 38,000 and 40,000 voter signatures. According to Rosie Fitzpatrick, they got a late start because they were waiting to learn if it was in their interest to go countywide or to target individual townships. Within a week of the Byrne decision, however, STOP spokespersons like Gail Dunham were telling reporters like Alex Rodriguez of the *Sun-Times* that she and her comrades expected to do better than the three-to-one drubbing CTIG had suffered in McHenry two years earlier. Taxes, high taxes, were to be the engine that drove the DuPage petition machine that summer as taxpayers sought, STOP hoped, to stop extravagances such as the new $900,000 town hall Naperville Township was then constructing.[26]

Township officials and other supporters rallied in defense of grassroots government by founding DuPage Citizens for Township Government and issuing a flyer that provided facts, figures, and a quotation from Illinois Congressman Henry Hyde: "Township government is a hands-on government you can rely on. It's responsive, affordable and easily accessible. No bureaucracy. No red tape. In these days of downsizing government it's a good example of how government should work." "A group of very concerned citizens" (presumably) from Bloomingdale Township cited, in its light blue campaign brochure, the dozens of services that its township provided, from line dancing lessons at the senior center to foster care for youth to mosquito controls, and warned how the demise of townships would eliminate the county board and its chairman. The group asked if voters could "imagine the chaos" dissolution would cause.[27]

August arrived. Dunham, in a guest essay for the *Elmhurst Press,* reported STOP operatives in DuPage were working hard to meet the deadline for signatures and requested "just a little" help in getting the issue of dissolution "to the voters in November." Help failed to arrive, however. "We needed an army of petition circulators," Fitzpatrick told me by telephone. "We hardly had a platoon." On 19 August she announced that some two hundred volunteers had been soliciting, but STOP had not bothered to gather all petitions in one place for a signature count because leaders knew they stood short of the 37,900 minimum and far short of the cushion needed to combat challenges "from politicians, using lawyers often paid for with taxpayer money." She told Mike Burke of the *Daily Herald* that STOP hoped to mount a more successful drive "either next year or in 1998." Bryan Smith, in one of his first official commentaries on abolition as the new TOI leader, told Burke such "antitownship groups organize from time to time, but are generally unsuccessful because township services are needed and run efficiently."[28]

"Overwhelmed" was the word Fitzpatrick used to describe for me her feelings and that of other STOP leaders at the end of the summer in DuPage. "Disappointed" is how Bob Anderson described to me his feelings about Fitzpatrick, Dunham, Hamm, and other leaders there. "They put zero effort into gathering signatures" and wasted the victory STOP won in Judge Byrne's court. "DuPage would have been the perfect county," he said, "because no rural areas remained." Anderson expected a lot more than STOP got from Fitzpatrick, a veteran campaigner for political office.[29]

Rosie Fitzpatrick erred in her press release when she reported that STOP would not be filing any of its petitions. Actually the election commission did receive one STOP petition, containing 15 names, on the deadline Monday. The submitter was Wayne Johnson of West Chicago, a man about whom STOP principals knew nothing. Johnson told me that he only filed in order to test in court the questionable practices of the election commission. When, in his opinion, the commission, convening as "DuPage County Officers' Electoral Board," failed to live up to the laws controlling it, he sued, in the circuit court for the Eighteenth Judicial Circuit, asking that the issue of dissolution go on the 5 November ballot. Theresa Amato, in an effort to force such bodies to follow the law, helped Johnson argue that the board failed to notify him of its initial (27 August) hearing of his case. More pertinent to this chapter, she wanted to go on record regarding her beliefs about the referendum process facing abolition-minded citizens.[30] In a reply, filed 28 October 1996, asking the circuit court to reverse a ruling of the board unfavorable to Johnson's cause, Amato wrote these heartfelt words—more for the general assembly, I believe, than for the court in Wheaton.

> Placing the petition [referendum] on the ballot may appear to enrich Mr. Johnson's cause unjustly, but any other result is equally or more unjust and laden with the pernicious ramifications set forth above and in Plaintiff's [Johnson's] Memorandum. Furthermore, the entire electoral process is rife with injustices where valid petitions are knocked off or kept on the ballot for picayune deficiencies in the mandatory requirements in either the petitions or the objections, and one party or another—petitioner or objector—is unjustly enriched thereby. A convoluted, ill-defined, adversarial code allows for such results. Now, when the same morass threatens to keep an insufficient petition on the ballot for a serious deficiency—lack of notice and due process—a procedural flaw that is arguably much more fundamental than pagination or binding which routinely keeps good people and questions off the ballot—and when these deficiencies are created not by the unaware citizen filer, but by the sophisticated entity in charge of administering this system, the system is ridiculed. Agreed. Nonetheless, what is sauce for the goose is sauce for the gander, and it is high time that the General Assembly fixes this mess. Otherwise, is it not a bit disingenuous to contend that for access to the ballot

> average citizens must follow a byzantine code with religious adherence on their first try but that the sophisticated government boards in charge of this electoral gomorrah are entitled to multiple bites at the apple to cure their failures, with impunity, at the price of a substantial delay and disadvantage for whomever happens to be the aggrieved party?[31]

Despite Amato's impassioned appeal and Johnson's 15 signatures, of which 6 were apparently invalid, the referendum did not reach the voters of DuPage that fall. Amato continued to litigate for and otherwise support the rights of local citizens to serve as watchdogs over their governmental officials until she left in 2001 to become president of Citizen Works, a national nonprofit organization in Washington, D.C.[32]

McHenry County

State law prohibits for four years the resubmission of a countywide abolition referendum, so Bob Anderson and the McHenry County wing of STOP chose to take advantage of section 5 of article VII of the Illinois Constitution allowing dissolution of a single township—McHenry Township. Ignoring protownship rhetoric from the editors at the *Northwest Herald,* calling him "a two-time loser" and suggesting he and his supporters "get a new hobby," the barber from Wonder Lake set out, largely on his own and mainly door-to-door, to gather over two thousand signatures.[33]

Bob and Judy Anderson held a news conference at the township warming shelter, north of the City of McHenry, just before beginning their 1996 campaign. He admitted that some accused him of beating a dead horse, but he felt he had several others willing to share the burden of getting names on petitions. Anderson then proceeded to work the streets and roads of his township for 84 straight days, averaging 3 hours per day and covering (he guessed) 200 miles on foot. STOP needed 2,100 signatures; he collected over 2,600 of the 2,743 they submitted on the 19 August deadline.[34]

Within days of the nineteenth, the McHenry Township Electoral Board rejected STOP's petitions. Supervisor Albert Adams, Clerk Tom Birmingham, and Trustee Bruce Novak looked at 27 of the more than 2,700 signatures and projected 800 invalid signatures overall. They listened to John Heidler, a citizen who had assembled a team of 45 volunteers and had spent more than a day checking the signatures against the voter registration records. Heidler said he and his team were questioning almost 1,100 names. Losing either 800 or 1,100 meant Anderson's total valid signatures would fall below the magic number of 2,098. So the board denied him ballot access, citing invalid signatures and failure to follow procedures to the letter of the law. By that time, Anderson and attorney Michael Walkup had walked out of the hearing room, complaining about "'an obvious conflict of interest in having the township board determine whether or not a question should be on the ballot that would dissolve the township.'" Walkup told the *Tribune* if the three township officials failed to turn the final responsibility for a decision over to the county election board, Anderson would sue them.[35] The board refused; he sued, primarily arguing that elected township officials had no business serving as an electoral board on a challenge in which the officials had a vested interest.

Robert G. Anderson v. McHenry Township ended up establishing legal precedent in Illinois, but it took nearly two years to complete all the proceedings. On 5 September 1996, Anderson's attorneys, Michael Walkup and Joel Hurewitz, filed suit with the Circuit Court of the Nineteenth Judicial Circuit asking for a review of the decision to deny ballot access and for a writ of mandamus to provide a spot on the 5 November ballot. When the court received the election case, 8 of 10 judges on the Nineteenth Circuit recused themselves. The ninth was on vacation; and the tenth, Judge James Franz, did not suit one of the defendants, so the case went to Judge Jack Hoogasian of the Lake County Circuit. Judge Franz, by the way, was the man who, in 1994, had ordered the CTIG referendum to the countywide ballot.[36] In late October 1996, with only a couple days remaining before the county clerk had, by law, to publish the official notice of the upcoming election, Judge Hoogasian "affirmed the Electoral Board's decision and denied the writ of *mandamus.*" The judge found no "violation of due process" when the three elected officials ruled on a situation that affected them personally: "While certainly ripe for legislative intervention, Illinois statutes do not require a township electoral board to recuse itself on the issue of dissolution of the township where the board consists of township officers." He further decided that the sampling of signatures, to which Anderson objected, "was not arbitrary." Joel Hurewitz later commented to me that he guessed Judge Hoogasian had not "ruled on the issue of the permissibility of the random sampling; rather only that it supported the Board's decision." Hurewitz agreed that it supported the decision; but he did not "think they had any authority to conduct a random sample, and if they did have the authority, their procedure was not random."[37] Anderson appealed the Hoogasian decision to the Appellate Court of Illinois, Second District, and continued, as he had for several weeks, going from house to house distributing the some three thousand flyers he had printed to help make the dissolution case. The appellate court denied the Anderson request of an accelerated docket—meaning the issue of abolition would not be in voters' hands on election day—but agreed to rule later on the merits of his suit.[38]

In June 1997, Judge Lawrence Inglis of Gurnee, writing for two concurring justices, reversed the Hoogasian ruling and ordered the appointment of a different electoral board "composed of impartial members" to hear challenges to the signatures Anderson and STOP/CTIG had collected in 1996 because "petitioner was deprived of his right to a fair and impartial tribunal to hear the objections to this petition." Henceforth, attorneys could cite *Anderson v. McHenry Township* as a precedent for asking members of an electoral board to recuse themselves when an issue affecting them personally came before them for action. Anderson told the *Tribune*'s Charles Mount that he was "'very surprised'" to have won. "'I would have bet,'" he said, "'it would have gone against us.'" Cyndi Wyss of the *Northwest Herald* had to admit admiration for the tenacity of the Wonder Lake barber. "Mom and Dad always taught me to fight for what I believed was right," she wrote in an opinion piece. "Apparently, Anderson had a similar upbringing."[39]

Six months later, John Bolger, Case Ellis, and James Hecht began a series of hearings as the circuit court-appointed replacement electoral board. The three Woodstock attorneys ordered the county clerk to perform a detailed check of the 1,087 challenged names on the 1996 Anderson petitions and then voted to certify the Anderson petitions for the 17 March 1997 ballot, because County Clerk Schultz had found only 588 invalid signatures, leaving Anderson a cushion of 57 signatures. Attorney Hurewitz contended Schultz had excluded too many names while challenger Heidler's attorney, Patrick Coen, still questioned the validity of another hundred names. A few days later, the electoral board met and agreed to reduce the Anderson cushion to only 39 signatures, because of Heidler objections. Anderson told the press afterward that he would not, even if eventually victorious in the hearings, have enough time to campaign effectively prior to 17 March; and he preferred the upcoming November election. Finally, in February 1998, the electoral board ruled Anderson had fallen four signatures short of the 2,098 he needed and ruled he had no right to the ballot in either March or November. Anderson said he would immediately begin planning another petition drive in McHenry Township.[40]]

Rock Island County

STOP operatives estimated they needed 9,100 signatures in 1996 to place the countywide dissolution of townships on the November ballot in Rock Island County. They brought in 9,913. Among the signatures were several from local elected officials, including one from a sitting South Moline Township trustee, Bill Fisher, who signed not because he necessarily favored dissolution but because he thought the people deserved a chance to have their say. On the other hand, he told listeners at a press conference on the courthouse lawn in Rock Island that fellow township officials did not like his stance on the issue. STOP's 12 July 1996 press release carried Fisher's name along with those of the sheriff, a circuit court judge, and various active and former city council members from the Quad Cities who had signed STOP's petitions.[41]

Michael Richardson needed a lot of help collecting the nine to ten thousand signatures to which STOP aspired, and went out of his way to find kindred petition-passing souls, in part by dressing and behaving conspicuously. For instance, at a meeting of Ross Perot's United We Stand group in Moline one night in June, wearing a tie with an American flag on it, Richardson got into an "old-fashioned...name-calling" fracas with the assessor of South

Moline Township, Pat Wendt, a man Richardson called a triple dipper and zebra that had still not "'changed its stripes.'" Unhappy taxpayers, Richardson told me, were the people he wanted to reach, in the hope they would assist STOP with its petition project. On other occasions, Richardson, with his beard and craggy countenance, donned a top hat and dark suit in order to remind potential petition signers/circulators of the nation's sixteenth president (figure 23). Dressed as Honest Abe, Richardson showed up in July at a play in Rock Island's Lincoln Park with his clipboard and began moving among the seated audience before that evening's performance of *The Persians.* Don Wooten, a local celebrity and director of plays in the park, put his hand on Richardson's shoulder and asked him to leave peacefully, a sequence that John Kanthak's sources reported, was a "sight worth seeing." Into the next play, *The Birds,* Wooten introduced a character dressed like Lincoln who pestered other actors with a pen and clipboard. Eventually, those who were the objects of the actor's solicitations netted him and hauled him, in Wooten's words, "offstage to great applause."[42]

Figure 23 Michael D. Richardson in Rock Island, October 1996. Richardson sometimes dressed in the fashion of Abraham Lincoln to attract attention during petition drives.

There existed, of course, an easier way to get countywide abolition of townships on the ballot than sweating out the summer in a top hat and suit or confronting an irritated director. All STOP needed to do was convince the county board to declare a referendum for the fall election. Perhaps taking encouragement from an editorial in the *Argus* urging county leaders to let the people decide, Randy Popp, in mid-July, asked the board to do the right thing and order the issue to the ballot but got no response. Richardson, just prior to the petition deadline, tried talking the county board's Legislative Committee into asking the full board to put the issue on the ballot. By then, STOP estimated it had eight thousand signatures. Committee members briefly debated with Richardson, expressing their preference for maintaining townships and refusing to act on his request. A couple of days later, the *Argus* said, "County residents deserve the opportunity to let their feelings be known. They are, after all, the ones paying the bills to keep Rock Island County's townships going. They should also be the ones to decide if they want them eliminated." In lieu of a unilateral move by the board, the *Argus* urged readers, if they had not already signed a STOP petition, to "please do so."[43]

Over a thousand more did sign by the deadline. On the Saturday before that Monday, Richardson told the *Argus* that STOP had 9,400 names on the line and expected to file just over ten thousand, which would have represented "a roughly 10 percent buffer against petition challenges." Experience, he said, had helped. This time they tried hard to interrogate signers "'to be sure they're qualified.'" STOP also had deputy registrars in tow "'getting people registered to vote right on the spot.'" Popp told the *Quad-City Times* on Sunday night they were at 9,800 and planned to "continue collecting signatures until the deadline." The 9,913 they actually submitted gave them only a 9 percent cushion; and the next night Richardson stood before the full county board requesting, in light of STOP's delivery of the goods, automatic, board-decreed ballot placement, which he noted would save county employees countless hours in the face of the challenges sure to come. Again, STOP got no response.[44]

A week later, several township officials and others challenged 2,263 of the 9,913 signatures. Thus began another lengthy litigation situation. Richardson testified before the county electoral board about the fact that the county clerk was not carefully purging voter lists, even of the deceased, which meant STOP had to produce 10 percent of an inflated number. According to Richardson, "'When Gabriel blows his trumpet on Judgment Day, the dead of Rock Island County will rise up and go to the polls.'" County Clerk Leibovitz, presiding officer of the electoral board, blamed the retention of county dead, as part of the registered-voter total, on the "motor voter law." The Illinois State Board of Elections had told county clerks to leave the dead on their lists pending full resolution of the federally inspired measure to encourage voter registration. The electoral board did reduce the STOP target to 9,023 valid signatures, prompting an objection from township supporters.

STOP lost the next battle, when the electoral board ruled 1,445 signatures invalid, dropping the valid total below even the new target of 9,023. As he had been saying publicly and privately since 1991, Richardson believed that the TOI-inspired law requiring abolitionists to gather signatures totaling at least 10 percent of the registered voters was unconstitutional.[45]

On 17 September 1996, Richardson appealed, requesting the circuit court vacate the board's decision and order the township question to the November ballot. Among his points in the four-page petition was the fact that the Township Code forced abolition-minded petition passers to acquire the signatures of 10 percent of a "vague and uncertain" number of registered voters. "'All we want,'" he told the *Quad-City Times,* "'is a real number to meet, not a false number inflated with three years of deaths and moves'" by voters. Judge Martin E. Conway, Jr., sat through a half day of testimony in the first hearing before his court and noted that Richardson had failed to raise properly the "constitutionality issue" in his original court petition; but Conway, perhaps making a concession for an outsider to the legal profession "who represented himself in court," allowed him "to amend his petition." On the first of October, Judge Conway ruled STOP 552 signatures shy and its referendum not eligible for certification as part of the upcoming election. He also ruled unconstitutional that part of section 25-5 of the Township Code that read "as determined on the date the petition is filed" because it was, as Charles Keeshan of the *Moline Dispatch* wrote, "not fair to expect petition circulators to gather signatures without knowing how many names they need[ed] to place the issue on the ballot." Conway "likened it," reported Keeshan, "to asking someone to kick a field goal but not allowing them to see the goal posts."[46]

Legal maneuvering occupied the rest of October. Richardson filed an appeal with the Third Appellate Court, asking that a lower signature requirement prevail, one based on the pre-1991 law and calling for 10 percent of those who had voted at the previous presidential election. According to Richardson's calculations, STOP would therefore need only 7,375 signatures; and they still had 7,429 valid ones, according to the electoral board. Following additional oral arguments, Judge Conway reversed himself, saying STOP had presented more valid signatures than necessary and ordering the Rock Island County Board certify the referendum question. When County Clerk Leibovitz wondered how the county was to give three thousand absentee voters, whose ballots he had already issued, the chance to vote on the referendum, Richardson said Leibovitz should have prepared for this eventuality. Learning that the county board might avoid placing the referendum on the ballot because the judge had written "should...certify" instead of (as the statute read) "shall certify," Richardson gave notice in letters to all county board members that he intended to sue each of them if they failed to certify the referendum. Petition challengers, through attorney Thomas Kilbride, appealed the Conway ruling to the appellate court, asking for reversal of the ballot order and affirmation of the electoral board's original denial of ballot access. The county board did refuse to certify the referendum, and several members admitted that Richardson's threatening letter had insulted them. Richardson apologized for the harsh wording but intended to file a federal lawsuit naming the 15 board members who voted against certification. The Third Appellate Court stayed the Conway order to place the referendum on the ballot and transferred the constitutionality issue to the Illinois Supreme Court. Justice James Heiple reaffirmed a stay that the appellate court had that day, 30 October, issued. Richardson admitted defeat but vowed to see resolution of the constitutional issue he had raised and to pursue a ballot spot in 1998, based on the 1996 signatures collected.[47]

Michael D. Richardson v. Rock Island County Officers' Electoral Board et. al. came before the Illinois Supreme Court in the fall of 1997. Attorney Richard A. Cowen of Chicago rose first, at the 18 September session, and used his 20 minutes to ask the court to reverse Judge Conway's ruling and affirm the original electoral board denial of ballot suitability. Paralegal Michael D. Richardson rose to speak as the appellee; but Chief Justice Charles Freeman asked, "Are you an attorney?" "No," said Richardson. Freeman thanked him, assured him the justices would read his brief, and informed him that the Illinois Supreme Court does not listen during oral arguments to anyone but attorneys. When Richardson requested a clarification, the justices rose and exited stage rear. The supreme court ruled against Richardson two months later. Writing for the court, Justice Heiple said, "mootness" then prevailed. The election of 1996 had come and gone, without the referendum on the ballot; and the general assembly had addressed, in Public Act 90-112 (see below) the presumably unconstitutional aspect of the Township Code, to which Judge Conway had negatively reacted. Having nothing left upon which to offer relief, the court said, "This case is moot." Richardson immediately declared victory, telling the press that STOP would have its referendum on the ballot after all. He filed the following April a complaint for mandamus with the circuit court asking for a 3 November 1998 ballot slot. Judge Conway denied Richardson's complaint, in part because Richardson had failed to challenge the legality of state election codes.[48]

Richardson's days as an antitownship activist, for all intents and purposes, ended on 18 September 1997, as we watched the justices exit the hearing room. The fact that he was not an attorney, never a significant liability in so many previous hearing rooms, had finally caught up with the talented paralegal. Even Judge Conway implied that Richardson's lack of attorney status was a consideration in his last ruling on the case. Tired, financially strapped, and beaten at almost every turn of the judicial process, Richardson went through the motions for the next few months, filing a request for investigation of Justice Freeman with the Illinois State Judicial Inquiry Board and threatening a federal lawsuit. STOP made a feeble attempt to bring about the demise of individual townships in 1998, but those who had heard him before and read his fiery rhetoric in biennia past knew the township side had won in Rock Island County. In October of 2000, as he was preparing to leave Illinois for a new life in New England, he still could not "understand why STOP did not make the ballot in 1996." "I just lost heart," he said.[49]

Madison County

Michael Richardson kept a running inventory of where antitownship turmoil might erupt. Madison County, across the Mississippi from St. Louis, often came up as possible hot spot; and Mel Jones of Collinsville, a veteran taxpayer advocate, was the man Richardson expected to lead there the charge to the ballot box against township government. He had not expected David Partney of Granite City to pick up the activist clipboard.

Businessman and son of a former mayor, Partney was running himself in 1996 for the mayoralty. Partly to gain publicity for that campaign but perhaps because of genuine concern over waste and nepotism in Granite City Township (coterminous with the municipality), Partney instigated a petition drive calling for the elimination of that one township. He cited for me figures purporting to show four times as much tax money there went to pay the township to administer welfare as went to the general assistance recipients in 1995. Marilyn Vise of the *Belleville News-Democrat* focused one of her articles on the fact that Granite City Township was paying the mother of Assessor Darlene Laub to clean the township hall and had hired her sister-in-law as a computer programmer. Township Supervisor Bernadine Hagnauer put her daughter-in-law on the township payroll as a bookkeeper. A "long-standing tradition" was the way Vise characterized nepotism at the township's headquarters.[50]

He got a late start, but Partney estimated he needed only 1,850 signatures. By the second week of August 1996, facing the same petition deadline as the STOP initiatives (about which he had never heard), he felt he was doing "'pretty good'" at gathering the signatures himself. Then he fell ill and ended up in the hospital. To obtain the last few hundred names on his petitions, he hired two temporary helpers; and on Monday, 19 August, he turned in approximately nineteen hundred signatures.[51]

Township officials naturally prepared to challenge him. Laub said they would be inspecting the 79 pages carefully. Partney, much like his northern counterparts, at the close of the petitioning period, cautioned township officials not to stand between the people and the polls on this issue. He also planned to consult his attorney about whether township officials should be able, as the law provided and as occurred in McHenry Township, to sit in judgment of his hard-won signatures.[52]

The day after he deposited his petitions with the town clerk, he picked up an endorsement from the *Belleville News-Democrat.* Discounting his potentially self-serving motive (making an issue of Granite City Township while running for mayor), the editors at Belleville (in St. Clair County) said he was giving voters a chance they wished "more area residents could have." Editors went on to chastise Laub and Hagnauer for their familial hiring practices and to point out that, in the editorial opinion of the *News-Democrat,* the bulk of the people would not "even notice that the township was gone—except to see their property tax bill go down."[53]

Assessor Laub's chief deputy, Deborah K. Gray, filed the objection to Partney's petitions, having found, in the legal fine print, his fatal flaw: the two young ladies he had hired to complete the signature drive. Misty Hammond and Kristin Willis did not register to vote in Madison County until they had almost finished their temporary job for Partney. They had, they testified before the township officials composing the electoral board, tried to register on Tuesday, 13 August; but employees of Town Clerk Judy Whitaker, a member of the electoral board, had told them they could not register for three days, because the office had no registration forms. While waiting to register, they continued collecting signatures, all of which Whitaker, Hagnauer, and senior Trustee Casmer Skubish threw out, along with 115 that had bad addresses or other problems. Those 115 alone would have been enough to bring Partney's signature total below the 1,793 the board said he needed. In the end, the Hammond-Willis delay cost Partney 471 signatures.[54]

"Once again," said the Belleville editors, "a committee of politicians had the authority to reject a ballot measure that would directly affect them." The editorial went on to recount for readers how in 1995 candidates for the Belleville City Council had their petitions go uncertified by the very council members the outsiders were trying to unseat. Unlike the editors at the *Northwest Herald,* who a few days earlier had praised the decision of a township electoral board ruling on it own future, those at the *News-Democrat* warned against leaving the foxes "in charge" and wondered if the legislature might want "to address this loophole."[55]

That Belleville editorial ran on Tuesday before the Granite City Township Electoral Board revealed its sealed findings. On Saturday next, the *News-Democrat* editors scolded the electoral board again, wondering why it should matter who circulates a petition and noting that was just the sort of technicality "professional politicians use all the time to block the people from having a bigger voice in government." Rewrite the law said the editors "with the goal of getting issues on the ballot, rather than providing excuses to keep them off."[56]

Savvy in the ways of local politics, Partney had attorney Morgan Scroggins of Edwardsville on call from the outset of the abolition action. His attorney, Partney told me, had interceded during the summer when county officials tried to convince Partney that he needed signatures from 10 percent of the voters in Madison County—in order to call a referendum in one township. He needed Scroggins again when he sued the electoral board on13 September 1996 in Madison County Circuit Court (Third Judicial Circuit) claiming bias on the part of the three hearing officials and challenging the rule requiring at least 10 percent of the township's voters sign the petitions.[57]

A St. Clair County associate circuit judge, Richard Aguirre, heard the case on 3 October after, in his words, "The Madison County judges wanted no part of it." Scroggins pointed out that Hagnauer, Whitaker, and Skubish had financial stakes in the electoral board's verdict and should not have been part of the hearing. Township attorney William Schooley said, "anyone serving on the electoral board would have reached the same conclusion as to the certifiability of the of the ballot request.[58] The next day Judge Aguirre ruled against Partney. Facts demonstrated to the judge that the activist had too few valid signatures, and neither the parties to the suit nor the judge could find "any precedent upholding a challenge [of] an electoral board's composition on judicial review." Partney, said the judge, had not demonstrated that the statutory requirement of 10 percent was "an unreasonable limitation" of the electorate's access to the ballot via the petition route.[59]

David Partney never sought assistance from Richardson or Anderson. Certainly they would have been able to save him the trouble of his second futile signature solicitation, in the winter of 1996–1997.[60] Earlier, however, when it really mattered and when he stood a good chance of performing the first hostile takedown of an individual Illinois township since the 1970 Constitution made it possible, his isolation from the rest of the abolition movement hindered him. Why do I feel he missed an excellent opportunity? First, Granite City (the municipality) had helped fund a 1993 study in which the investigators found Granite City Township "costly, redundant and unnecessary." Melville Public Advisors recommended at that time what Partney was doing in 1996: elimination of the coterminous township.[61] Granite City Township was thoroughly urbanized and had little or no rural tradition upon which protownship forces could play. Second, Partney was a seasoned and fairly savvy politician, and he apparently had the personal financial resources to devote himself to a civic cause like abolition of his township or running for mayor. He had a respected daily newspaper carefully covering his movements and editorializing on behalf of his plan to try and save taxpayers the cost of running a township that was coextensive with a municipality. Third, tax issues had received a lot of play in the 1990s, and taxpayers were seeking ways to control outlays. Finally, Fortress Astoria had its sights trained on the Chicago area, Rock Island County, and Saline County. Partney's little operation attracted no obvious interference from TOI.

Saline County

One other non-STOP abolition effort popped up in 1996—in Saline County—about as far as one can get in Illinois from Rock Island (270 crow-flight miles) and Wonder Lake (310 miles). Michael Richardson learned about Saline, after the fact, from Carl Edwards of neighboring Gallatin County, and told me in a letter, dated 12 February 1997. Bryan Smith, however, had gotten wind of the Saline challenge early on, in time to dispatch a veteran road warrior the 190 miles from Astoria (or perhaps the 275 miles from Leyden Township in Cook County). Yes, Smith's choice for rallying the troops in Saline County, only a few days into a well-deserved retirement, was George Miller, the same man who had headed southward into Gallatin County in 1988 to help sidetrack Edwards's plan to terminate townships there.

Roads were the big issue in Saline County, just as they had been 20 years earlier with the Saline County Taxpayers Association (chapter 6). "SAVE TAX $ Consolidate Township Roads" was the bumper-sticker slogan that Bill Bethel and Ben Brinkley adopted in 1996 to call voters' attention to the efforts of their County Commission Committee (CCC). They began, as had Anderson and Richardson, in May; and the positive reaction surprised them. Voters complained of the county's lack of a centralized road "'system to prioritize the roads and bring them all up to a certain level,'" Brinkley told the *Harrisburg Register.* He reminded readers that Williamson County (next door to the west) had turned to commissioners in the Depression and that people in several other nearby counties remained convinced of commission superiority over government by a larger county board with an extra layer (the townships). Formerly a schoolteacher and union negotiator, Brinkley (a Harrisburg resident) denied that anyone among the CCC faithful had promised to blacktop every road in the county within five years. Bethel, who lived south of Harrisburg in rural Independence Township, said he used the argument that few tourists, in their fancy cars and sport utility vehicles, were going to venture onto Saline's numerous gravel-surface roads, to visit its considerable scenic attractions, until locals invested in hard-surface improvements.[62]

The Miller mission culminated on Monday, 12 August, one week before the petition deadline. He spoke that night to an audience of township officials, county board members, other county officeholders, and ordinary citizens at the Galatia Village Hall, northwest of Harrisburg, and donated the first fifty dollars to prime the pump for a round of advertising. "It is safe to call Miller biased on the issue," said Scott L. Hines of the *Register;* but Miller was not present just to spout vague praise of the status quo. He or someone else had put together figures, which he shared, about costs of township roads versus county roads and about other economic matters. Meeting organizers circulated a photocopy of one of the CCC's petitions, a petition that they were eager to challenge—because it appeared to contain several signatures in the same handwriting, the circulator had left it at a business and had not been present to witness all signatures, and someone knew one of the signatories actually resided in Hamilton County. Brinkley and Bethel would have benefited from the how-to workshop that STOP had held in Elgin earlier in 1996.[63]

The CCC turned in petitions, containing between 1,800 and 1,900 names. Bethel thought they had a sufficient cushion beyond the minimum of

1,612 that someone had calculated by taking "10 percent of the register[ed] voters from the last general election." What someone should have calculated was one-tenth of those registered to vote that August Monday. No matter what the actual target, however, Bethel and Brinkley were in for a fight.[64]

This fight ended quickly. By the time it had reviewed only half the signatures, the electoral board had lopped two hundred names from the 1,867 submitted. Eventually, the board eliminated 299, leaving the CCC 44 signatures short of the 1,612. Township officials in Saline County had not even bothered to find a token objector or two, as their brethren often had done elsewhere in the 1990s. Each objector, save one, was a township official. The exception was a county board member. All, of course, stood to lose their jobs if the CCC had had its way in the challenge round, the courts, and the general election. Speaking for the objectors, after the electoral board ruled, Galatia Township Supervisor Leroy Bird said, objectors "'merely wanted the election laws upheld.'" Brinkley said he had a hard time believing Bird, that the letter of the law rather than the spirit had prevailed, and "'No petition could pass under such scrutiny.'"[65]

Bethel, Brinkley, and the CCC did not pursue ballot certification through litigation that fall. The errors they had made, as novices, in collecting petition signatures, left them far short of the magic number. They never hired an attorney, nor did they again pursue another countywide referendum, as Brinkley said they would the Tuesday the electoral board ruled. When I spoke with Bill Bethel, in the summer of 2002, he told me that they still hoped to relieve Saline taxpayers of their township burden by first electing enough Republicans to the county board and then getting the board to call a vote on townships.

The Smith Agenda

The era of Bryan Smith at TOI was off to a great start. When he took office as executive director on 1 August 1996, three campaigns to rid entire counties of their townships were underway, as were two to expunge single townships in two other counties. Within just over a month of his ascension to the directorship, challenges from township officials and allied individuals had succeeded in keeping all five from reaching the voters. By election day that year, township attorneys had successfully countered appeals for ballot certification in the four counties where litigation followed signature solicitation. Ultimately, in the two cases that dragged on through 1997 and into 1998, the Smith side prevailed as well. So the winning streak against abolitionists that predecessor Troy Kost had begun in the 1940s and successor George Miller had continued, Bryan Smith extended with a five-for-five rout of STOP, David Partney, and the CCC. Not satisfied, however, with being an instant winner, Smith set out to overhaul TOI in several ways and to prevail upon

TOI's Springfield allies to help make the abolition attempts of people like Anderson, Bethel, Partney, and Richardson even less likely to occur.

Fortress Springfield

In its "Recruitment Profile" for Miller's replacement, the PAR Group listed 25 "*ideal* attributes" of the new executive director, one of which was that he or she would need to "be committed to residing in the Springfield metropolitan area." TOI officials did not expect their new leader to commute from Springfield to Astoria but rather to reside in the general vicinity of a relocated headquarters in the state capital. In order to make that move, TOI needed help from Jerry Kats.

At the end of chapter 6, I profiled briefly three men who had by the late 1980s begun to make a difference in the realm of Illinois townships: Michael Richardson, Bryan Smith, and Jerry Kats. In June of 1986 Kats became the guiding force behind what has since become TOI's successful and profitable Risk Management Association (TOIRMA). Kats takes a lot of credit for the fact that the view out the front door of TOI headquarters now includes the capitol building, the dominant planet in the TOI solar system, instead of several Astoria storefronts.

TOIRMA provides townships with insurance alternatives to coverage they could buy on the commercial market. By the end of TOIRMA's first 12 months and Kats's first 12 as program director (later executive director), he had brought 181 townships into the fold. At the beginning of 1991, the number stood at 896; and, when the TOIRMA Operating Committee decided to declare the first member dividend later that year, the membership figure easily exceeded 900. Now more than 80 percent of Illinois townships belong, plus a few road districts from commission counties. Keeping costs low, with tactics like outsourcing some of member services, spending little on marketing, and having a favorable claim history (in part because townships have no fire or police forces, unlike their municipal cousins), TOIRMA accumulated a significant reserve fund. "The problem," Kats revealed, "was in getting rid of excess money," which by the fall of 2000 stood at forty million dollars. Annual dividends to townships and other member bodies was one way to build good will and to reduce the reserve nest egg. Another way was to invest in TOI, the mother ship. Investment took at least three forms. First, TOIRMA would pay TOI an annual "sponsorship" fee of several hundred thousand dollars ($400,000 in 2000). Second, TOIRMA was largely responsible for funding the establishment of TOI's new education arm (see chapter 9). Third, TOIRMA helped TOI establish Fortress Springfield.[66]

The move to Springfield did not occur immediately. Kats said he first talked George Miller into purchasing, largely with TOIRMA surplus funds, a building on South Fifth Street in Springfield, just east of the capitol. Then TOIRMA bought that building from TOI and paid for extensive renovations.

Finally, TOIRMA sold TOI an improved structure at a loss and arranged for a generous, interest-free repayment schedule. TOI is also able to earn rent by leasing out the other offices in its handsome building. TOI's legislative director worked solo out of the 408 South Fifth site for several years. Smith began listing, in *ICTO,* the Springfield address below the 118 West Broadway address in Astoria. By March of 1997, the South Fifth fortress was above West Broadway on the *ICTO* masthead page. Ribbon cutting occurred at the new headquarters 22 June 1997, and a legislative reception on 29 October that year helped "to reaffirm the partnership between state and local government." A notice in the January 1998 *ICTO* directed "all calls and correspondence to our Springfield office." Thus ended an Astoria affair that had begun during World War II when Troy Kost stepped up to become the head man of TOI. Bryan Smith could now rise at 5:30 a.m. rather than two hours earlier for his commute from Decatur. Legislators were now only a few blocks away from the executive director's large and nicely appointed office (figure 24).[67]

Figure 24 Bryan E. Smith, executive director of the Township Officials of Illinois, in front of the organization's Springfield headquarters.

Magazine Facelift

The youthful executive director had plans too for the magazine, as he settled in and got ready to occupy a new command post. First to change, in the February 1997 issue, was the masthead page, which would henceforth occupy a more prominent position, provide more information, and feature a thumbnail of the cover story. A new title logo appeared above a modified table of contents. For the April 1997 issue, a major overhaul of the front and back covers occurred. Slick paper replaced the cheaper stock that had protected *ICTO* for decades. The title logo from the title page began appearing on the front cover along with a four-color photograph of a different township official each month. The back cover no longer carried the photograph of and information about a missing child. Instead, TOIRMA has claimed the back of every subsequent issue for advertising. The most obvious change has been a new title, *Township Perspective,* since April 1999; but the careful reader will have noted many things that make the modern magazine a better advocate for the townships than was *ICTO*.

Silver Bullet

Every other year, beginning in 1988, township officials somewhere in Illinois had faced an abolition threat, culminating with the five in 1996. Raising the signature bar in 1989 to 10 percent (from 5 percent in most counties) of the vote at the previous presidential election and in 1991 to 10 percent of the registered voters had not provided the desired deterrent to determined township critics. Court action had, in 1996, rendered unconstitutional (in DuPage County) the supermajority aspect of Public Act 89-365. Consequently, huddling with attorneys and legislators in the fall of 1996, TOI leaders devised what they hoped would be the answer to their longstanding problem, in other words, a silver bullet. "With continuing anti-township efforts cropping up in different parts of the state, " Smith wrote in his October 1996 *ICTO* column, "it is important to have individuals who understand 'the township story' in the general assembly."[68]

Legislative Director Ed Stasiewicz introduced the proposed panacea to TOI membership in his December 1996 report. Noting the problems of the supermajority law, he said TOI's Legislative Committee believed they might have better luck "addressing signature requirements on petitions to abolish." They wanted to keep the existing rule in effect requiring 10 percent of the voters in the county to sign but also force abolitionists to secure signatures from 10 percent of the voters in each township, with no more than half coming from incorporated areas. If a township did not have a sufficient number of voters in its unincorporated areas to provide half of the 10 percent, the committee wanted at least half the total population of the unincorporated areas to sign petitions, with the balance coming from incorporated areas.[69]

Rep. Ann Hughes, house sponsor of the bill that became constitutionally questioned Public Act 89-365, also served as sponsor of H. B. 480 in 1997,

"An Act to amend the Township Code by changing Section 25-5." She introduced it on 4 February 1997, and voted in favor of passing it out of committee, on a 17-0 recommendation. All of the rural versus urban confusion that Stasiewicz had described was gone. In its place was the simple fact that, for a countywide referendum, an Anderson or a Richardson would now need at least 10 percent of the voters in each township to sign, no matter where the voters lived, just so they lived in that township. An amendment from the house floor corrected the problem that Richardson had argued before Judge Conway and would try to argue before the supreme court in 1997 regarding the difficulty of knowing how many voters would be on the rolls come petition deadline day. The amendment specified that "registered voters" would be the number of "registered voters...on the date registration closed before the regular election next preceding the last day on which the petition may be filed." This statutory change, by the way, was what Justice Heiple had noted in his narrative dated 20 November 1997. The bill passed the full house, 110-0-2 on 19 April 1997.[70]

Awaiting TOI's silver bullet in the senate was a newly appointed senator, a man who had been the supervisor of Jackson Township, in Will County, for a quarter of a century. Lawrence M. Walsh, a farmer from south of Joliet, served as senate co-sponsor with fellow Will County senator, Ed Petka of Plainfield. Walsh offered the motion on 7 May that led to the bill's second reading and voted along with 55 other senators to approve on 8 May. No senator voted no. June's *ICTO* included a short story about and photograph of Walsh, noting how H. B. 480 would give "added protection" to the township cause and make impossible the calling of a referendum with "all of the signatures coming from a single township." Governor Edgar signed Public Act 90-112 into law 14 July 1997 (effective 1 January 1998); and the September *ICTO* carried a photograph of the event, with 19 township officials and TOI leaders forming a backdrop.[71]

Reactions to the new law followed predictable patterns. Bob Anderson told the *Northwest Herald* that he just wanted the people across his county to debate and vote, a scenario the new law was going to make virtually impossible. Ersel Schuster said it was about time her small township got some protection from large townships. The supervisor of one of those large townships, Pat Floeter of Algonquin, agreed the law could save small townships that depended on road district snowplows. Editors at the *Northwest Herald* called the latest Hughes law "clear and common-sensical." Citing the fairness principle, editors said getting 10 percent of the voters in every township should not prove to be "an unreasonable hurdle." Theresa Amato and the Citizen Advocacy Center took a close look at Public Act 90-112 and decided not to challenge it in court because there existed no glaring fault—such as the supermajority provision of Public Act 89-365. STOP had now to do some serious contemplating of the evolving abolition landscape.[72]

STOP Survived

On the STOP wish list for 1997 were successful outcomes in the McHenry Township and Rock Island County lawsuits and a statute to clarify the thorny transition issue. I have already discussed the court cases, neither of which led to ballot spots for STOP, though one led to new case law regarding electoral boards and the other to a change in the way a county or township determines the basis for calculating the number of signatures necessary to call a referendum on township abolition. What about transition? TOI's legislative director had said once that TOI would "support and work towards legislation clarifying what happens to township property and the county form of government should township government be dissolved in a county."[73]

Perhaps 1997 would be the year when the general assembly finally did the job that framers of the 1970 Constitution had authorized. Taking further encouragement from the decision of the Rock Island County Board to study transition issues locally and an editorial in the *Moline Dispatch,* posing many questions to which a transitioning county would need answers, Michael Richardson, on 20 January, wrote to a pair of legislators: Joel Brunsvold and Ann Hughes. He reminded Brunsvold, his district's representative, of the abortive 1991 effort Brunsvold had made (H. B. 68) and asked him to work with Rep. Hughes on a new version. His letter to Hughes, took a less diplomatic tone, however, as he asked her to work with Brunsvold but could not resist chiding her for the "unconstitutional H. B. 438" and warning her that her "political allies in Township Officials of Illinois…[had] a history of opposing transition legislation," like H. B. 68. Richardson clearly had little hope of help from Hughes, but he did receive a reply from Brunsvold promising another try. Randy Popp met with another local representative, Mike Boland, about the transition push and reported Boland uncommitted but interested.[74]

By the middle of March 1997, Brunsvold was saying publicly that he was "considering" the revival of H. B. 68 in order to clarify at least the county board/county commission aspect of a transition away from the status quo. "'Township organizations,'" he told John Kanthak, "'have been arguing that the county board becomes a board of commissioners….They want people to think that, if they go down, everyone goes down with them.'" Brunsvold felt it highly unlikely that "anyone could convince a court a 123-year old law governing the now-extinct boards of supervisors would override the 1970 Constitution which created the current county boards." He planned to talk to the chairman of the Rock Island County Board, Paul Mulcahey, before resurrecting the legislation."[75] This interview sounds a lot like an effort by a lawmaker to test the political winds and to make the legislation's real advocates feel he had at least tried.

Rep. Brunsvold quickly found how the transition winds were blowing between Fortress Astoria and Springfield. He and Senator Denny Jacobs of East Moline met 7 April to discuss the topic with the Legislative Committee of the Rock Island County Board. Jacobs said most of his colleagues in the general assembly felt they should not even be discussing transition. Brunsvold blamed "influence from the Township Officials of Illinois" for keeping "bills dealing with post-township issues from seeing the light of day." TOI had succeeded in killing everything in committee, and Brunsvold doubted any legislative action would occur until abolition "'actually happen[ed] somewhere first.'" Paul Mulcahey said, in response, that action after the fact would be too late for the poor county that had to navigate the uncharted waters of transition on its own. Abolitionists and the media pushing the abolition agenda, according to Senator Jacobs, ought to stop and look at the consequences of such drastic action.[76]

Battered but not yet ready to concede defeat to a foe that had won almost every contest, STOP leaders sought any sign of hope for their cause in the coming election year, as 1997 drew to a close. Dennis Hamm, calling himself the Chief of Stuff, put together in October the first (and likely only) *S.T.O.P. Watch,* a newsletter that looked and read a lot like *TSG news.* Hamm's quotation from Mark Twain characterized the mood of those then still a part of this declining organization: "Success is the ability to go from failure to failure without losing your enthusiasm." STOP held a rally, on 19 October 1997, at the CAC offices in Elmhurst. Richardson predicted abolition would be on the ballot in a "dozen" Illinois counties the next fall. He was only off by 11 counties.

Celebration

TOI and Illinois's township officials prepared in 1998 to begin a multiyear, sesquicentennial celebration of the period 1848–1850, during which had occurred a key series of events, culminating with the activation of townships in 24 counties. Bryan Smith mentioned the coming jubilee in his April 1998 column and devoted his entire November 1998 column and the front cover that month to the milestones that township government was reaching. He asked officials to plan local events that would give residents a chance "to see first-hand what their township government is all about" in the months to come.[77]

Besides the sesquicentennial, township advocates had much more to celebrate in the last biennium of the millennium. Their grassroots government had weathered several petition drives, one countywide referendum, and substantial criticism from newspapers and activists. In place was a new law (Public Act 90-112) that rendered further countywide referenda much more difficult. Occupying attractive and well-placed Springfield offices was a dedicated crew of TOI officials, under the direction of an articulate and effective executive director. Generating surplus cash, some of which was available to TOI for appropriate uses, was TOIRMA with its original champion the helm. Serving in the general assembly were many men and women who either were currently or had been at one time township officials—men and women upon whom the TOI legislative team could count to be out front pushing the TOI agenda to clean up statutes or introduce relevant new legislation—most of which had nothing to do with thwarting abolition-minded folk. TOI even had a new, professionally prepared promotional video to help officials tell the story: *Your Township Government: They're The People Next Door,* which the associate director of TOI, Heather McAnarney, had helped produce.

Party Poopers

Apparently not in celebratory mood, however, were still a few diehards. And there was a newcomer to the antitownship scene, in Lake County.

Shields Township

Among the original townships, in one of the 24 original township counties, is Shields, on the Lake Michigan side of Lake County. Shields takes in roughly half of Lake Forest, half of North Chicago, the Great Lakes Naval Training Center, most of Knollwood, and all of Lake Bluff. Beginning in 1997, a determined resident of Lake Bluff, often on her own, challenged the elected officials of Shields Township, especially the supervisor, Charles B. Fitzgerald, IV.

Mother, grandmother, and educational consultant with a doctorate in education, Janice Schnobrich, who is also active in the League of Women Voters, began to take an increasing interest in Shield Township when, in 1997, she learned how much her township's officials were earning and how much the township was spending on rent for its office complex in Lake Forest. She started by writing letters to Fitzgerald and began (in May 1997) attending township board meetings to ask questions about township business. Schnobrich wrote also several letters to the editor of the *Lake Forest Lake Forester,* pointing out the shortcomings of the township's leaders and prompting replies from the supervisor as well as news articles covering the township's business. Eventually she learned of STOP, made contact with Bob Anderson, and gave some thought to an abolition effort, because, as she closed a September 1998 letter, "Shields Township has become a most expensive embarrassment and a political liability to its residents." The Lake Forest-Lake Bluff League of Women Voters took up her cause, put township government (locally) on its study list, and presented the study's results at a League meeting on 27 September 2000.[78]

Schnobrich has done what any able Illinois citizen should do, in the interest of all: question the ways of governmental officials if something seems amiss, go public in a reasonable manner with the findings, and seek a better alternative—even if the only alternative is running for office or attempting to change the government through petition and referendum. At the least, citizens like Schnobrich can induce elected governmental officials to do a better job doing the public's business.

Rock Island County

When STOP convened its final rally, on 17 May 1998 in Elgin, Michael Richardson still harbored hope that Judge Conway, back in Rock Island, would order the question of a countywide referendum to the November 1998 ballot, as a carryover from *Richardson v. Rock Island County Officers' Electoral Board.* When Conway denied that request, Richardson called a news conference outside the federal courthouse in Rock Island, nearly a month into the 1998 petitioning window. On the mind of the man who had been in the face of township officials for more than a decade were two strategies: sue in federal court and challenge townships one by one. He did neither.

Later he told me he regretted not going into the building that formed the backdrop for his last of several dozen news conferences and suing for what he felt the township officials had stolen from him—a day on the ballot to allow the people of Rock Island County to decide the fate of their townships. His treatment by the supreme court, as a man whose credentials did not measure up, and the final defeat in Judge Conway's court had left him much less confident than he felt he would have needed to be for a federal lawsuit.[79]

Because of TOI's silver bullet, Public Act 90-112, and the fact that Rock Island County still had several rural (township-friendly) townships, he knew the only other possibility of reaching the ballot was to attack the more vulnerable urban townships. "If the townshippers want to make us do this township-by-township, then we will just do it township-by-township," he said in the final news release. But his heart was never really into a fifth summer of petition circulation and another autumn of litigation, partly because he felt it was unnecessary (and was even then hoping "to get the judge to grant another hearing on the issue" of a ballot mandate) and partly because he was tired of the whole thing. He did a bit of signature chasing that summer in his township, South Rock Island; but a colleague there failed to deliver. Randy Popp backed out of doing a drive in Rock Island Township, and they could not find anyone to work Coal Valley or get nearly enough effort in Moline and South Moline. The only consolation he could mention, as we talked by phone in late September 1998 about the status of the abolition movement in Rock Island, was what Bob Anderson had accomplished that same year.[80]

McHenry Township

As Bob Anderson's 1996 McHenry Township signature count reached its final stages, early in 1998, a staff writer out of the Crystal Lake bureau of the *Chicago Tribune,* called me for some background on township referenda that I had uncovered in my research. We also talked, off the record, about the barber from Wonder Lake. People, she said, had warned her that

Bob Anderson was "a sort of kook," but she had found him rational and able to defend his position articulately. "He is a barber," I recall her saying, "and look what he has done so far." Barber, bus driver, public servant, STOP co-founder, Bob Anderson likes people. In the circulation of petitions, whatever his cause, he gets to renew old acquaintances and make new ones. Refusals only spur him on to the next person. When I asked, after the decade had ended, what for him was the low point in all those campaigns, hearings, and lawsuits? He really had to think, because to him, he was always successful—in victory or in defeat. Even his critics had to pay him an occasional compliment, as did Cyndi Wyss at the *Northwest Herald,* when she wrote, "Think what you like of this Wonder Lake barber and bus driver. If nothing else, he is one heck of a living civics lesson."[81]

Gone from the battlefield in 1998 was the township supervisor who had gotten the whole McHenry thing started by insisting on building the warming shelter. Replacing Albert Adams was Donna A. Schaefer, whom Anderson characterized for me as a "good elected official." She did seem to want her constituents to have a say in their future, though she said she had to question Anderson's ultimate strategy. Schaefer told the *Northwest Herald,* in July, as Anderson sought 2,450 signatures, that she could see him being dissatisfied with individual elected officials and working for their removal; but she thought his struggle to eliminate a whole governmental unit "'short-sighted.'"[82]

Editors at the *Northwest Herald,* perhaps sensing a new mindset at the McHenry Township offices about allowing the people a choice, and sensing no "dramatic change in the electorate's mood," issued an endorsement of a referendum, something that would cost the public almost nothing. "What would cost money, however, would be a court fight waged by township officials in an attempt to keep Anderson's referendum off the ballot." Anderson, Richardson, and many other abolition advocates had been saying this, of course, for years. The *Northwest Herald* did still wonder why Anderson could not "satisfactorily" answer questions about the transfer of essential services should the voters do the unthinkable.[83] Where had the Crystal Lake editors been when Anderson and Richardson were practically begging Ann Hughes, surely a subscriber, to work in Springfield for transition legislation?

True to her earlier word, Schaefer and her township board did not challenge Anderson as he completed his summer's crusade and filed his petitions containing three thousand personally collected signatures on Monday, 17 August, with Town Clerk Bruce Novak. Any other McHenry Township voter could, however, have brought an objection; but Schaefer, who greeted Anderson at the township offices on the seventeenth, said she hoped no one would. Otherwise, she implied (in a *Northwest Herald* article), Anderson would keep petitioning indefinitely. She and her board wanted the people to put a stop to Anderson's biennial abolition exercise, his "'rite of election.'" Schaefer did challenge Anderson to a debate. He accepted, and the *Northwest*

Herald asked who could blame her for wanting to debate: "She has a referendum-tested product that people like."[84]

Even before he and Schaefer got to the debate, Anderson had a tangle of sorts with Jack Schaffer, new chair of the McHenry County GOP Central Committee, over a poll that Schaffer took on Saturday, 13 September, as precinct committeemen met to plot general election strategy. Anderson objected to the poll before the meeting, which Schaffer convened at the McHenry Township offices. "'To me,'" said the barber/precinct committeeman, "'this is a fight between Concerned Taxpayers for Integrity in Government and McHenry Township, not the county Republicans.'" He could not understand why committeemen from all over the county needed to go on record for or against an issue that affected only one of the county's townships; and he sensed the same sort of political pressure that had permeated the 1994 referendum campaign, when his party had taken a protownship stance. Anderson could not attend but asked Schaffer to read a letter to those who did, a letter in which he charged that such a vote would tell the public that township government could not stand without party support. Abolition got 4 votes; status quo got 48.[85]

Both sides waged a printed-matter campaign as the 3 November election approached. CTIG/STOP had an 11 by 17 inch, one-fold flyer; at least one newspaper ad; and yard signs. Citizens for McHenry Township (CMT) used several different flyers, including at least one with bulk postage paid; yard signs; and one or more display ads in the *Northwest Herald.*

The CTIG flyer (print run of three thousand) featured the octagonal STOP sign as a theme, noting that STOP was CTIG's "Statewide Organization." Anderson emphasized the "rural" roots of township government, duplication of service within municipalities (as they overlapped the township), and the political patronage and nepotism themes. McHenry Township, said the flyer, had the county's highest paid supervisor, although Schaefer was only part-time. Anderson tried to address the thorny transition issue by citing the legislature's constitutional mandate to provide for such. He explained how general assistance and property assessment would devolve upon county and state offices if voters agreed with his point of view on abolition. To the township roads and bridges he devoted even more space on the flyer, explaining that the township and its coterminous road district were "two separate units of government" and that snowplows would still be plowing the township's roads no matter what the referendum's outcome. He predicted future attempts to consolidate road districts for economy purposes. In closing, Anderson asked for financial contributions.[86]

CMT (with a mailing address identical to the place where Supervisor Schaefer lived in McHenry) did not solicit contributions of cash, only a vote against dissolution. One piece of CMT literature used familiar phrases like "The People Next Door" and "Government Closest to the People." It also listed township services like roads (over a hundred miles), senior citizen companions, mosquito abatement, paint recycling, and welfare, but did not mention tax assessment. Another flyer laid out the potentially grim results of dissolution for those living in the township's several municipalities. Road maintenance and snow removal would surely suffer because county crews would focus on "the southeastern section of the county." Citizens should not plan to complain to the county commissioners after dissolution because only 5 of the 24 current board members (including Schaefer) were from McHenry Township. Children of McHenry, then involved in sports leagues using township parks, would "have to find another place" to play when the parks were gone. Seniors would not have anyplace to go either, once the various township services for them ended. On the reverse of that flyer, CMT reprinted a township-friendly editorial from the *Northwest Herald.*[87]

Debate night for Anderson and Schaefer came on 28 October, under the sponsorship of the Woodstock-McHenry League of Women Voters. Each debater was to have 10 minutes for an opening statement, and then the audience would get a chance to ask questions. Between one and two hundred souls crowded into the McHenry City Hall, mostly staunch township supporters, employees, and officials. Anderson, looking out on the audience, saw none of his former CTIG comrades as he made his initial remarks about what he considered outdated, old-boys, and replaceable government. Schaefer downplayed his opening as having "nothing new to add." She referred to herself as the "'poster child for township government'" and ticked off all the services that citizens might lose if her side failed to prevail next Tuesday. She said the road district would disappear right along with the township, if voters failed to heed her message. Anderson said the road district would remain to plow the roads and perform all the usual road and bridge duties. Voters could make history by forcing the general assembly to do its duty and finally pass a law or laws making way for a smooth transition. Mitch Martin of the *Tribune* heard her say that the taxpayer would suffer "'from this big transfer of responsibility.'"[88]

Few bothered this time around to write letters to the editor. Anderson did, however; and his letter ran on 26 October. Bryan Smith wrote a rebuttal, warning of "long legal battles" that would add to the taxpayer burden in order to resolve them. "When the truth is told," Smith concluded, "it is apparent township government is still one of the best bargains around. Township government is the real thing."[89]

Editors at the *Northwest Herald* never wavered from the paper's initial position opposing governmental reform involving township abolition. "No question on the Nov. 3 ballot is more ill-conceived than the referendum to dissolve McHenry Township. Bob Anderson is apparently devoting his life to eliminating township government in Illinois." Send him a message, said the editors, by a "lopsided" margin. Anderson told me he took some consolation that week because his nemesis newspaper actually asked him for his side of the story.[90]

More people voted for dissolution than had signed the petitions, but not nearly enough said yes to make it happen. He got 3,182 votes; while Schaefer's side, with far more resources, the advantage of a system that had worked for nearly a century and a half, and the unwavering support of the *Northwest Herald,* counted nearly nine thousand (8,924) votes. Crystal Lake editors took one last 1998 shot at Anderson, a few days after his second three-to-one referendum defeat, suggesting, "Maybe now Bob Anderson will sit down and shut up." Editors could not imagine more convincing proof and warned that trying again "would be insulting the intelligence of the voters." He would, furthermore, risk "making himself a laughingstock" if he were to try again, and in the official opinion of the *Northwest Herald,* should just "spare us all and drop it."[91]

Did the barber from Wonder Lake just drop it, sit down, shut up, and find a new hobby? No, he did not. In fact, the *Northwest Herald* had already reported his vow the day after the election to approach his newly elected (to fill the Hughes seat) representative, Democrat Jack Franks of Woodstock, about sponsoring a transition plan. Franks promised to "study" the issue. At the 1999 annual meeting of McHenry Township, Anderson stood to offer Resolution 99-002, "A resolution to support investigation of the benefits of the merging of McHenry Township with one or more surrounding townships." Electors have few real powers at annual township meetings, and authorizing such a study was not among them; but he kept his reform agenda rolling and picked up press coverage in the process. Schaefer said Anderson might be correct about achieving some economy through scale, but people prefer their local government in the hands of those they know.[92] Anderson would have even more ideas for streamlining the polity of his area in the millennium to come.

A Busy Decade

No decade of the twentieth century, save perhaps the 1930s, saw as many threats arise for the township system, as did the 1990s. Desperation and the Great Depression drove the reformers of that earlier decade; and one county, Williamson, succumbed to the lure of the county commission. Ardent local reformers, catching the early 1990s wave of taxpayer distress and thriving on the challenge of bringing down some small part of the township megalith, fueled the reform movement of the century's last decade. No counties lost their township governments nor did individual townships dissolve under hostile fire, and TOI saw to it that abolition at the tail end of the decade was much tougher than it had been at the outset. Figure 25 depicts the battlefield of the 1990s.

Rock Island County played a role in every other year, but Michael Richardson and Randy Popp never succeeded in winning a ballot slot much less a referendum victory. They missed their best chance in 1990, at a time when the strongly blue collar Rock Island-Moline metropolitan area seemed

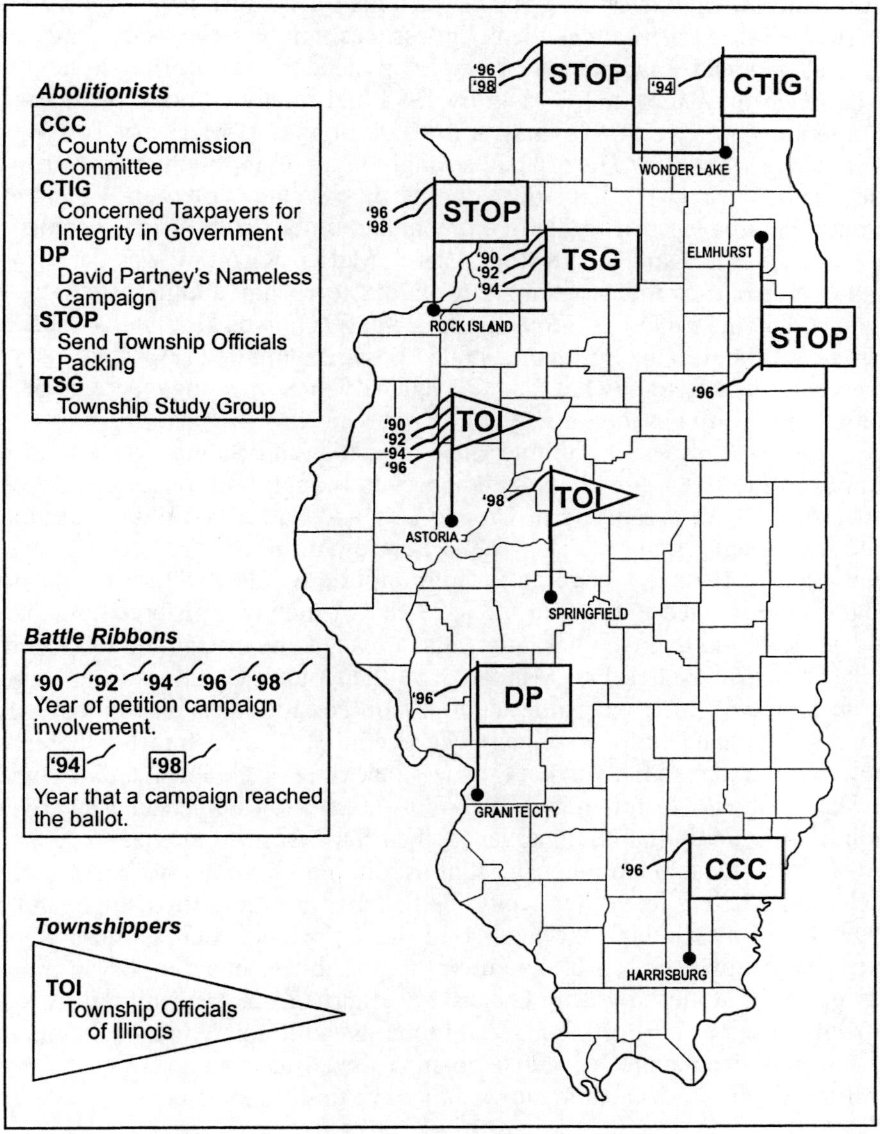

Figure 25 Township Wars: Illinois in the 1990s.

ripe for taxpayer revolt, they enjoyed excellent newspaper support, and the general assembly had yet to tip the landscape to favor the township forces.

Conservative and prosperous McHenry County produced three petition drives and the only two referenda to reach the voters during the 1990s. At the heart of all three drives was the relentless, seemingly tireless Bob Anderson who, at personal financial risk, hired a top election law attorney to help his issue reach the ballot in 1994 and by 1998 had finally convinced the townshippers to just give him his day at the ballot boxes of McHenry Township. Even if the *Northwest Herald* had strongly favored abolition, however, that endorsement might not have been enough to convince voters in Anderson's county that this layer of local government had outlasted its long welcome.

Along with DuPage and Rock Island, McHenry County was part of the first-ever multi-county coalition of Illinois township abolition activists—under the banner of STOP. Peaking in 1996, STOP was hardly a whisper by the end of the decade, but demonstrated how like-minded reformers can cooperate to reduce somewhat the odds against which the single activist or local reform group has to contend.

Abolition efforts in Granite City Township and Saline County had no connection to STOP or to one another, even though both occurred in 1996. As it turned out, organizers in Granite City and Saline could have used advice from battle-tried veterans. David Partney, in urban, blue collar Granite City, seemed to have lots going for him until he fell ill. A STOP connection might have warned him about the necessity of having registered circulators and advised him to collect far more signatures than he thought he would possibly need. Bill Bethel, Ben Brinkley, and the County Commission Committee in Saline County, with other commission counties at their doorstep, also had a good chance to reverse just over a century of township rule there; but they sadly underestimated what they were tackling in the local and statewide power of the township network. Bob Anderson and Michael Richardson would have gladly shared, in advance, their hard-won experiences.

Most township counties in Illinois did not have an Anderson or a Richardson telling them they would be better off without townships, and as 1999 gave way to 2000, the likelihood that reformers such as these would step forward became less likely but not impossible to imagine. Bryan Smith, having brought the Township Officials of Illinois from rustication in Astoria to sophistication in Springfield, could look ahead to many local celebrations of the sesquicentennial of the townships that organized in 1850, even in counties that did not catch the township wave until years later.

Notes

1. *Ann Hughes: Candidate for House of Representatives District 63,* n.d., <www.votesmart.org/state/Illinois/legis/Lower/63/b024445.html> (11 October 1996); Illinois, *Blue-*

book of the State of Illinois: 1995–1996 (Springfield, 1995), 97; "Members of our General Assembly with a Township History," *ICTO,* May 1995, 5–7.

2. Kurt Begalka, "Township Fans Check Signatures," *Northwest Herald,* 24 August 1994, A/1.
3. Ed Stasiewicz, "Township and Road Districts Legislative Reporting," *ICTO,* March 1995, 6; *House Journal,* 89th G. A., 1995, 1162, 2326, and 2556; *Senate Journal,* 89th G. A., 1995, 1242, 1872, 2039, 2201, and 2358–59; *Laws Eighty-ninth General Assembly, 1995,* 3811.
4. Miller, "From the Executive Director," *ICTO,* September 1995, 4.
5. Bob Anderson to Senator Dick Klemm, 19 December 1994, personal copy; Bob Anderson, "Anti-township Effort," *Daily Herald,* 10 February 1995, 1/13; Lavelle, Juneau & McCollom, Ltd. to Robert G. Anderson, 5 March 1995, personal copy of billing statement; *Concerned Taxpayers for Integrity in Government,* March 1995; Bob Anderson to Members of the General Assembly, 20 March 1995, personal copy.
6. "More Scenes from the 88th Annual Conference," *ICTO,* December 1995, 18.
7. [Joni Jacobsen], "Constitutionality of H.B. 438," 23 July 1995, personal copy.
8. Joni Jacobsen to Jim Edgar, 3 August 1995, personal copy.
9. Theresa Amato, "Citizen Advocacy Challenges Constitutionality of Act Designed to Inhibit Petition Challenges to Township Government," 5 October 1995, personal copy.
10. Bob Anderson, "Press Conference—Immediate Release," 5 October 1995, personal copy; Rosanna "Rosie" Fitzpatrick, "For Immediate Release," 5 October 1995, personal copy; Michael Richardson, "STOP: Send Township Officials Packing," 5 October 1995, personal copy.
11. Dave McKinney, "Citizens Group Challenges Law that Makes it Difficult to Abolish Townships," *Daily Herald,* 5 October 1995, 1/12; Charles Mount, "Suit Takes Aim at Townships," *Chicago Tribune,* 6 October 1995, McHenry County/1 & 4; Kurt Begalka, "Group Attacks Township Law," *Northwest Herald,* 6 October 1995, C/1–2; "Still Tilting at Windmills," *Northwest Herald,* 7 October 1995, A/16; Brett Johnson, "End of the Road: Group Aims to Put a Stop to Townships," *Elmhurst Press, Hinsdale The Doings,* 11 October 1995, 5; Karen Kaegebein, "Township Abolitionists Will Go with STOP Effort," *Naperville Sun,* 6 October 1995, 10; "David Wickert, "Group Sues to Ease Township' s End," *Pantagraph,* 7 October 1995, A/3.
12. Mike Burke, "Hearing Delayed in Suit Challenging New Township Law," *Daily Herald,* 15 December 1995, 1/4.
13. Miller, "From the Executive Director," *ICTO,* December 1995, 4.
14. Michael D. Richardson, letter to the author, 31 October 1995; Jonathan Turner, "Richardson Will Seek Evans' Seat," *Argus,* 30 December 1995, A/2.
15. *Send Township Officials Packing, et. al. v. DuPage County Election Commission, et. al.,* 95 MR 750 (DuPage County), "Petition to Intervene" and "Plaintiffs' Memorandum in Opposition to Petitioners' Petition to Intervene"; Kurt Begalka, "Judge: Township Officials of Illinois OK to Defend Lawsuit Challenging Townships," *Northwest Herald,* 9 January 1996, D/3.
16. *Send Township Officials Packing, et. al. v. DuPage County Election Commission, et. al.,* 95 MR 750 (DuPage County), "[Defendants'] Motion to Strike and Dismiss," "Defendants Wayne Township and the Township Officials of Illinois Memorandum in Support of Motion to Dismiss," "Memorandum in Support of Intervening Defendants' Motion for Summary Judgment," and "Memorandum in Support of Plaintiffs' Motion for Summary Judgment."

17. *Send Township Officials Packing, et. al. v. DuPage County Election Commission, et. al.*, 95 MR 750 (DuPage County), "Opinion and Order."
18. Mike Burke, "Ruling Makes Changing Government Easier," *Daily Herald,* 4 June 1996, 1/7; "In the News: In the Suburbs," *Chicago Daily Law Bulletin,* 4 June 1996, 1; Lynn Van Matre and Art Barnum, "Challenge to Townships Gets a Boost," *Chicago Tribune,* 4 June 1996, DuPage County/1–2; Kurt Begalka, "Township Abolition Law Ruled Unconstitutional," *Northwest Herald,* 4 June 1996, 1–2; Cheri Bustos, "Pro-township Law is Ruled ' null and void,' " *Times,* 4 June 1996, M/3; Steve Zalusky, "Township Foes Win Round 1," *Elmhurst Press,* 5 June 1996, 1/10; John Kanthak, "STOP Wins Victory," *Argus,* 6 June 1996, 1. TOI and Wayne Township appealed the Byrne decision in late June but withdrew it from the supreme court' s consideration when STOP failed to place a referendum on the ballot that year in DuPage County. *Send Township Officials Packing, et. al. v. DuPage County Election Commission, et. al.*, No. 81402, "Motion to Withdraw Appeal," (Ill. Sup. Ct. 1996).
19. George H. Miller to Township Friend, 10 June 1996, personal copy.
20. The PAR Group, "Recruitment Profile, Executive Director, Township Officials of Illinois (TOI)," (Lake Bluff, Illinois: n.d.); Miller, "From the Executive Director," *ICTO,* July–August 1996, 4; Michael D. Richardson to [PAR Group], 31 March 1996, personal copy of Richardson' s letter of application for position of TOI executive director.
21. Miller, "From the Executive Director," *ICTO,* July–August 1996, 4.
22. Bryan E. Smith, interview by author, 4 October 2000; Smith, email correspondence with author, 22 April 2002. Smith did not have to resign as supervisor but felt he needed to devote full-time effort to the executive director job.
23. Paul Gale, "Townships Stand Strong," *Sterling Gazette,* 8 March 1996, A/1.
24. Ray Quintanilla, "Township Foes are Strife of the Party," *Chicago Tribune,* 27 February 1996, 1/1 & 11; Robert G. Anderson, telephone interview by author, 20 May 1996; George Houde, "Woodstock Man Reignites Campaign to Abolish Townships," *Elgin Courier-News,* 28 March 1996, A/1 & 6; "Residents Remain behind Townships," *Northwest Herald,* 10 April 1996, 1–2; Nina Bernardi, "Anti-township Leader to Present Case to Dundee GOP," *Elgin Courier-News,* 24 April 1996, A/1; Send Township Officials Packing, "News Release," 8 April 1996.
25. Steve Zalusky, "Is STOP Making Sense?" *Elmhurst Press,* 7 February 1996, 1/7; Steve Zalusky, "How Many Layers of Government Qualify as ' One Layer Too Many' ?" *Elmhurst Press,* 7 February 1996, 1/7; Steve Zalusky, "Townships: Transfer of Services Won' t Be Easy," *Elmhurst Press,* 14 February 1996, 1/7; "6,722 Units of Government is Far Too Many," *Elmhurst Press,* 14 February 1996, 1/8; Sarah Greider, "LWV to Examine Township Viability," *Edwardsville Intelligencer,* 3 April 1996, 1; Cathy Hensley, "Illinois Group Wants to Sink Townships," *Times,* 9 April 1996, M/1.
26. Alex Rodriguez, "DuPage Group Fights Long Odds to Abolish Township Government," *Chicago Sun-Times,* 10 June 1996, Suburban News/8.
27. DuPage Citizens for Township Government, "Township Government: Serving You and Your Family, summer 1996, campaign literature; A Group of Very Concerned Citizens, "If You Were Asked to Sign a Petition," summer 1996, campaign literature. DuPage Citizens for Township Government, according to William Grady of the *Tribune,* did not just ask for contributions to pay for brochures and other protownship items. Instead, the leadership assessed each of the county' s township officials 3.5 percent of their township salaries and collected about fifteen thousand dollars. William Grady, "Township Officials Still in Business," *Chicago Tribune,* 19 September 1996, 2/8.

28. Gail Dunham, "It' s Time to STOP Township Government," *Elmhurst Press,* 9 August 1996, 1/9; Rosie Fitzpatrick, telephone interview by author, 28 August 1996; Send Township Officials Packing, "Press Release," 19 August 1996; Mike Burke, "Signature Shortage Puts a Stop to Anti-township Petition Drive," *Daily Herald,* 20 August 1996, 1/4 (DuPage edition).
29. Robert G. Anderson, interview by author, 8 December 2000. A Democrat in a heavily Republican county, Fitzpatrick needed any boost she could get for her own political career. The township issue, as was the case with Patrick Quinn in the late 1980s, provided a perfect opportunity for press attention and perhaps the break she needed. On the other hand, few of the abolitionists of the 1990s were as doggedly dedicated as Bob Anderson. Fitzpatrick ran for DuPage County recorder in 1996 and Bloomingdale Township supervisor in 1997—losing on both occasions.
30. Wayne Johnson, telephone interview by author, 20 October 1996; *Wayne Johnson. v. William J. Toerpe, et. al.,* 96 MR 783 (DuPage County), "Petition for Judicial Review of Decision of the DuPage County ' Officers' Electoral Board and Other Relief." Amato told me that she was basically neutral about the advantages and disadvantages of township government, but she did not want an electoral board to be able to hold a hearing on challenges to petitions without notifying the person who submitted the petitions. She and Johnson did succeed in getting the DuPage Electoral Board to recognize Johnson as the "principal proponent" of the 1996 STOP petition, which he had submitted. Theresa Amato, telephone interview by author, 18 June 2002.
31. *Wayne Johnson. v. William J. Toerpe, et. al.,* 96 MR 783 (DuPage County), "Plaintiff' s Reply in Support of His Motion to Reconsider and Supplemental Petition for Judicial Review."
32. "Citizen Advocacy Center Names Terry Pastika as Executive Director," *Everyday Democracy,* Spring 2002, 4. *Everyday Democracy* is the newsletter of the Citizen Advocacy Center (CAC).
33. "Support Strong as Ever," *Northwest Herald,* 11 April 1996, A/6.
34. Robert G. Anderson, telephone interview by author, 4 September 1996.
35. Jeanette Lach, "Panel Abolishes Anderson Petitions," *Northwest Herald,* 27 August 1996, A/1; Dave Barnes, "Township Foe Calls Hearing Improper," *Chicago Tribune,* 27 August 1996, McHenry County/1–2.
36. Jeanette Lach, "Lake County to Hear Anderson Case," *Northwest Herald,* 11 September 1996, C/3.
37. *Robert G. Anderson v. McHenry Township et. al.* 96 MR 196 (McHenry County), "Order"; Joel B. Hurewitz, correspondence with author, 14 October 1997.
38. Stephen Lee, "Township Foe Vows Appeal as 2nd Drive Stalls," *Chicago Tribune,* 23 October 1996, 1/1; *Robert G. Anderson v. McHenry Township et. al.,* 225 Ill. Dec. 56 N.E.2d 1133, 57 (2d. Ill. App. Ct. 1997).
39. *Anderson v. McHenry Township,* 59; Charles Mount, "Appeals Court Revives Bid to End Townships," *Chicago Tribune,* 18 June 1997, 1/1 and 19; Cyndi Wyss, "Tenacity Keeps Township Dissolution Notion Afloat," *Northwest Herald,* 18 June 1997, A/3. Identical reviews of the appellate court' s ruling appeared in two Chicago area legal publications. See, "Election Law—Conflict of Interest" in *Chicago Daily Law Bulletin,* 9 July 1997, 4, or *Chicago Lawyer,* August 1997, 36.
40. Jeanette Lach, "Schultz Given OK to Check Issue' s Validity," *Northwest Herald,* 17 December 1997, Community/1; Dave Barnes, "Question Seeking End to Township is Put on Ballot," *Chicago Tribune,* 12 January 1998, 2/1; Dave Barnes, "Activist Faces Electoral

Board, Clock in Fight to End Township," *Chicago Tribune*, 3 February 1998, McHenry County, 3; Dave Barnes, "Drive to Abolish Township' s Government Comes Up 4 Signatures Short on Petition," *Chicago Tribune*, 14 February 1998, 1/5.

41. John Kanthak, "Trustee Says Township Opponents Should Get Vote," *Dispatch*, 7 June 1996, A/7; Send Township Officials Packing, "News Announcement," 12 July 1996.
42. Seth Hettena, "Crowd Fails to Stand United at Moline Township Meeting," *Times*, 12 June 1996, M/3; John Kanthak, "Political Theater Disrupts Genesius Guild, "*Dispatch*, 25 July 1996, C/1; Don Wooten, undated note to author, circa 1 September 1996. Wooten provided his perspective for the public an opinion piece he wrote for the *Argus*. Don Wooten, "A Park Tussle? Hardly," *Argus*, 18 August 1996, A/4. Randy Popp, who had once been a "Rootin for Wooten" supporter in Wooten' s successful run for the state senate, took exception in a letter to the editor to the disruption of Richardson' s right to pass a petition. Randy Popp, "Petitioning at Public Park a Citizen' s Right," *Argus*, 2 August 1996, A/4.
43. "Without Challenge, ' Democracy Stagnates' ," *Argus*, 15 July 1996, A/4; John Kanthak, "STOP Must Go Ahead with Petition Drive," *Argus*, 17 July 1996, C/2; John Kanthak, "Anti-township Activist Gets Cold Shoulder," *Argus*, 15 August 1996, C/3; "Voters Should Decide Township Question," *Argus*, 16 August 1996, A/4.
44. John Kanthak, "Voters May Get Their Say," *Argus*, 19 August 1996, A/1; "Anti-township Group Gets Signatures," *Times*, 19 August 1996, M/1; John Kanthak, "STOP Asks for Help," *Argus*, 21 August 1996, C/2.
45. John Kanthak, "Dead Eligible to Vote," *Dispatch*, 5 September 1996, A/1–2; Cathy Hensley, "Township Fight Goes Before Electoral Board," *Times*, 6 September 1996, M/2; John Kanthak, "Township Vote off Ballot," *Dispatch*, 13 September 1996, A/1–2.
46. *Michael D. Richardson v. Rock Island County Officers' Electoral Board et. al.* 96 MR 195 (Rock Island County), "Petition for Judicial Review of Electoral Board Decision"; Cathy Hensley, "RI Township Ballot Issue Gets Appealed in Court," *Times*, 18 September 1996, M/3; Doug Schorpp, "Handling of Township Referendum Comes Under Fire," *Times*, 28 September 1996, M/1. Richardson had filed an unconstitutionality claim with Attorney General Jim Ryan on 23 September 1996. *Michael D. Richardson v. Rock Island County Officers' Electoral Board et. al.* 96 MR 195 (Rock Island County), "Notice of Claim of Unconstitutionality"; *Michael D. Richardson v. Rock Island County Officers' Electoral Board et. al.* 96 MR 195 (Rock Island County), "Opinion and Order"; Charles Keeshan, "Judge Rules STOP Petitions 552 Short," *Dispatch*, 2 October 1996, C/1. Public Act 89-235, which had gone into effect 4 August 1995 (to correct the relevant portion of Public Act 88-62, effective 1 January 1994 [see chapter 7]), altered section 25-5 by reintroducing the standard of 10 percent of the legal voters in an area as determined on the date of petition filing. *Laws Eighty-ninth General Assembly, 1995*, 2950.
47. *Michael D. Richardson v. Rock Island County Officers' Electoral Board et. al.* 96 MR 195 (Rock Island County), "Appellant' s Brief"; Carolyn Hardin, "Judge: STOP is a Go," *Dispatch*, 24 October 1996, A/1–2; Michael Richardson to William Armstrong, 25 October 1996, personal copy of one of the letters Richardson mailed to each county board member. Richardson closed by writing: "It is my sincere personal hope that the next time I correspond with you it will be in the form of a thank you letter rather than as plaintiff to defendant." *Michael D. Richardson v. Rock Island County Officers' Electoral Board et. al.* 96 MR 195 (Rock Island County), "Motion for ' Emergency' Stay of Judgment and Expedited Hearing"; Doug Schorpp, "Township Issue Goes Off Ballot," *Times*, 29 October 1996, A/1–2; *Michael D. Richardson v. Rock Island County Officers' Electoral*

Board et. al. 96 MR 195 (Rock Island County), "Order"; *Michael D. Richardson v. Rock Island County Officers' Electoral Board et. al.* No. 82136, "Order," (Ill. Sup. Ct. 1996); John Kanthak, "STOP Stopped for Now," *Dispatch,* 1 November 1996, A/1–2.

48. John Kanthak, "Richardson Muzzled in STOP Court Case," *Argus,* 26 September 1997, C/1–2; *Michael D. Richardson v. Rock Island County Officers' Electoral Board et. al.* 179 Ill. 2d 252, 688 N.E.2d 633 (Ill. Sup. Ct. 1997); "Election Law—Signatures," *Chicago Lawyer,* January 1998, 27; John Kanthak, "Court: Referendum Moot," *Argus,* 21 November 1997, A/1 and A/5; John Kanthak, "Attorney: Township is Off Ballot," *Argus,* 25 November 1997, C/1; *Michael D. Richardson v. Rock Island County Officers' Electoral Board et. al.* 96 MR 195 (Rock Island County), "Third-Party Complaint for Mandamus"; *Michael D. Richardson v. Rock Island County Officers' Electoral Board et. al.* 96 MR 195 (Rock Island County), "Order."
49. Michael D. Richardson, interview by author, 16 October 2000.
50. David Partney, telephone interviews by author, 17 October 1996 and 29 January 1997; Marilyn Vise, "Partney Assails Nepotism," *Belleville News-Democrat,* 18 August 1996, A/1 and 10.
51. Marilyn Vise, "Granite City Candidate Seeks to End Township," *Belleville News-Democrat,* 12 August 1996, B/1; David Partney, interview by author, 17 October 1996.
52. Marilyn Vise, "Granite City Residents to Consider Township," *Belleville New-Democrat,* 20 August 1996, A/1 and 3.
53. "No Need for Townships," *Belleville News-Democrat,* 20 August 1996, A/4.
54. Brian Brueggemann, "Township Officials Vote on Petitions," *Belleville News-Democrat,* 30 August 1996, B/1 and 4; Marilyn Vise, "Signatures Considered Invalid," *Belleville News-Democrat,* 4 September 1996, A/1 and 4; Charles Bosworth Jr., "Granite City Officials Rule Petition is Illegal," *St. Louis Post-Dispatch,* 4 September 1996, B/4.
55. "Foxes Left in Charge," *Belleville News-Democrat,* 3 September 1996, A/4.
56. "Let the Voters Decide," *Belleville News-Democrat,* 7 September 1996, A/4.
57. Marilyn Vise, "Election Panel Sued over Petition Ruling," *Belleville News-Democrat,* 14 September 1996, B/3.
58. Richard Aguirre, telephone interview by author, 31 March 1997; Brian Brueggemann, "Ruling Expected before Election," *Belleville News-Democrat,* 4 October 1996, B/1.
59. *David Partney v. Granite City Township and Judy Whitaker, Town Clerk* 96 MR 536 (Madison County), "Order."
60. Marilyn Vise, "Election Rule Leaves Signatures Useless," *Belleville News-Democrat,* 14 January 1997, B/1–2.
61. "The Township Dinosaur," *Belleville News-Democrat,* 13 September 1993, A/4.
62. "Push for Change Began in May," *Harrisburg Register,* 13 August 1996, 1 and 10; Bill Bethel, telephone interview by author, 17 June 2002.
63. Scott L. Hines, "Seasoned Veteran Fighting to Keep Township Government," *Harrisburg Register,* 13 August 1996, 1 and 10.
64. Scott L. Hines, "Referendum Ballot Petitions Being Reviewed," *Harrisburg Register,* 22 August 1996, 1.
65. Gregory Norfleet, "Names Removed from Petition," *Harrisburg Register,* 4 September 1996, 1 and 11; Gregory Norfleet, "Township Question off the Ballot," *Harrisburg Register,* 5 September 1996, 1 and 12.
66. Jerry H. Kats, interview by author, 26 October 2000; "TOIRMA Returns $100,000 to First Year Member Townships," *ICTO,* July–August 1991, 32.

67. Kats interview; "TOI Hosts Open House," *ICTO,* September 1997, 39; "TOI Hosts Legislative Reception," *ICTO,* December 1997, 30–31; "NOTICE," *ICTO,* January 1998, 5.
68. Bryan E. Smith, "From the Executive Director," *ICTO,* October 1996, 4. Hereafter, I cite items he wrote for the magazine as Smith.
69. Ed Stasiewicz, "TOI Legislative Report," *ICTO,* December 1996, 4.
70. Illinois, General Assembly, House, Bills, 1997, State Archives, Springfield; *House Journal,* 90th G. A., 1997 session, 35, 151, 275, 1,256, 2,650, and 2,653; *Michael D. Richardson v. Rock Island County Officers' Electoral Board et. al.* 179 Ill. 2d 252, 688 N.E.2d 633 (Ill. Sup. Ct. 1997).
71. *Senate Journal,* 90th G. A., 1997 session, 628, 1,729, 1,762, 1,854; "Jackson Township Supervisor Becomes State Senator," *ICTO,* June 1997, 15; Illinois, General Assembly, *Laws of the State of Illinois, Ninetieth General Assembly, 1997* (Springfield, 1998), 2338–39; "Governor Edgar Signs House Bills 248 and 480 into Law," *ICTO,* September 1997, 38. H. B. 248 made sure that road districts would continue to receive revenue from the motor fuel tax in Illinois.
72. Karen Rivedal, "New Law Gives Safety Net to Townships," *Northwest Herald,* 1 August 1997, C/1–2; "Fair to the Whole County," *Northwest Herald,* 4 August 1997, A/6; Theresa Amato, telephone interview by author, 18 June 2002.
73. Michael D. Richardson, telephone interview by author, 2 December 1996; Ed Stasiewicz, "Township and Road District Legislative Reporting," *ICTO,* January 1996, 6.
74. "Out of the Bunker," *Dispatch,* 19 January 1997, A/4; Michael Richardson to Joel Brunsvold, 20 January 1997, personal copy; Michael Richardson to Ann Hughes, 20 January 1997, personal copy; Joel Brunsvold to Mike Richardson, 27 January 1997, personal copy; Citizens' Township Review Committee, minutes of meeting from 20 February 1997.
75. John Kanthak, "Brunsvold Gives Thought to Townships," *Dispatch,* 12 March 1997, C/1;
76. John Kanthak, "State Won' t Help Plan End to Townships," *Dispatch,* 8 April 1997, C/2.
77. Smith, "Set Your Goals Now," *ICTO,* April 1998, 4; Smith, "Today, More Than Ever, Township Government Should be Celebrated," *ICTO,* November 1998, 4.
78. My first meeting with Jan Schnobrich occurred 14 March 1998 at her home in Lake Bluff. Since then she has periodically kept me apprised of her township-related activities. Here, in chronological order are some items from the *Lake Forest Lake Forester* related to the Schnobrich initiative. Dan Rafter and Kyle Leonard were reporters for the paper. Dan Rafter, "Shields Twp. Salaries Questioned," 17 April 1997, 5; Janice Schnobrich, "Townships Have Outlived Usefulness," 17 July 1997, 19–20; Dan Rafter, "Township Road Spending Attacked," 21 August 1997, 5 and 23; Dan Rafter, "Township Forms Committee to Study Spending on Roads," 28 August 1997, 9; Dan Rafter, "Township Officials Defend Memberships," 2 October 1997, 11; Jan[ice] Schnobrich, "No Accountability in the Township," 1 January 1998, 16 and 19; Dan Rafter, "Accountability Questioned at Township Level," 15 January 1998, 5 and 11; Janice Schnobrich, "Reader Questions Township' s Duties," 26 March 1998, 22; Janice Schnobrich, "Township Closed for Good Friday," 23 April 1998, 20; Charles B. Fitzgerald, IV, "Fitzgerald Sets the Record Straight," 30 April 1998, 16 and 18; Kyle Leonard, "Township Battles Continue," 11 June 1998, 9 and 11; Charles B. Fitzgerald, IV, "Help Improve Process, Don' t Just Tear it Down," 18 June 1998, 21 (a guest essay); Janice Schnobrich, "Where Does All the Money Go?" 18 June 1998, 22; Janice Schnobrich, "Critic Urges Township Vigilance," 10 September 1998, 24.
79. Michael D. Richardson, interview by author, 16 October 2000.

80. Send Township Officials Packing, "News Release," 17 June 1998; Marcy Norton, "STOP Hopes to Take Townships One at a Time," *Dispatch,* 18 June 1998, C/1; Michael D. Richardson, telephone interview by author, 28 September 1998.
81. Cyndi Wyss, "Energizer Bunny Has Nothing on Anderson," *Northwest Herald,* 8 July 1998, A/3.
82. Jeanette Lach, "Activist Puts Township to Voters," *Northwest Herald,* 7 July 1998, D/1–2.
83. "Put Township Issue to Rest," *Northwest Herald,* 9 July 1998, A/8.
84. Jason J. King, "Township Set for Abolition Vote," *Northwest Herald,* 24 August 1998, B/1–2; Steve Stanek, "Township Foe Seeks Referendum," *Chicago Tribune,* 18 August 1998, McHenry County/1 and 4; "Township Relishes Fight," *Northwest Herald,* 26 August 1998, A/8.
85. Jeanette Lach, "GOP Mulls Township Question," *Northwest Herald,* 11 September 1998, C/11; Dave Barnes, "GOP-backed Township Government Still Has a Detractor," *Chicago Tribune,* 17 September 1998, McHenry County/7.
86. Concerned Taxpayers for Integrity in Government, "Operation Tax Relief," autumn 1998, campaign literature.
87. Citizens for McHenry Township, "Stay in Control. Vote No on Nov. 3rd. Punch 176" and "Why Do I Need the Township? I Live in the City!" autumn 1998, campaign literature.
88. Jeanette Lach, "Anderson, Schaefer Argue Ballot Issue," *Northwest Herald,* 29 October 1998, C/1; Mitch Martin, "Township Abolition Issue Heads to Voters," 29 October 1998, McHenry County/1 and 9.
89. Bob Anderson, "Dissolve and Save," *Northwest Herald,* 26 October 1998, A/9; Bryan E. Smith, "The Real Thing," *Northwest Herald,* 30 October 1998, A/10.
90. "Don' t Dissolve Township," *Northwest Herald,* 25 October 1998, A/8; Robert G. Anderson, telephone interview by author, 24 November 1998.
91. "Ringwood, McHenry Township Reject Dissolution," *Northwest Herald,* 4 November 1998, A/1; "Listen to the Voters," *Northwest Herald,* 8 November 1998, A/8.
92. Jeanette Lach, "Township Issue Shifts to Franks," *Northwest Herald,* 5 November 1998, C/8; Thad Rueter, "Merger Proposal Raises Skepticism," *Northwest Herald,* 17 May 1999, B/4. Mergers, of course, were then and are now possible, between and among townships, and had occurred in Ogle and Sangamon counties within the past two decades.

9 Closure

THE YEAR 2000 BROUGHT to an end the 36-month sesquicentennial that Bryan Smith had declared in 1998 and, for most of us, despite purists' protests, the beginning of the new millennium. Here in this final chapter I focus first on the year 2000, turn next to a summary of the previous century and a half, and lastly look beyond 2000 at what might lay ahead for Illinois's civil-political townships.

2000

At TOI, 2000 was a year of celebration, innovation, press confrontation, and determined opposition (to yet another attempt at countywide abolition). The sesquicentennial celebration culminated in April with a gubernatorial proclamation. Innovation came later in the year with the launch of a new TOI education program, a program that can only help township officials do a better job serving the public interest. Two rounds of press confrontation kept Smith and other township advocates busy that year, in June and especially in December. A citizen challenge to the township way of governing, this time in southern Illinois, demonstrated the value of TOI-inspired legislation from the 1990s.

Commemorating 150 Years

In the months preceding Township Government Month, *Township Perspective* carried reports about various townships had gotten into the sesquicentennial spirit and offered suggestions as to how other officials could mark the occasion. Naperville Township (DuPage County), for instance, had refurbished a public park and renamed it Grassroots Overlook. One Cook County township, Palatine, reported several 1999 events including participation, with a township float, in the Palatine Jaycees' Independence Day parade. Another Cook County township, Bremen, had sponsored a logo contest among township school children. Rick Davis and Julie Pittman devoted column space in the magazine to suggesting ways of informing the citizenry that something special was happening at the grassroots level. Davis, the producer of TOI's latest video, *Your Township Government: They're The People Next Door,* offered tips on the preparation of a video scrapbook, which might contain still shots of township documents, old photographs, and contemporary photos of relevant local sites. A show featuring transparencies (slides), Davis said, would be fine, while a simple, narrated videotape show would play even better. Pittman, the woman who had replaced Heather McAnarney as the associate director of TOI late in 1999, encouraged township officials to think about asking other governmental bodies to pass congratulatory resolutions, working with local newspapers on suitable township-oriented stories, or burying a time capsule. TOI, Pittman noted, was in the process of developing radio scripts that townships could appropriate.[1]

Despite the fact that only the original townships in the 24 counties adopting in 1849 and organizing in 1850, like the 8 in Boone County or the 11 in Marshall County, truly had the time in service to celebrate 150 years of existence in 2000, Smith invited all 1,433 survivors to the party. In proclaiming April 2000 as Township Government Month, Governor George Ryan, no doubt working from a script that TOI had provided him, cited the pre-Revolution origins of America's governmental townships and their function from the outset as a setting for citizen involvement in governmental decision-making.[2]

In his April column, Smith chose to dwell not on the proclamation but rather to acknowledge several more townships for their methods of marking the anniversary and to praise (though not by name) men like Hezekiah Morse Wead (chapter 2), Onslow Peters (chapters 2 and 3), and Elijah M. Haines (chapters 3 and 4) for their roles as "forefathers" of township government in Illinois. "Please take time," Smith wrote, "to celebrate the best form of government we have today."[3]

Near year's end, the executive director addressed the thousand or so delegates and guests attending the annual educational conference in Springfield. There he recounted how townships had, during the celebratory period, convened retro town meetings, conducted essay contests, held picnics and running/walking events, and created special t-shirts to commemorate the 150 years. If township government was to thrive, however, to be "the government of the people next door…for 150 years to come," changes in the way TOI educated its member officials would be crucial. "Nothing," Smith said, "not a building, not a family, not a government can survive for 150 years without constant attention to detail, without complete care and maintenance." Training and retraining were, as they should be, an aspect of routine personnel maintenance.[4]

Educating Township Officials

TOI (and its predecessors) had, of course, been serving an educational role through annual statewide meetings and later via the magazine, manuals, regional meetings, and topical seminars/training sessions. Until the latter half of 2000, however, the statewide organization had neither a staff slot devoted to education nor an education office. With the help of Jerry Kats and discretionary TOIRMA funds, Bryan Smith set out to remedy the situation.

TOIRMA, according to Kats, had a longstanding interest in the dissemination of job-related information to the officials of its member townships and road districts. To accelerate the education process already in existence through TOI, for insurance purposes and to foster improvement of the township product, Kats offered TOIRMA resources for the establishment and continued operation of a dedicated TOI education office, separate from Fortress Springfield. He and Smith asked Jill Gleason and Steven Stanger to submit a joint proposal that would lay the groundwork for the expanded education mission of TOI. Gleason had worked 17 years as a reporter for the *Bloomington Pantagraph* before leaving to become the meeting coordinator and later the executive director of the Illinois Property Assessment Institute, in Bloomington. Stanger had been (and would continue to be) the full-time, elected assessor for West Deerfield Township (Lake County). He had written since 1994 a column for the magazine entitled "Assessing the Facts," and knew Gleason as the result of having served as an instructor for the Illinois Property Assessment Institute.[5]

The Gleason-Stanger proposal called for the establishment of a TOI education office in McLean County with Gleason as the director of education and Stanger as lead advisor. From there the education branch of TOI would initiate and coordinate programs, according to priorities that a new TOI committee would set. Gleason would write a monthly column for the magazine to solicit interest in coming educational initiatives and to acknowledge officials for completing programs past. Among short-term goals for the education center were the scheduling of grant-writing workshops and forming of a speakers bureau to enlighten the public about township government. Longer term they hoped to help coordinate computerization grants for rural townships and conduct videoconferences for distance learning in order to serve those who could not attend formal class sessions. Kickoff for the education office was 1 September 2000, and by October that year Gleason and her small staff had occupied new quarters on the northeast side of Normal.[6]

I have no means of formally assessing TOI's 2000 education initiative, but there are several positives worth noting. The program remains in operation although Jill Gleason (now Jill Werner), as of the fall of 2002, resigned the directorship. Bryan Smith has reaffirmed his commitment to continuing the education initiative, telling me in a telephone conversation, "Education is part of our mission."[7] During the years after opening the Normal office, TOI employed additional educators, each taking responsibility for a different region of the state. The fact that TOI recognized a shortcoming and sought to overcome it with enhanced educational effort speaks well for an organization that in the 1990s too often had to depend on its legislative friends or employ scare tactics to counter the efforts of abolition-minded citizens, most of whom were only demanding a better product from elected officials. TOI's education program seems headed in the right direction; and, given the fact that countywide abolition of townships is much less likely to occur than it was in, say, 1990, better attitudes and better service from township officials may be the most for which those dissatisfied with "the people next door" can hope.

The Copley Series

Celebration of 150 years of grassroots government called attention to townships in many positive ways but also partly prompted a pair of critical investigative efforts by the press. One series, from the Copley News Service, appeared in the summer. The other, from the Associated Press, surfaced right before Christmas. In between June and December, disgruntled Franklin County taxpayers tried to rid their hard-pressed area of townships altogether.

Writers for Copley played upon the fact that Illinois, despite massive consolidation of rural school districts after World War II, still has more governmental units that any other state. Copley called its three-part series "Illinois Government: Layer upon Layer." Dean Olsen, in the first part, which ran over the weekend of 3–4 June 2000, focused attention on governmental congestion in the area around Chicago; but he did not mention townships specifically. That task fell to Jennifer Nelson, whose companion piece appeared the same weekend as Olsen's and recapitulated the 1994 McHenry County abolition effort of Bob Anderson and CTIG. On Monday, 5 June, the second Copley segment, by Nelson, took a look at special taxing districts in particular and townships peripherally, while Nick Reiher's article that same day dealt almost exclusively with townships. "Township government in Illinois," he began, "is celebrating its 150th anniversary this year. And it seems as if township officials have had to defend themselves for 149 of them." In the view of former state senator Dawn Clark Netsch, however, townships seemed likely to continue with business as usual because of special sacred cow status in the eyes of the legislators and the public. Copley's final series segment ran on Tuesday, the sixth. In it Nelson wrote about how difficult it was to reduce the number of taxing bodies or even for them to save money by cooperating on purchases of goods and services.[8]

At least two related, post-series editorials appeared; and both urged an end to Illinois's townships. "We see no reason," said the *Springfield State Journal-Register,* "why most of the road maintenance, tax assessment and social service functions of townships could not be assumed at the county level, as they are in many states." That editorial did leave some wiggle room for townships in certain circumstances because, "given the different challenges in sprawling suburban areas near Chicago and sparsely populated rural counties, there is no one-size-fits-all model of government that will work everywhere." At the *Waukegan News Sun,* editorial staff suggested a reduction in the number of school districts and called for the outright abolition of "an anachronism," the civil-political township.[9]

Franklin County

Neither James Muir nor Charles Horton needed advice from upstate editors when it came to the matter of doing something about what they considered excessive property taxes. These two and several thousand others asked in 2000 for the elimination of the township assessors and all the township officers in this county just north of Williamson, the last Illinois county to send township assessors packing.

At the heart of the Franklin County controversy was the *Benton Evening News,* a family-owned and operated daily, for which Muir worked as city editor, chief reporter, and featured columnist. From early May 2000, when he wrote a short news item concerning the mailing to taxpayers there of the quadrennial property assessments, until he wrote, in December, about the court ruling on the abolition effort, Muir was on duty, serving not only as the eyes and ears of *News* readers but also as an opinion leader, in the spirit of the newspapermen whose prose peppered earlier chapters of this book. Charles Horton, an unknown to Muir until May, became a key player in the petition drive after Muir recruited him to serve as taxpayer spokesperson.[10]

Franklin taxpayers did not appreciate what they discovered when they opened their quadrennial property tax assessment envelopes. All townships showed an increase in equalized assessed valuation, mostly at the expense of ordinary citizens on their homes and family farms. Coal mines, once a great source of local tax revenue, had all but closed by 2000, putting the bulk of the burden on homeowners, farmers, and retailers. In some cases, Muir said, taxes had gone up 30 percent on homes whose owners had seen similar increases the previous year. Horton and Muir both told me about how unfair the system seemed because an assessor in one township might underassess to the point where the owner of an obviously expensive home would pay less in taxes than would the owner of a modest place in another township in the county.[11]

Taxpayer protests started to flow into Benton's courthouse. Even *News* publisher Danny Malkovich, brother of celebrated film actor John Malkovich, protested his revised tax assessment. Muir editorialized a few days later that there was nothing "equal" about these so-called equalized assessments. Tax assessors were not, he said, doing their jobs properly, so "some type of initial step must be taken to take the pressure off Franklin County taxpayers. The current system simply cannot continue." Calls and visits to his desk at the *News* offices on East Church Street convinced him that momentum was building to make some "changes," he wrote on 10 May. In his editorial of 13 May, Muir asked the "commissioners" to do something about "the highway robbery that is commonly referred to as Franklin County real estate taxes." Franklin County had not had county "commissioners" since 1885, but Muir was probably wishfully thinking about the possibility of revisiting that nineteenth century state of affairs.[12]

Among those stopping by to see Muir was Charles Horton, a man whose taxes on his relatively new home in Browning Township had gone up 73 percent in four years. Horton had not previously known Muir but wanted to thank him for the stance he had taken in print. Muir must have seen something special in the retired metallurgical engineer because Muir later called Horton to ask if he would agree to speak up at an upcoming county board meeting. Horton resisted, in part because of his friendship with his township supervisor; but Muir persisted until Horton agreed to read a statement to the board and the crowd expected to pack the meeting room the night of 16 May and then serve on a committee to study the problem of tax assessment and tax levies.[13]

Townships were, of course, only one among several governmental units levying taxes against taxable real property in Franklin County that year. Tax protesters could have picked another target, such as the hospital district; but they picked townships. Why? First, townships provided the tax assessors. Actual or perceived discrepancies in assessment practices, from township to township within Franklin, bothered many of the dissatisfied. Second, the average property owner there was then paying 10.4 percent of her or his tax bill to support the township framework, according to research Muir conducted and reported. Comparable percentages in other Illinois counties were hovering around 2 or 3 percent of the total property tax bills. Third, and perhaps of greatest significance, was the stance that Muir took publicly. Between 9 June and 26 June he published a series of five highly critical research-based articles about townships, beginning with "Township Government: Needed Unit of Government or Outdated Bureaucracy." The last element of the series appeared just before the big tax protest meeting of 29 June and contained the first reference in print by Muir to the idea of reversion to commission government for Franklin. At the meeting, which Muir chaired and about six hundred (Muir's estimate) attended, township abolition was among several possible targets on the table. Nevertheless, the Property Tax Awareness Committee (P-TAC), an outgrowth of the meeting, decided, with Muir's encouragement, to pursue a petition-initiated referendum on abolition of all 12 Franklin townships, even though, by the time they got organized, they had hardly a month in which to gather the necessary signatures.[14]

"Townships were a bad choice," Horton told me almost a year later. For one thing, P-TAC was the first antitownship group to confront TOI's 1997 silver bullet, Public Act 90-112 (H. B. 480). Nobody had ever had to collect signatures from at least 10 percent of the voters in each township of an Illinois county in order to call a referendum on abolition of townships. Signatures needed, according to a Muir article, ranged from 682 in Benton Township to just 30 in Northern. For another thing, P-TAC, lacking any knowledge of STOP or other recent efforts in Illinois to take down townships, greatly underestimated, Horton confided, the determination of both local officials and TOI to save the township way of life. Third, much of the three-month window for passing petitions had closed before they (about ten men and women) began on 27 July to solicit public approval through the petition process. Fourth, P-TAC operated under the assumption throughout the short campaign and inevitable challenge period that the supermajority rule, from Public Act 89-365 (H. B. 438), would apply if the issue actually went to the voters in November. Muir emphasized that point again and again in his *News* items and no doubt behind the scenes with Horton and P-TAC as well. "According to the ILCS [Illinois Compiled Statues]," Muir wrote on 19 July, "if the required number of signatures are obtained, 75 percent (9-of-12) of the townships must pass the referendum." Yes, the statute still stood, as revised in 1995 after enactment of the supermajority bill from Rep. Ann Hughes; but Judge Byrne had already declared the supermajority portion of the bill unconstitutional for DuPage County (chapter 8). My own feeling, in part the result of email conversations with practicing attorneys, is that another judge in a case involving township county other than DuPage would also recognize the divergence between the statute and the 1970 Constitution and rule that a simple majority vote would be sufficient to allow for discontinuance of townships countywide. An observer has to wonder how much more aggressive P-TAC might have been in gathering signatures and pushing its case for reform through the system had Horton, Muir, and others believed that a simple majority in November almost certainly would have prevailed.[15]

Despite their silver bullet and abolitionist ignorance of the DuPage ruling regarding the supermajority law, TOI and local township officials took the P-TAC threat seriously. Franklin County, in the year 2000, had all the makings of a real test of township staying power. While much of Illinois and the nation had enjoyed a prosperous 1990s, Franklin County had suffered mine closures and seen no significant emergence of replacement jobs. People from all over the county had expressed distrust of their township assessors and apparently felt the 10 percent township cut of the property tax pie was too hefty. Comparisons with neighboring Williamson and other Little Egypt counties were inevitable and might favor the commission-county model. Franklin had a scrappy city editor and a tax-protesting newspaper publisher, both of whom seemed determined to bring down townships. And the county had Charles Horton, a man who had stepped up reluctantly to accept a leadership role in the protest movement and was clearly not someone that township proponents could easily vilify—as they had Bob Anderson and Michael Richardson.

Bryan Smith came to Franklin County for a meeting in Zeigler, at the Six Mile Township building, in late July. Several dozen township officials joined him, as did reporter Jim Muir. Six Mile's supervisor, Rannie Snider, set the stage for their distinguished guest from Springfield by reminding fellow officials about all the valuable services they had been providing; and he entertained comments from several local townshippers who agreed that the 601 miles of township roads were likely to be a strong selling point for retention of townships and their road districts. Muir reported that Smith then rose to offer a series of suggestions for combating P-TAC, suggestions such as the inclusion of all the county's townships in any battle plans and the need to educate the public about the good that townships performed. Smith, according to Muir, spent considerable time outlining ways for townships to cut costs through sharing of officials (like the assessor), consolidating road districts, or even consolidating townships. At meeting's end, officials chose Snider to head the newly formed Friends of Township Government (FTG).[16]

Derek Johnson, Benton Township supervisor, swung into action with a lengthy and cogent *News* guest column, which appeared on 5 August. A main premise of his argument (to retain townships) was his prediction that abolition of all townships would save few taxpayer dollars and might end up costing even more. Salaries of the 13 general assistance employees in Franklin, for instance, came from the Illinois Department of Human Services. Would the state pay for similar personnel in a commission county? He did not venture a guess but left the reader to conclude that the state might withhold in such a circumstance. Road maintenance was another sticky point for Johnson. As we have seen in other counties, road district employees routinely earn less than county road department employees in the same county. For that reason and others, Johnson argued that cost per mile of road maintained was then and would still be less under township control than under county control. As for assessors, Johnson noted there were already three multi-township assessment districts in the county; and other townships might wish to consider joining or forming new ones.[17]

Johnson's other main premise for disagreeing with the P-TAC solution to Franklin's taxpayer dilemma deserves special note, in light of new legislation that offered a change in a mode of governing counties that had been in place since statehood. Here is what Johnson wrote in his guest column:

> If township government is abolished, Franklin County would automatically change to a commissioner form of government. Three (3) county commissioners would be elected to govern the county....These three commissioners are elected at-large, meaning they all could be from one town or different areas of the county....Do the residents of Franklin County desire three individuals to control the entire county?

H. B. 3132, which became Public Act 91-846 when Governor Ryan signed it into law on 22 June 2000, allowed commission county voters to opt for five instead of three commissioners and to opt, whether choosing to go with three or five, to have them serve specific districts of the county. Therefore, even if township laws from the nineteenth century were to prevail over laws from the twentieth century calling for a county board based on the principle of one-man, one-vote districts, voters could bump their commission up to five members and force them to run in districts (modified every ten years after the Census of Population) rather than run at large. Charles Horton told me that P-TAC was aware of this change in Illinois law and ready to use me that P-TAC was aware of this change in Illinois law and ready to use the legislation to argue against the consequences that Johnson was implying.[18]

While P-TAC struggled to meet the 21 August filing deadline, the *News* recorded the progress and Muir continued to editorialize. Horton told Muir, for a story that ran on Tuesday, 8 August, that he remained "very optimistic that [they] would get the matter on the ballot." Muir, a couple days later in his regular column, wrote a scathing comparison of the competing groups. One was "all volunteer, non-partisan…[with] no paycheck or political favor at stake." The other's "main objective was not to figure out how to possibly lower their tax levies and help the struggling taxpayers of Franklin County, but was focused on how to save the collective 'keesters' of those drawing a paycheck from township government." By Sunday, 13 August, P-TAC figured it had sufficient signatures in seven of the dozen townships; and Horton thanked voters for their interest in the issue. Scott Hines talked to Horton later that week and reported eight and maybe nine townships were in the safe column with respect to signatures. Horton emphasized to Hines that the real debate ought to be between filing date and election day, not before the board had certified the referendum for the ballot. The *News* on 18 August carried two letters whose writers spoke against dissolution. One came from the supervisor of Frankfort Township, the other from the executive director of Franklin County Senior Services, an agency whose director claimed dependence on township support for transportation and facilities.[19]

Finally, on the morning of the twenty-first, P-TAC leaders dropped off petitions containing more than 3,700 signatures. They had averaged nearly a thousand signatures per week, since beginning solicitation in late July. Among the signers were the chair of the county board and some township trustees. Horton, like Bob Anderson in McHenry County, really enjoyed the petition process, finding the people extremely receptive to his argument. The smaller (in population) townships proved, however, a challenge even to Horton. "You might think 30 signatures would be easy to obtain," he told me, but township proponents worked hard in sparsely populated Barren, Eastern, and Northern townships to keep people from signing. After all, the silver bullet meant that P-TAC only had to fall short in one township, no matter how peripheral, to have the whole operation fail.[20]

When objections followed, Horton seemed more or less resigned; Muir went sarcastic. The first objection, a rather simplistic one, challenged the validity of all signatures because some folks had not written the name of their county beside their name and address. Horton responded that "Franklin County" appeared in four places on each petition sheet and noted that only residents of Franklin County were to sign. In a prepared statement, Horton also questioned the intent of anyone who would seek to take the decision about form of government out of the hands of the people. "'As far as I know,'" Muir quoted from the press release of Horton, "'there has been no voter resistance, but all resistance has come from township officials.'" A more substantive set of objections arrived on Monday, the twenty-eighth. These came from officials representing most of the 12 townships and cited "wrong addresses, invalid signatures and multiple signatures."[21]

Muir, who had (according to Horton) endured threats to himself and his family during the abolition effort, used his column to ridicule township officials for their poor citizenship. His diatribe sounds to me much like the letter that Michael Richardson sent to the Rock Island County Board in 1996 when he got word that the board might fail to certify his referendum for the ballot despite a court order to do so (chapter 8). Muir seemed to recognize, as he composed his 30 August column, that unless someone could convince FTG to relent, all the energy he and others had put into the campaign would prove futile. Richardson threatened to sue anyone on the board who voted against certification. Muir warned township officials that nearly four thousand voters wanted their referendum and that the voters would always prevail.[22]

One of the reasons Muir would have taken such a hard stance in his column was the FTG tactic he reported elsewhere in the *News* that day. Township officials were apparently asking selected voters to recant and request removal of their signatures from the submitted petitions. According to Muir, certain officials, especially those from townships where P-TAC had the slimmest of signature cushions, noted the names of those who had signed, went to them in person or by telephone, and asked them to fill out an affidavit attesting to the fact that they no longer wished to be on record in favor of an abolition referendum. "In some cases," said Muir, "the number of affidavits filed in a township coincided with the number of names needed to push the total number of signatures below the 10 percent needed to put the referendum in the hands of the voters in the general election." The attorney for the state's board of elections, said Muir, had told him that Illinois law forbade removal of signatures once circulators had filed their petitions. An FTG attorney later confirmed that opinion.[23]

Readers should by now be conversant with the makeup of the local election board (county clerk, state's attorney, and clerk of the circuit court) for such petition challenges and be able to predict the sequence of events that transpired when the board convened in Benton on Wednesday, 6 September 2000. Attorney (and Benton Township Trustee) David Garavalia argued for strict adherence to Illinois statutes, as he proceeded township by township. He pointed out on several occasions how objections, if sustained, would drop township signature totals below the 10 percent figure that the law required. Rebecca Whittington, a Benton resident practicing law out of Carbondale in nearby Jackson County, pleaded for a fair and liberal interpretation of issues like omitted middle initials in signatures and the use of a valid post office box instead of actual street address on petitions. Garavalia did withdraw the objection claiming every voter should have written "Franklin County" at the end of the signature lines. Parties agreed at hearing's end that the 10 percent rule (the silver bullet) would prove the deciding factor, with Goode, Frankfort, and Six Mile townships appearing to control the fate of the referendum.[24]

Goode, Frankfort, and Six Mile did the trick for TOI and FTG, but failure to reach 10 percent in any one would have sufficed. Horton expressed his disappointment; supervisors Tom Broy of Frankfort, William Crocker of Goode, and Rannie Snider of Six Mile said they had no comment. Muir devoted his column the next day to bullies. In it he recalled how the bully in his fourth grade class finally got his comeuppance when the favorite target of his ridicule punched him in the nose. "Bigger, stronger and with many more assets, Franklin County township officials outmuscled the motley crew at P-TAC and in turn took the opportunity away from taxpayers to have a say in the form of government they wanted."[25]

One of the supervisors whose township had come in under the 10 percent standard wrote to the *News* to explain his actions. Tom Broy said he did not oppose having people vote on the issue. He just felt some voters had signed the petitions out of frustration with "an unfair tax system" and needed more time to cogitate about the differences between the commission and township models of county government. Besides, he wrote, "Voting against townships is voting for bigger government."[26]

Rebecca Whittington, who, according to Horton, charged P-TAC a "reasonable" $8,000 for her legal services, filed an appeal of the electoral board ruling with the circuit court. She alleged therein several errors on the part of the board but focused mainly on the fact that many of the sustained improprieties in street addresses stemmed from the failure of the county clerk (an electoral board member) to update addresses all over the county. People had given their true addresses, argued Whittington, in the petitions; but Clerk Dave Dobill's official records still showed an old address. When the two addresses failed to match, the board disqualified those signatures.[27]

Judge David Underwood of the Second Circuit Court, on Wednesday the twenty-seventh of September, granted FTG a fateful continuance, which guaranteed no ballot slot that November for P-TAC. Why the continuance? FTG had changed attorneys; and neither John Hoffman, the new attorney for the township side, nor any member of his Carbondale law firm could appear before the judge prior to 11 October. Muir, of course, had to point out the brilliance of that FTG maneuver:

> I mean, think about it; you changed attorneys midstream for a hearing in which all the evidence had already been presented (when an attorney's job is mostly done). But by pleading for time to bring your new counsel up to speed, you got the court date set back so far that even if the final decision went against you it would be moot: the election would already be history.[28]

Final arguments on the petitions took place on 29 November, and a ruling came one week later. Hoffman praised the electoral board for its efforts to determine fairly the validity of the names and addresses on the petitions, noting that nothing thus far precluded the disgruntled taxpayers from "mounting another petition drive." Whittington chose to remind the court that many of the apparently bogus addresses had turned out to be accurate—once the county clerk's staff updated the address roster. Judge Underwood found in favor of the electoral board, a board that (Muir reported) consisted of the man running the office with the address problem and two who were not running for re-election at the time of the board ruling. Muir's article about the final decision concluded with a reminder to readers that every township official had to run for his or her seat again in the coming April elections. Readers who might wish to challenge an incumbent or seek a vacated seat had to "pick up an intent-to-run form before Dec. 30."[29]

A few months after the judge upheld the board ruling, I sat down with Charles Horton to obtain his perspective on the whole 2000 affair. He seemed convinced that P-TAC would have gotten the measure on the ballot if he and his circulators had started earlier—and thereby had time to gather more cushion signatures. Horton maintained his neutrality on the matter of township abolition, implying that he might have voted to retain in the end. To Jim Muir he ascribed whatever success P-TAC achieved in its short life. Muir kept up the relentless press pressure, helped develop strategy, and took a lot of the heat that Horton and his petition passers would have otherwise had to bear. Imagine, I thought as we talked, what Bob Anderson might have been able to accomplish if a Jim Muir had worked at the *Northwest Herald* in the 1990s. For Horton, his biggest surprise throughout the Franklin effort was how strong the local Democrat machine had proven itself and how resistant township officials had been toward affording the public its right to decide at the ballot box. Politicians, however, never publicly attacked Horton; and one or more judges even called to encourage him and provided off-the-record advice. P-TAC had only enough funds to pay it attorney fees and nothing for advertising that, for instance, could have explained the implications of Public Act 91-846 for changes in the traditional county commission or to provide comparisons with neighboring nontownship counties. Best of all, for Horton, P-TAC did succeed in giving citizens a new view of local politics. People found they could (almost) effect change in the system, and some asked Horton to try again. Talking to him that day in April 2001, however, I could tell his heart was not in it for an encore. The last thing he said was there was no way they could have carried the day at the election in nine or more of Franklin's townships. He still thought they needed a supermajority. As for Jim Muir, he left the *News* early in 2001 for a position with the *Carbondale Southern Illinoisan.*[30]

The Associated Press Series

Whereas November 2000 had ended on a positive note for Bryan Smith, with the confirmation by Judge Underwood of the fact that P-TAC had failed to gain the ballot, December of township government's year 150 did not end on such a high note for TOI's youthful leader. John Kelly, Christopher Wills, and host of local reporters and editors from Carbondale to Waukegan saw to that. Kelly and Wills co-authored the three main articles in the 20–22 December Associated Press (AP) series, which more than a dozen Illinois newspapers carried. Many editors, after receiving access to the AP articles a month in advance, assigned reporters the task of generating local stories that followed up on or expanded what Kelly and Wills found during their many months of digging. Several editors wrote accompanying editorials, which turned out to be almost universally negative about townships government. In January of 2001 a number of out-of-state papers carried a condensed version of the three main AP articles, and one Illinois paper continued the pressure for reform with a critical spread of its own.

The AP called its series "Illinois Townships: Bargain or Boondoggle?" Subscriber newspapers were free to title the articles as their editors saw fit and to cut or reorder parts. The *Bloomington Pantagraph,* for example, named the first part "Township Governments Can Be Inefficient, But Few are Noticing" while Rock Island's *Argus* went with "Townships: Are They Worth It?" Decatur's *Herald and Review* ran the same story on the Web as "AP Investigation Questions Need for Government Form."[31]

Part 1 provided historical background; lists of township officer duties; and comments from Smith, various township officials, and observers like a professor and a legislator. The most negative aspect of the first AP segment dealt with the surpluses many townships were holding in their banks. Newspapers received a table containing information, by township, on revenue, spending, and (1999) surpluses reported. Local reporters perused the table for local examples of large surpluses and then prepared stories about those townships. The *Argus* reproduced that portion of the AP table dealing with four nearby counties. The *Pantagraph* created a table listing 17 townships that the AP said had at least three years of surplus funds on deposit. Officials defended the practice of holding large surpluses by noting that they were anticipating large purchases, like a backhoe, or projects, like a bridge, and that they needed a reserve for unanticipated events, like a health emergency of a general assistance client.[32]

Part 2 began with the story of a Franklin County resident that had received general assistance funds from Six Mile Township. Kelly and Wills talked with Jim Muir, Charles Horton, and others as the Franklin abolition effort wound down to its conclusion. They talked with Bob Anderson and his township's supervisor, Donna Schaefer, about their views on the topic of townships. Services provided and returns on the taxpayers' dollars were a big focus of the second series part. The *Herald and Review* did a story about Macon County's Bryan Smith, while Hampton Township's health clinic was the topic of an *Argus* article. Sonya Klopfenstein wrote a companion piece for the *Peoria Journal Star* in which she compared township services with county services.[33]

Part 3 dealt largely with abolition and the difficulties facing anyone seeking that drastic remedy for township shortcomings. Charles Horton told the AP that he believed townships survived because they had stacked the deck against their own people, the taxpayers, while Bryan Smith attributed persistence to the fact that the people really wanted to retain their townships, as they had made known again and again. Along with part 3, the AP distributed an article explaining how one would go about trying to abolish all a county's townships, given the laws making the process extremely difficult. The *Argus* packaged part 3 alongside a Stephen Elliott farewell to Michael Richardson, who was soon to leave for New England. In Champaign County, the *Champaign News-Gazette* ran its surplus stories on the same date as part 3.[34]

At least five Illinois papers also offered editorial commentary. From none of the editors did Executive Director Smith get much of a Christmas present. "Seventeen of the state's 102 counties," said the *Bloomington Pantagraph,* "have already abolished township governments. Many more need to follow suit. And the Legislature needs to do everything it can to make the process of consolidation or elimination easier." What the editorial staff meant to write was that, while 2 of the 102 had abolished townships after brief trials, the other 15 had done relatively well without ever adopting this Yankee innovation. Although editors at the *Carbondale Southern Illinoisan* admitted rural residents might receive better service from their townships than a more centralized (county) source, "on balance...township government appears to have outlived its usefulness and has become an unnecessary burden upon the taxpayer." In the *Chicago Tribune* Smith might have read how "It's time to give Illinois townships a decent burial. Pols who defend these obsolete artifacts need to first consider one question: If townships are so all-fired necessary, why do 17 counties already function perfectly well without them?" In Peoria, *Journal Star* editors wrote, "With taxpayers demanding less government, it's indefensible that the most bloated government around continues to operate in their backyards. Lawmakers should shut them down." And in the *Springfield State Journal-Register* ran an editorial that said in part: "Seventeen counties in Illinois...get along fine with no township government....It is time for the remaining 85 counties to take a hard look at doing away with townships and centralizing their functions."[35]

On into 2001 fallout from the AP series continued to drift past the public eye. Part of that public was out of state, reading a condensed version of the Kelly-Wills stories and viewing a photograph of Bob Anderson in his barber chair or one of Michael Richardson in the spare bedroom that he had devoted to abolition and his other causes. Editorial staff at the *Journal Star* brought prose contributions from Anderson and Bryan Smith to their "Viewpoints" page on the first Sunday of 2001, along with a piece from another township spokesman and one from a Peorian who agreed with Anderson's proposed remedy. Both of the other *Journal Star* guest commentators that

January Sunday were Peoria lawyers. Timothy L. Bertschy, a past president of the Illinois Township Attorneys Association and in the employ of townships for over two decades, asked readers to ignore all the anecdotal township tales that were circulating about misdeeds and the like. Instead the thoughtful voter would ask if whatever was to replace the township would be better or worse. He opted for worse, of course. Picking up the pen to help Anderson make the opposite case was John C. Parkhurst, the same Parkhurst who had chaired the Local Government Committee during the 1970 Constitutional Convention (chapter 6). He confessed again to having caved, in 1970, under the pressure from township officials but urged amending now the constitution "to eliminate this waste and inefficiency." Anderson posed a simple question at the end of his op-ed piece, asking whether Illinois residents of a township county would institute township government today if none existed.[36]

Township Perspective carried reactions to the AP series well into the year 2001. Smith presented a vigorous defense of townships in his February column. Noting the critical *Chicago Tribune* editorial of 20 December (see above), he chose instead to quote from a more favorable 1994 editorial out of the same source, an editorial that appeared right before the McHenry County abolition referendum. Personally and professionally committed to township longevity, Smith, as this column demonstrated, sometimes seemed truly perplexed by the claims and charges of the antitownship crowd. Barbara Weyrick, assessor for Lyons Township in Cook County and chair of TOI's Education Committee, used her regular column in April and May of 2001 to examine the generally defamatory Associated Press series, etc. She accused Kelly and Wills of having their minds made before they began their research for the series. Her hope was that the information she was presenting in the magazine would offer officials ammunition to counter the criticism that was still circulating. Weyrick commented, in April, on the prominence that Kelly and Wills gave Anderson and Richardson, "old familiar names," and wondered why no "fresh faces come [had] forward to challenge townships." Maybe she missed the part about Charles Horton.[37]

Looking Back at What Came Before

Illinois could well have ended the last year of the township sesquicentennial with 12 fewer townships, with 1,421 instead of 1,433. Franklin County had the makings of trouble for the Illinois township, as an example to which the determined Bob Anderson or the next Michael Richardson could then have pointed and said, "If Franklin County can successfully abandon townships, then our county ought to be able to do as well or better." Yet the Franklin dozen persisted, as they had for over a hundred years, doing the people's business, in part, perhaps in large part, because of the silver bullet and petitioner perception of the supermajority law.

Petition Signatures and Referendum Votes

Among the more confusing aspects of this research has been the task of tracking, from year to year, the number of signatures necessary to call a referendum and the number of referendum votes needed to secure a change in government at the grassroots level in Illinois. In an attempt to summarize these moving targets, I have created three tables, two dealing with signatures and one with votes.

Most of the 438 countywide referenda over the issue of township organization, roughly 80 percent, were attempts to establish the township mode of local government rather than to discontinue it. One of the principal reasons why counties like Perry could vote year after year on the issue of adoption was the miniscule number of signatures required by law to place the question on the ballot from 1851 until 1995 (table 1). Legislators made it easier to adopt than to abolish, because township organization quickly became the norm rather than the exception, once the courts had resolved thorny issues in the 1850s and the tide of diffusion gained momentum just before and right after the Civil War. On a good day, around the courthouse square of almost any Illinois county, a proponent of this Yankee innovation could gather 50 signatures to present to the county commissioners. Only in the 1990s, long after the last (1957) attempt by a county to adopt townships, did TOI acquiesce to criticism about how much more difficult it was to collect the signatures needed to call a referendum to discontinue than one to adopt. Miller agreed to a modification of the statutes to call for a minimum of 10 percent of the legal voters in a county on petitions calling for adoption.

Table 2 reveals how much more effort legislators have put into the matter of signatures on petitions to discontinue than they have on those needed to call for a referendum to adopt township organization. Voters have had the legal authority to petition for discontinuance since the Township Law of 1851 went into effect. From 1851 until the early 1870s, the signatures of the same 50 men who had asked the county commissioners, in one year, to place a referendum to adopt on the ballot would have been sufficient on another petition, the next year, calling on the new board of supervisors to place the issue of township discontinuance on the ballot. By the time the Township Law of 1874 left Springfield, with the 50-signature requirement on petitions to adopt intact, the minimum number needed for calling a discontinuance referendum had jumped to 20 percent of those voting in the previous presidential election. In an Illinois county, for instance, in which 10,000 had voted a couple years previous, the increase would have been from 50 signatures to 2,000.[38] During the 1930s, legislators, under pressure to help taxpayers cope with the Great Depression, eased the discontinuance petition standard for most counties by dropping the minimum to 5 percent of those voting at the last presidential election. Counties in which 75,000 more had voted at that last election remained on the 20 percent standard, however.

Table 1 Signature Requirement: Petitions to Adopt Township Organization

Period	**Minimum Number of Signatures (or Percentage of Voters) Needed on Petitions to Call for Referendum to Adopt Township Organization**
1849	No signatures required. Township Law of 1849 provided for referendum in every county.
1850	No provision for referendum in counties that failed to adopt in 1849.
1851–1995	50 signatures required.
1995–present	10 percent of a county's legal voters must sign petitions.

Sources: Illinois, General Assembly, *Laws of the State of Illinois Passed at the First Session of the Sixteenth General Assembly (1849)* (Springfield: 1853), 190–224; Illinois, General Assembly, *General Laws of the State of Illinois Passed by the Seventeenth General Assembly at Their [Regular] Session* (Springfield: 1851), 35; Illinois, *General Assembly, Laws of the State of Illinois, Eighty-ninth General Assembly, 1995* (Springfield: 1996), 3,811.

Big changes in the signature situation regarding discontinuance took place from the late 1980s on into the 1990s, and mainly affected in a negative manner those wishing to challenge the township's hegemony in any of the 85 township counties. Public Act 85-1146, in 1989, standardized the requirement for petition signatures related to discontinuance of township organization at 10 percent (dropping the 5 and 20 rule that had existed since 1933). That 1989 law also specified a period of just 90 days for gathering the signatures in an effort to discontinue, but left unchanged the unlimited period for a petition campaign to adopt townships. A couple years later (1991), the Eighty-sixth General Assembly, at the request of TOI, stiffened the discontinuance signature target with Public Act 86-1486, which called for basing the 10 percent figure on the number of county voters registered on the day of petition filing. Legislation in 1993, effective 1 January 1994, accidentally reduced the number to the 1989–1991 level (10 percent of those voting at the last presidential election) while a corrective measure passed in 1995, returning the requirement to the 1991–1993 level. The most recent ratcheting upward of the signature hurdle on the discontinuance side of the ledger, the silver bullet, made mandatory the collection of signatures sufficient to exceed 10 percent of the voters in every township of the county. Public Act 90-112, however, did replace petition filing day, as the date for calculating the number of voters, with a day in the recent past—in order to provide all parties with a clearer idea of the base number of registered voters against which officials would measure the number of signatures collected.

Table 2 Signature Requirement: Petitions to Discontinue Township Organization

Period	**Minimum Number of Signatures (or Percentage of Voters) Needed On Petitions to Call for Referendum to Discontinue Township Organization**
1849	Not applicable.
1850	No provision for discontinuance of township organization.
1851–1874	50 signatures required.
1874–1933	20 percent of those voting in the last presidential election.
1933–1989	20 percent of those voting in the last presidential election in counties in which more than 75,000 voted and 5 percent in counties in which fewer than 75,000 voted.
1989–1991	10 percent of those voting in the last presidential election.
1991–1993	10 percent of legal voters as of the date of filing.
1994–1995	10 percent of those voting in the last presidential election.
1995–1997	10 percent of legal voters as of the date of petition filing.
1998–Present	10 percent of the legal voters in each township, based on the number registered to vote in each township on the date registration closed before the regular election next preceding the last day for petition filing for the referendum in question.

Sources: Illinois, General Assembly, Laws of the State of Illinois Passed at the First Session of the Sixteenth General Assembly (1849) (Springfield: 1853), 190–224; Illinois, General Assembly, General Laws of the State of Illinois Passed by the Seventeenth General Assembly at Their [Regular] Session (Springfield: 1851), 78; Harvey B. Hurd, ed. and comp., The Revised Statutes of the State of Illinois. A. D. 1874 (Springfield: Illinois Journal Company, 1874), 1068; Illinois, General Assembly, Laws of the State of Illinois Enacted by the Fifty-eighth General Assembly at the Regular Biennial Session (Springfield: 1933), 1108–9; Illinois, General Assembly, Laws of the State of Illinois, Eighty-fifth General Assembly, 1988 (Springfield: 1989), 4058; Illinois, General Assembly, Laws of the State of Illinois, Eighty-sixth General Assembly, 1990 (Springfield: 1991), 3,998; Illinois, General Assembly, Laws of the State of Illinois, Eighty-eighth General Assembly, 1993 (Springfield: 1993), 888; Illinois, General Assembly, Laws of the State of Illinois, Eighty-ninth General Assembly, 1995 (Springfield: 1996), 2950; Illinois, General Assembly, Laws of the State of Illinois, Ninetieth General Assembly, 1997 (Springfield: 1998), 2338.

For most of the century and a half that has elapsed since township government became an Illinois option, discontinuance has required fewer votes than adoption (table 3). According to the township legislation of 1849, *People v. Brown* in 1850, and the revised township laws in 1851, adoption required the approval of a majority of a county's voters, whether they voted or not on the referendum. Discontinuance, first an option in 1851, took merely the approval of a majority of those voting at the referendum. *People v. Warfield* (1858) applied that standard of "majority" to subsequent adoption referenda as well, a standard that still exists. Both adoption and abolition faced the same definition of majority until the 1870 Constitution lowered the standard further for discontinuance to a simple majority of those who marked ballots for or against keeping townships. Thus, it was easier for Johnson County to disadopt in 1876 than it had been to adopt in 1870. The Township Law of 1874, which on 1 January 1994 became the Township Code, following passage and signing of Public Act 88-62, confirmed the discrepancy between vote requirements for adopting or discontinuing. Then, in 1995, came H. B. 438 and Public Act 89-365, the supermajority law, calling for majorities favoring discontinuance in at least three of every four townships, provided those townships contain at least half the population of the county. Almost a year later, Judge Robert E. Byrne of the DuPage Circuit Court declared the supermajority section of the law unconstitutional in his jurisdiction; but the law remains on the books until other challenges to the township system lead to other court action. For the moment, however, except in DuPage County, the laws in force seem to make discontinuance much tougher to achieve than adoption.

Table 3 Vote Requirement: Countywide Referenda to Adopt or Discontinue Township Organization

Period	**Votes Required to Adopt Township Organization**
1849–1858	Majority of voters in county.
1858–Present	Majority of voters voting at the election in which the referendum occurs.
Period	**Votes Required to Discontinue Township Organization**
1849–1851	Not legally possible.
1851–1870	Majority of voters voting at the election in which the referendum occurs.
1870–1995	Simple majority of those voting on discontinuance.
1995–Present	Simple majorities in at least three-quarters of the townships provided those townships contain at least half the county's population. Simple majority applies in DuPage County.

Sources: Illinois, General Assembly, *Laws of the State of Illinois Passed at the First Session of the Sixteenth General Assembly* (Springfield: 1853), 190; *The People v. Brown et. al.*, 11 Ill. 479–43 (Ill. Sup. Ct. 1850); Illinois, General Assembly, *General Laws of the State of Illinois Passed by the Seventeenth General Assembly at Their [Regular] Session* (Springfield: 1851), 35–36 and 78; *The People v. Warfield,* 20 Ill. 160–66 (Ill. Sup. Ct. 1858); Illinois, *Constitution of the State of Illinois* (1870), article X, section 5; Illinois, General Assembly, *Laws of the State of Illinois, Eighty-ninth General Assembly, 1995* (Springfield: 1996), 3811.

If the foregoing discussion of signature and vote requirements seems confusing, I ask the reader to consider how citizen-reformers must have felt over the years. Imagine how those possessing better information, perhaps county commissioners or county board members in township counties, would have had the advantage if there existed a desire to thwart change and maintain the status quo. Signature needs were simple to calculate when the statutes called for just 50 names but were not so simple once legislators tied petition signatures to number of voters. Confusion over what constituted a majority for township organization in the 1849 referenda kept several counties from organizing their townships in 1850, even though other counties in the same predicament went ahead and became part of the 24 charter township counties. Lack of knowledge about the Byrne decision in DuPage regarding supermajorities almost certainly discouraged members of P-TAC from more vigorous pursuit of petition signatures in Franklin County and may, as a result, have allowed a legitimate opportunity for countywide discontinuance to slip away in 2000.

Referenda Review

Franklin, if Charles Horton and Jim Muir had gotten their wish, would have been countywide referendum 439, the first since 1994. If P-TAC had obtained a few more signatures in a handful of townships, the year 2000 would have been the thirty-first calendar year in which one and only one county voted on whether to adopt or abolish all its civil-political townships. So the number remains at 30, compared to 51 calendar years with multiple countywide referenda, beginning with 1849. All the countywide referenda after Morgan in 1957 were efforts to abolish; but, for the twentieth century as a whole, beginning with 1900 and ending with 1994, 29 of the 47 attempts arose because citizens in those counties wanted to try this nineteenth-century innovation. A dozen of those 29 attempts to extend the township tide were in a single county—Perry. Overall, countywide referenda to adopt townships accounted for 349 of the 438—close enough to call it 80 percent. For every abolitionist over the township century and a half, there were four township champions. For every Horton there were four Haines. For every voter seeking to save a remnant of the Upland South there were four who sided with the Yankees.

P-TAC might have tried picking off Franklin County townships one at a time, choosing perhaps a scapegoat township wherein the taxpayers were blatantly bilious about their quadrennial assessments. If Muir and Horton had isolated one township and succeeded in gaining ballot access in the fall of 2000, that referendum would have marked only the fifth time that a portion of an Illinois county had voted on township organization. Based on results elsewhere of attempts to discontinue a single township, P-TAC would have failed, as did the League of Women Voters in 1972 and 1973 (Cook County) and Bob Anderson in 1998 (McHenry County). The only successful knock-out of some townships in a county involved the special circumstance of Chicago in the early 1900s. Voters there said shut them down, and the mayor complied.

Franklin County would, by voting in 2000, have joined neighbors Hamilton, Perry, and Williamson as counties with 10 or more countywide referenda on the matter of township organization (figure 26). Bumping Franklin from 9 referenda to 10 and into the highest category would have produced a contiguous cluster of seven counties, with the leading county on the map, Williamson, as the anchor of this core of uncertainty about the township way of doing the people's business, deep in the heart of Little Egypt. Even without Franklin in the top category, however, the southeast-northwest tidal line that I noted on other maps in this series stretches obviously from Johnson in the south to Cass on the north.

Two of every three Illinois counties voted more than once but fewer than 10 times on township organization. Franklin and the other 22 counties in the "5–9" category nicely reinforce the dominant southeast-northwest trend. The "2–4" category, with 47 member counties, holds sway in the central portion of the state but shows outliers in every other area as well, from the Ohio to Wisconsin, and Indiana to the Mississippi.

Most of the 19 counties that voted just once form a tight cluster in the northeastern quadrant, a cluster that includes about half the counties that voted to adopt in 1849. If not for Bob Anderson, of course, McHenry would be a member of that group, too. Outliers of the cluster of counties that voted only once included Pike and Stephenson, both of which adopted in 1849, and Alexander and Hardin, which voted in 1849 with no intention of adopting then or ever. Contiguous to the big cluster with one referendum are Kankakee and Ford, counties that did not exist in 1849. Once citizens of these counties did organize, in 1853 and 1859 respectively, they quickly opted for townships and the board of supervisors.

That leaves Effingham County, which adopted 6 November 1860. Effingham stands well south of the state's east-west centerline, amid counties in the "2 to 4" group and only a few miles from the margin of the "5 to 9" category. As I indicated in chapter 1, while reviewing the process that I followed to document these referenda, I am confident a few slipped through the net. Effingham may account for one or more of those undocumented occasions when voters decided the fate of township organization, although a thorough review of meeting minutes from the county commissioner era of Effingham for the 1850s and county board minutes in the 1860s revealed no other votes on the matter. Be that as it may, Effingham just looks out of place in figure 26.

The three counties in the "zero" category, on the other hand, are right where I would expect to find them—tucked in next to counties that never adopted. Scott is part of the cluster of four, with Calhoun, Menard, and Madison, just west of Springfield. Until the general assembly split off Scott

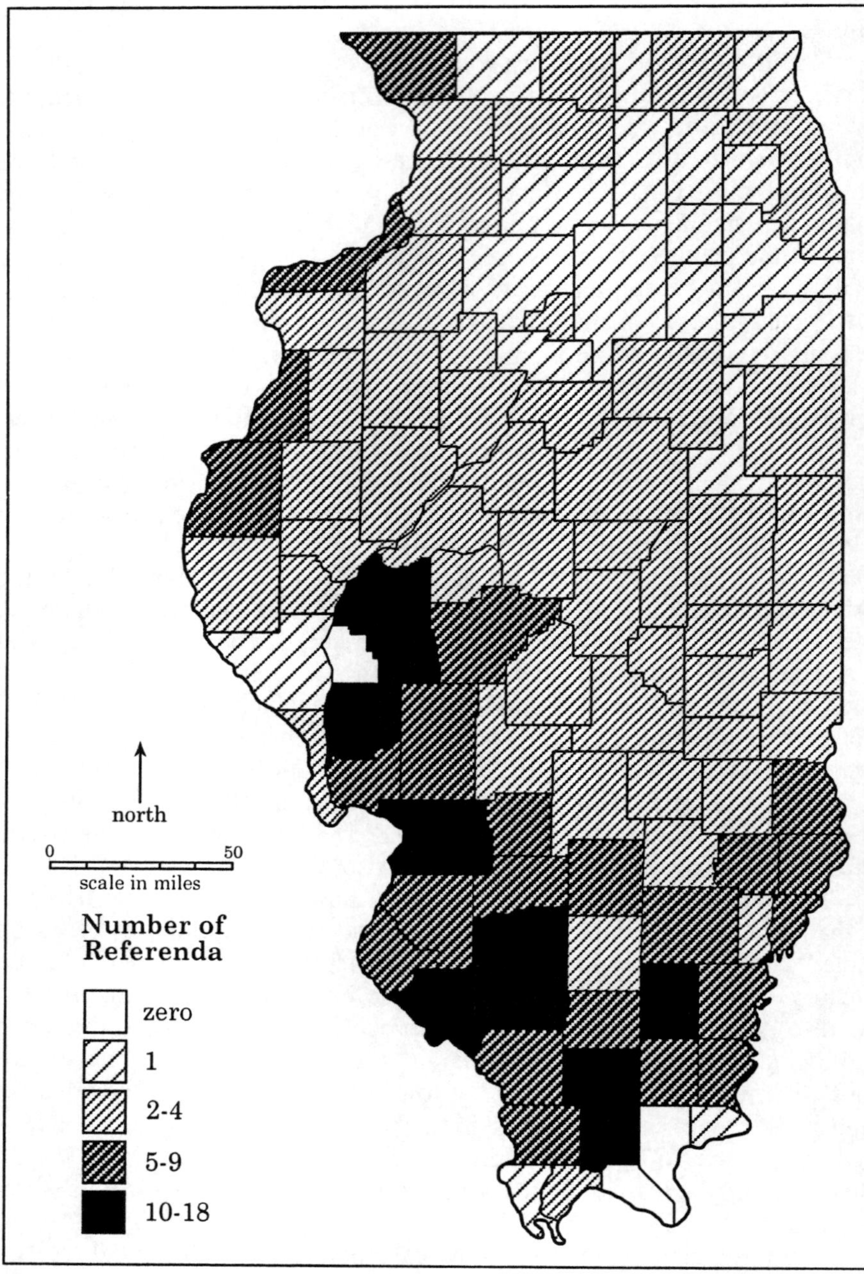

Figure 26 Countywide Referenda on Township Organization, by County: 1849–1994.

in 1839, it was a part of Morgan and earlier, the predecessor of both, Madison. Adjacent to one another, hard by the Ohio, and lightly populated (with folks still looking more to Kentucky in many instances than to Springfield or Chicago), Massac and Pope had little reason to even consider what Wead and Haines thought so critical to a civilized polity in the nineteenth century. Thus the 17 commission counties count among their number the three counties that never put township organization on the ballot and the two, Williamson and Morgan, that voted on it most often, as well as representatives from each of the other categories I have designated in figure 26.

Typology

Another way to summarize what has happened to Illinois's counties with respect to the acceptance, rejection, discontinuance, and reaffirmation of township organization is with a typology that takes into consideration the number of times a county's citizens voted countywide on the matter of township organization and decisions they rendered (table 4). Rather than code by numbers or letters, I have chosen to represent each class with the name of a county that seems to typify it.

The three counties of the Scott Class, Massac, Pope, and Scott, stood out in figure 26, and they stand out here for the fact that they never voted, even in 1849, on whether or not to have townships. All were among Douglas Meyer's Upland South counties for 1850 (chapter 3) and today retain strong rural connections, especially Pope and Scott. Massac, facing Paducah across the Ohio, has experienced modest population growth recently, with a total in 2000 exceeding 15,000. Pope and Scott stayed at about the same population between 1990 and 2000, with approximately five thousand souls each. Scott is my choice for class eponym because of it close relationship to Morgan County, which lends its name to the next group.

Morgan Class counties put township organization on their ballots at least once, without budging from the commission model. Morgan's proponents of township organization tried 17 times, failing each time to get a majority of the votes cast at the election. On most occasions the yes votes exceeded the no votes; but constitution framers, courts, and legislators demanded a majority of the county's voters (determined by the number voting at that election) for adoption to occur. Coming in behind Morgan in this class of 12 were Perry with 12 failed attempts and Randolph with 11.

All the remaining 87 Illinois counties adopted townships at least once. The 14 members of the Pike Class were among the 24 to adopt in 1849; but, unlike the other 10 members of the charter group, these 14 never checked signals and have been securely in the township column since the beginning. Pike gets the class-name nod because of how far it was in distance and culture in 1849–1850 from the other counties in its class, which clustered in the northeastern corner of the state.

Table 4 A Typology of Illinois Counties According to Decisions That Citizens Made Regarding Township Organization

Class	Decisions	Counties
Scott	To never vote on township organization.	Massac, Pope, Scott
Morgan	To vote at least once but never adopt.	Alexander, Calhoun, Edwards, Hardin, Menard, Monroe, Morgan, Perry, Pulaski, Randolph, Union, Wabash
Pike	To vote in 1849 and adopt that year.	Boone, Bureau, DeKalb, DuPage, Grundy, Kane, Kendall, Lake, LaSalle, Lee, Marshall, Pike, Stephenson, Will
Effingham	To vote only once, and to adopt, but not in 1849.	Effingham, Ford, Kankakee
Hancock/ Wayne	To adopt, perhaps in 1849, and then confirm at least once.	Adams, Carroll, Clark, Cook, Fulton, Hancock, McHenry, Macon, Ogle, Peoria, Tazewell, Vermilion, Wayne, Winnebago
Madison	To reject at least once, adopt, and never vote again.	Christian, Clinton, Coles, Cumberland, DeWitt, Edgar, Henry, Iroquois, Knox, Livingston, Logan, McDonough, Macoupin, Madison, Mercer, Montgomery, Piatt, Putnam, Warren, Whiteside, Woodford
Cass	To reject at least once, adopt, and confirm at least once after adoption.	Bond, Brown, Cass, Champaign, Clay, Crawford, Douglas, Fayette, Gallatin, Greene, Hamilton, Henderson, Jasper, Jefferson, Jersey, Jo Daviess, Lawrence, McLean, Marion, Mason, Moultrie, Rock Island, Saline, Sangamon, Schuyler, Shelby, Stark, St. Clair, White
Washington	To reject at least once, adopt, discontinue, and readopt.	Franklin, Jackson, Richland, Washington
Williamson	To reject at least once, adopt, discontinue, and reject.	Johnson, Williamson

I thought twice about creating the three-element Effingham Class because it is much like the previous class: all voted only once and adopted, never to vote again. On the other hand, none of the three can claim to be charter members of the township fraternity. Neither Ford nor Kankakee existed in 1849. Effingham voters decided not to vote on the issue of township organization that first year (1849), and (apparently) put off voting until there was, in 1860, sufficient voter support to accept on the first ballot. Ford's area, by the way, under administrative control of Vermilion County prior to Ford's organization as a separate county, would have enjoyed some benefits of township organization (e.g., the Vermilion County Board of Supervisors) in the 1850s.

The 14 counties in the next class, Hancock/Wayne, adopted the first time voters saw the issue on their ballots and then confirmed once (10 counties), twice (Clark), thrice (Ogle), or four times (Hancock and Wayne). In almost all cases, the first (or only) vote to confirm took place within three years of adoption. Eight of these checkups occurred in 1852, as early adopters from 1849, under the influence of the 1851 Township Organization Act reaffirmed their vows to township organization. McHenry, of course, waited a bit longer for its one checkup—from 1849 until 1994. Hancock's and Wayne's checkups all took place in the nineteenth century but spread over roughly a quarter of a century from their dates of adoption, which were 1849 for Hancock and 1859 for Wayne.

Once the voters in the 21 Madison Class counties tried the Yankee innovation, they liked it and never voted again on the matter. All these counties have in common also the fact that township proponents had to suffer through at least one failed adoption effort before their mates saw the township light. In Madison, the rejections prior to adoption went on and on, from 1849 to 1874, for a total of 13, including 5 years in a row right before the Civil War and 3 straight right after it ended. Madison's neighbors, Clinton and Macoupin, also members of this class, rejected the new method of local governance five times each before adopting in 1870 and 1873, respectively, just before Madison did likewise in 1875.

At 29 members, the Cass Class is the largest of the nine classes. Its namesake ranks near the top in terms of total countywide referenda with 14, which ties Cass with Madison for third place behind Williamson and Morgan. Cass, of course, was the last county to adopt, in 1923. Prior to adoption, Cass voters had considered the matter of township organization at least once in every decade, except the 1890s, since the 1848 Constitution and general assembly action gave them the option. Following adoption, voters there needed one more referendum in which to confirm that decision. All other counties in the Cass Class voted against adoption at least once, eventually adopted, and checked themselves at least once more down the line. A few of the 29 voted just the minimum for inclusion, three times; but most had multiple referenda before, after, or before and after adoption.

I discussed the circumstances of the Washington Class at length in chapter 4, as I tried there to convey the uncertainty factor they exhibited so well. Washington led the quartet with 13 referenda, followed by Franklin and Jackson with 9, and Richland at 6; but the most interesting aspect of all is that they adopted, disadopted, and readopted at least once. Franklin and Jackson readopted twice. None of the four has tested the township model at the ballot box since their most recent adoption, but P-TAC tried.

The last class, Williamson, contains the final two current commission counties. Both, after a series of rejections, gave township organization a try, and then reverted to commissioners. Proponents tried unsuccessfully four times in Johnson and twice in Williamson to revive townships. As I noted in chapter 4 for Johnson and chapter 5 for Williamson, however, the names of their defunct townships have persisted on the landscape in one form or another.

Reflections and Projections

Many inquired, during the course of this research endeavor, about where I stood on the matter of whether Illinois citizens in township counties would be better or worse off if township government were to disappear. In the gathering of information and the writing of this monograph, I tried to remain neutral, to keep an open mind, and to weigh all the evidence objectively. To a large measure, I believe I have succeeded, and as a result have maintained avenues of good communication between the major players and myself. When my own township's supervisor (now retired), Mary Caisley, introduced me to Executive Director George Miller at a 1992 TOI conference, she told him that I was "on their side." I suppose she was reacting to the negative press that townships had been getting in the early 1990s, to efforts in 1991 by Rep. Frank Giglio to secure passage of legislation that would have provided for the individual dissolution of Cook County's townships, and to the nearly successful countywide petition drive in 1990 of Michael Richardson and Randy Popp. I attempted to set the record straight about the fact that I was an academic interested in the diffusion and persistence of townships from a research perspective and not intent on prescribing the demise or in advocating the continuation of these tiny governmental entities. As I gained the confidence of Richardson, Bob Anderson, and others on the abolition side, I tried to keep an objective frame of mind, even when my natural instincts for siding with the underdog began to surface. I also had to bear in mind the fact that both sides knew I would likely publish my findings. Helping me meant, to some degree, helping their cause, their place in the record of Illinois politics.

So where do I stand on the Illinois version of the civil-political township? They were, I believe, a wonderful innovation when the Yankees at the 1847 Constitutional Convention practically forced them on the state. I believe that county option was and still is appropriate, however. Some counties, I believe, adopted township organization in order to keep up with neighboring counties. Some would have been better off leaving their commissioners in place. Once township supervisors and other aspects of the township system had operated for a few decades, inertia and uncertainty (about what would replace townships) made reversion to the commission model of county government difficult to achieve. It took the Great Depression for Williamson to revert successfully, and residents of that Little Egypt county seem to be missing townships not at all. My feeling is that voters get what they deserve in the case of local government. If they are willing to pay the usually small township premium in their property taxes, then they can have government that is a little closer to them, to the people. If not, they will get basically the same services from their county or municipality or special taxing district—probably for a little less in taxes. For me personally, as a geographer, townships have provided ideal local study units, areas with which residents and others nearby can identify.

I do have a problem, however, with some of the methods that the township side has employed since 1990 to defeat abolition efforts and make future efforts more difficult. The harassment by township supporters of those who questioned the township system seems inappropriate. I disagree with the need for a supermajority law, Public Act 89-365, or for the silver bullet, Public Act 90-112, requiring 10 percent of the voters of every township sign petitions calling for a countywide discontinuance referendum. Both laws have stood and may continue to stand between voters and the place where voters ought to have the right to effect changes—in the voting booth. TOI and local officials have been wrong to fight so strenuously to keep dissolution off the ballot at the county or township level. To me sometimes, it looks as if the government closest to the people does not trust the people. And that distrust is unfortunate.

At an academic gathering in Springfield, I once said that I expected challenges to the township way of governing to "bubble up" from time to time as the 1990s ended and a new millennium began, either as a result of dedicated citizen reformers or due to some issue that aroused the populace of a township county or a single township. That prediction preceded by just a few months the passage in 1997 of H. B. 480, which became Public Act 90-112. Since then, Bob Anderson did manage to get abolition of his township on the ballot, and Charles Horton came close to putting P-TAC's countywide referendum before the voters of Franklin County.

Here I am going now to revise my 1996 prediction. We have not, despite the legislative protection TOI leadership has leveraged and despite the educational initiative TOI has been funding (which I applaud) since 2000, seen the last challenge to the township way of life in Illinois. True, there may be fewer crusading mavericks with which to contend; but out there somewhere in the future an issue will arise that brings the people of a township or a county to the point where they hardly even need a champion of abolition to induce them to sign petitions calling for a referendum on discontinuance. Even without proper transition legislation in place, legislation from which a couple dozen post-1970 Constitution general assemblies have shied, the voters somewhere may just say that whatever comes next cannot be worse than what they now have. Once that first county or single township successfully negotiates the transition process, with legislative and court oversight, the door will have opened to others to ask thoughtfully which model of local government they really want. Then the best system, townships or commissioners or maybe something else, will prevail. Until that time, 85 Illinois counties will make the best of their townships. The diffusion of this innovation my have ended; but the outcome of that process, that significant change in governing at the grassroots level in Illinois, persists more than 150 years after 24 counties said yes in 1849 to township organization.[39]

Notes

1. "Park Name Commemorates Township Anniversary," *Township Perspective,* January 2000, 46; "Palatine Township Celebrates Township Anniversary by Helping Residents," *Township Perspective,* February 2000, 25; "Bremen High School Senior Wins Top Honors for Township Logo," *Township Perspective,* March 2000, 35; Rick Davis, "Strut Your Stuff: An Idea for the Sesquicentennial," *Township Perspective,* January 2000, 58–59; Julie Pittman, "From the Associate Director: Some Sesquicentennial Suggestions," *Township Perspective,* February 2000, 6–7.
2. "Governor Ryan Declares April as Township Government Month," *Township Perspective,* April 2000, 19.
3. Smith, "Happy Anniversary, Township Government!" *Township Perspective,* April 2000, 4–5.
4. Smith, "'A New Beginning' for Township Government Brings Need for Better Educated Officials," *Township Perspective,* December 2000, 24–25.
5. Jerry H. Kats, interview by author, 26 October 2000; Jill Gleason and Steven Stanger, interview by author, 11 October 2000.
6. Jill Gleason, telephone interview by author, 18 October 2000; Jill Gleason, "New Education Program Will Provide Learning Opportunities for Officials," *Township Perspective,* September 2000, 18–19; "TOI Education Program Addresses All Aspects of Official's Duties," *Township Perspective,* October 2000, 54–55; "The ABCs of Becoming a Better Township Official," *Township Perspective,* January 2001, 9–19.
7. Bryan E. Smith, telephone interview by author, 8 November 2002.
8. Dean Olsen, "Overcoming Rivalries," *Waukegan News Sun,"* 3 June 2000, A/1 and A/4; Jennifer Nelson, "Many Attempts to Downsize Have Failed," *Waukegan News Sun,* 3 June 2000, A/4; Jennifer Nelson, "Special Taxing Districts—Who's in Charge?" *Waukegan News Sun,* 5 June 2000, A/1 and A/4; Nick Reiher, "Townships a 'Whipping Boy' in Talk of Cutting Government," *Springfield State Journal–Register,* 5 June 2000, 4; Jennifer Nelson, "Reducing Number of Governments Faces Resistance," *Waukegan News Sun,* 6 June 2000, A/1 and A/4.
9. "Change Inevitably Part of Government," *State Journal-Register,* 13 June 2000, <www.sj-r.com/opinion/00/06/13/editorial.htm> (24 December 2000); "Too Much Government Not Serving Illinoisans," *Waukegan News Sun,* 8 June 2000, A/6.
10. James L. Muir, telephone interview by author, 12 October 2000; Charles Horton, interview by author, 26 April 2001. Muir proved to be an elusive source, following our October conversation. Much of what I say about his role comes from the Horton interview

and the many items that Muir wrote for the *Benton Evening News* before leaving to accept a position elsewhere. I cite hereafter as *News*.

11. Jim Muir, "Quadrennial Assessments on Way to Subscribers," *News,* 4 May 2000, 1; Muir and Horton interviews.
12. Jim Muir, "Assessment Protests Hit 150 Mark," *News,* 4 May 2000, 1; Jim Muir, "Assessments Raise More Questions Than They Answer," *News,* 10 May 2000, 4; Jim Muir, "An Open Memo to the Franklin County Board," *News,* 13 May 2000, 4.
13. Horton interview; Jim Muir, "County Board Hears Tax Protest," *News,* 17 May 2000, 1; Jim Muir, "Tax Protest Panel Named," *News,* 23 May 2000, 1.
14. The five-part Muir series unfolded as follows: "Township Government: Needed Unit of Government or Outdated Bureaucracy?" *News,* 9 June 2000, 1; "The Facts and Figures on County Townships," *News,* 13 June 2000, 1; Smaller Townships Get Share of Funding," *News,* 16 June 2000, 1–2; "Ewing, Easter, Barren and Northern Townships at a Glance," *News,* 21 June 2000, 1 and 3; "Two Sides Taking Differing Views on Value of Township Government," *News,* 26 June 2000, 1. Muir did not cover the protest meeting for the *News.* That task fell to Scott Hines, writing for the American News Service: Scott Hines, "Inequities of Property Tax System Addressed," *News,* 30 June 2000, 1. Franklin's opinion leader on abolition did write, however, the article that announced the petition drive: Jim Muir, "P-TAC Wants to Abolish Township Government," *News,* 19 July 2000, 1. In the midst of the five-part series, Bryan Smith got wind of the uprising in Franklin County and sent a typical TOI letter to the editor in which he reminded readers of the long history of townships, from New England to Illinois, and of the attempt to cut governmental costs in Illinois decades earlier through reduction in the size of the house of representatives. History proved wrong, Smith wrote, those who thought that action "would save millions of tax dollars." Bryan E. Smith, *News,* 15 June 2000, 4.
15. Horton interview; Jim Muir, "P-TAC Wants to Abolish Township Government," *News,* 19 July 2000, 1; Joel Hurewitz, email correspondence with author, 22 August 2002; Theresa Amato, email correspondence with author, 29 August 2002. The two attorneys I consulted, admittedly, have played important roles, on the abolitionist side, in abolition-related litigation in northeastern Illinois. Hurewitz has worked off and on for Bob Anderson. Amato led the way in the DuPage suit by STOP against the election commission over the supermajority law. Hurewitz eventually became a supporting attorney in that case, along with Michael Walkup. Be that as it may, I believe their reflections on the situation merit mention here. Hurewitz wrote in his email message: "It's a stupid statute, because it can result in a minority countywide vote properly distributed in particular townships [rejecting the abolition] referendum....Moreover, I still think the statute violates the constitution." Amato confirmed for me that the Byrne decision still holds sway only in DuPage County, "but it could be 'persuasive' in other districts, as it would surely be raised by counsel or the court in which any other challenge is presented." I think, she went on, "any judge anywhere would rule it unconstitutional for the same reasons Judge Byrne did so."
16. Jim Muir, "Local Officials Fight Back," *News,* 26 July 2000, 1.
17. Derek Johnson, "More to the Township Story than is Easily Apparent," *News,* 5 August 2000, 4.
18. Johnson, "More to the Township Story, 4; Illinois, General Assembly, *Laws of the State of Illinois, Ninety-First General Assembly, 2000* (Springfield, 2001), 1552–53; Horton interview.

19. Jim Muir, "Initial Steps Taken in Putting Reassessment on the Ballot," *News,* 8 August 2000, 1; Jim Muir, "First Person," *News,* 11 August 2000, 4; Jim Muir, "P-TAC Closes in on Needed Signatures," *News,* 14 August 2000, 1; Scott Hines, "Clock Ticking on P-TAC Petition Drive," *News,* 17 August 2000, 1; Tom Broy, *News,* 18 August 2000, 4; Ruby H. Melvin, *News,* 18 August 2000, 4.
20. Jim Muir, "Abolishment Petitions Filed Today," *News,* 21 August 2000, 1; Horton interview.
21. Jim Muir, "County Officials File Objection to Petition," *News,* 26 August 2000, 1; Jim Muir, "P-TAC Chairman Not Surprised by Challenge," *News,* 26 August 2000, 1; Jim Muir, "New Objections to Township Vote Petitions," *News,* 29 August 2000, 1.
22. Jim Muir, "Township Officials Working Overtime Trying to Avoid Vote," *News,* 30 August 2000, 4.
23. Jim Muir, "Election Board Attorney: Some Petition Objections Ruled Invalid," *News,* 30 August 2000, 1; Jim Muir, " Election Board Hears Evidence," *News,* 7 September 2000, 1.
24. Muir, "Election Board Hears Evidence," 1.
25. Jim Muir, "Bid to Seek Vote on County Government, Abolition of Townships Sidetracked," *News,* 12 September 2000, 1; Jim Muir, "More Than One Way to Deal With a Bully," *News,* 13 September 2000, 4.
26. Tom Broy, *News,* 22 September 2000, 4. Muir could not resist responding to Broy, noting that Broy had sent similar letters to several other newspapers. According to Muir, the real reason that Broy had worked so hard for FTG and against P-TAC was personal. If the townships had given way to commissioners in Franklin County, the Broy family (Muir claimed) stood to lose Broy's supervisor salary ($23,992), his wife's salary for overseeing general assistance in Frankfort Township ($22,662), his son's wife's salary for helping with general assistance ($18,852), and all the health and other benefits they could claim. Furthermore, Broy would have lost his $800 per month stipend from his county board seat, unless he had become one of the new commissioners. Muir noted that general assistance recipients in Frankfort totaled seven in the year 2000. Jim Muir, "Township Supervisor Doesn't Get It about Equalization Factor," *News,* 27 September 2000, 4.
27. Jim Muir, "Denial of County Government Referendum Will Be Reviewed Wednesday," *News,* 23 September 2000, 1; Horton interview.
28. Jim Muir, "Petition Drive Dead; Continuance Granted," *News,* 27 September 2000, 1; Jim Muir, "For the Sake of the County, Remove Your Political Blinders," *News,* 7 October 2000, 4. Stalling tactics by the township side had been a major challenge for activists elsewhere in Illinois during most of the 1990s. Any delay after the filing of petitions in August for a November referendum hurts the side that is seeking change.
29. Jim Muir, "Final Arguments Heard Over Petition Drive," *News,* 29 November 2000, <www.bentoneveningnews.com/display/inn_news/news01.txt> (30 November 2000); Jim Muir, "Back to the Drawing Board for P-TAC," *News,* 6 December 2000, <www.bentonevening news.com/display/inn_news/news09.txt> (11 December 2000).
30. Horton interview.
31. John Kelly and Christopher Wills, "Township Government Can Be Inefficient, But Few are Noticing," *Pantagraph,* 20 December 2000, A/1 and A/7; "Townships: Are They Worth It?" *Argus,* 20 December 2000, A/1 and A/8; "AP Investigation Questions Need for Government Form," *Decatur Herald and Review,* 20 December 2000, <www.herald-review.com/0/tshipap1220-0.html> (20 December 2000). I cite hereafter as *Herald and Review.*

32. Kelly and Wills, "AP Investigation Questions Need"; "Townships Surpluses," *Argus,* 20 December 2000, A/8; Bob Holliday, "Surpluses Often Fund Big Road Projects," *Pantagraph,* 20 December 2000, A/7.
33. John Kelly and Christopher Wills, "Township Provides Hometown Service to Needy," *Pantagraph,* 21 December 2000, A/1 and A/4; Ron Ingram, "Decatur Man Answers Township Post Calling," *Herald and Review,* 21 December 2000, <www.herald-review.com/0/smith1221-0.html> (21 December 2000); Stephen Elliott, "Hampton Clinic a Rare Breed," *Argus,* 21 December 2000, A/8. Sonya Klopfenstein, "Townships Take Burden Off Counties," *Peoria Journal Star,* 21 December 2000, <www.pjstar.com/news/ssection/cop170a.html> (21 December 2000). I cite hereafter as *Journal Star.*
34. John Kelly and Christopher Wills, "Township Foes Face Long Odds in Bringing Change," *Pantagraph,* 22 December 2000, A/1 and A/8; "Want to Kill Townships? Here's How," *Pantagraph,* 22 December 2000, A/8; Stephen Elliott, "Townships' No. 1 Gadfly Leaving Battle to Others," *Argus,* 22 December 2000, A/1 and A/8; Mike Monson, "Many Area Townships Have Huge Surplus Funds," *Champaign News-Gazette,* 22 December 2000, <www.newsgazette.com/story.cfm?Number=8508> (23 December 2000); Stephanie McClelland, "Vermilion Townships Defend Surpluses," *Champaign News-Gazette,* 22 December 2000, <www.newsgazette.com/story.cfm?Number=8508> (23 December 2000).
35. "Townships Have Outlived Usefulness in Many Instances," *Pantagraph,* 23 December 2000, A/8; "Our Opinion: Townships," *Carbondale Southern Illinoisan,* 21 December 2000, <www.southernillinoisan.com/rednews/2000/12/22/build/opinions/OPI002.html> (22 December 2000); "Townships and Your Tax Dollars," *Chicago Tribune,* 20 December 2000, 1/22; "Abolish Illinois' Wasteful Townships," *Journal Star,* 23 December 2000, <www.pjstar.com/news/editorial/cop9753a.html> (24 December 2000); *Springfield State Journal-Register,* 24 December 2000, <www.sjr.com/opinion/00/12/24/editorial.htm> (24 December 2000).
36. John Kelly and Christopher Wills, "Townships Facing Criticism," *Huntsville* (Alabama) *Times,* 7 January 2001, A/6; Kelly and Wills, "Illinois Townships Fat with Overhead," *Janesville* (Wisconsin) *Gazette,* 7 January 2001, A/3; Kelly and Wills, "Illinois Townships Fat with Overhead, Surpluses," *McAllen* (Texas) *Monitor,* 7 January 2001, A/5. The *Journal Star* items, all appearing 7 January 2001 on page A/5 were as follows: Bryan Smith, "Township Governments Meet Special Needs of Their Taxpayers"; Bob Anderson, "Townships Escape Most Public Scrutiny; Tax Money not Wisely Used"; Timothy L. Bertschy, "Illinois' Smallest Government Provides High-quality Service at Low Cost"; and John C. Parkhurst, "Political Pressure Saved Townships When the New State Constitution Was Written." For carbon copies of the Smith and Anderson items, see these: Bryan E. Smith, "Townships: Local Concerns Are Their Special Interest Every Day," *Pantagraph,* 5 January 2001, A/11; Bob Anderson, "Township Government Wasteful, Redundant, Unneeded," *Rockford Register Star,* 21 January 2001, H/3.
37. Smith, "Township Critics Should Take a Look at the Job Being Done," *Township Perspective,* February 2001, 4–5; Barbara Weyrick, "Did You Know…?" *Township Perspective,* April 2001, 36–37; Barbara Weyrick, "Did You Know…?" *Township Perspective,* May 2001, 52–53.
38. Mystery surrounds this sudden jump in the requirement for petition signatures calling for a referendum in which discontinuance might be the decision. The requirement of "one-fifth of the legal voters of any county having adopted township organization, to be ascertained by the vote cast at the last presidential election" just appeared on page 1,068 in the

township organization portion of *The Revised Statutes of the State of Illinois.* As an authority, Hurd cited the Township Organization Act of 1861. The 1861 law, however, called for the same 50 signatures that had been the rule since 1851. Perusal of all public and private laws between 1861 and 1874 failed to reveal any modification during that period of the signature requirement with respect to petitions calling on the county board to place a discontinuance referendum on the ballot. Is it possible that some person or persons, in order to protect the township system, took it upon themselves to alter the existing statute when preparing the 1874 compilation of Illinois statutes? Could Elijah M. Haines, the great advocate of townships for everyone and a major political force during the 1870s, have had anything to do with such deceit? As late as 1904 and long after his 1889 death, the Haines compilation of statutes was still citing the 1861 law as the source of the requirement for "one-fifth of the legal voters." Harvey B. Hurd, ed. and comp., *The Revised Statutes of the State of Illinois. A. D. 1874* (Springfield: Illinois Journal Company, 1874), 1068; Illinois, General Assembly, *Public Laws of the State of Illinois Passed by the Twenty-second General Assembly, Convened January 7, 1861* (Springfield, 1861), 265; Elijah M. Haines, *A Compilation of the Laws of Illinois Relating to Township Organization* (Chicago: The Legal Adviser Publishing Company, 1904), 40.

39. Bob Anderson continued his crusade for less government in McHenry County into the new millennium. In the fall of 2002, he managed to place on the ballot in his McHenry Township and adjacent Greenwood Township the proposition of consolidating the road districts of the two minor civil divisions. By a vote of 3,625 to 8,574 in McHenry and 1,257 to 1,554 in Greenwood, he failed once more to effect reform. For background, see Joseph Lopez, "Road Question on Ballot," *Northwest Herald,* 4 September 2002, C/1–2. For the results, I turned to "McHenry County Totals," *Northwest Herald,* 6 November 2002, <www.nwherald.com/elections> (6 November 2002).

Select Bibliography

Abress, Monica Dwyer. *Quietly at Work: Township Government in America.* North Branch, Minnesota: Specialty Press, 2000.

Allen, H. K. Control of Expenditures in the Local Governmental Units of Illinois, bul. 61. Urbana: University of Illinois, Bureau of Business Research, 1940.

———. "Efficiency and Economy in Local Government—With Particular Reference to Illinois." Ph.D. diss., University of Illinois, 1936.

———. "Relative Merits of Township and Non-Township Organization." *Illinois Journal of Commerce* 17 (January 1935): 8–9 and 29–30.

Allen, Howard W., and Vincent A. Lacey, eds. *Illinois Elections, 1818–1990: Candidates and County Returns for President, Governor, Senate, and House of Representatives.* Carbondale and Edwardsville: Southern Illinois University Press, 1992.

The Anatomy of Our Townships. Chicago: League of Women Voters of Illinois, 1976.

Anderson, Joan G., and Ann Lousin. "From Bone Gap to Chicago: A History of the Local Government Article of the 1970 Illinois Constitution." *The John Marshall Journal of Practice and Procedure* 9 (1976): 697–818.

Banovetz, James M., and Ann M. Elder. *The Local Government Article of the 1970 Illinois Constitution.* n.p.: Illinois Commission on Intergovernmental Cooperation, 1987.

Bauer, William J. "Present County Government System—Best of All Possible Worlds?" *DePaul Law Review* 17 (1968): 524–31.

Braden, George D., and Rubin G. Cohn. *The Illinois Constitution: An Annotated and Comparative Analysis.* Urbana: University of Illinois, Institute of Government and Public Affairs, 1969.

Bradshaw, William L., and Milton Garrison. "Township Organization in Missouri." *The University of Missouri Studies* 11 (October 1936): 1–70.

Bromage, Arthur W. "Functions and Powers of Village and Township Governments." In *Rural Government: Proceedings of the Fourteenth American Country Life Conference,* Ithaca, New York, August 17–20, 1931. Chicago: The University of Chicago Press, 1932.

———. "Shall We Save the Township?" *National Municipal Review* 25 (October 1936): 585–88.

Brown, Lawrence A. *Innovation Diffusion: A New Perspective.* New York: Methuen & Company, 1981.

Burman, Ian D. *Lobbying at the Illinois Constitutional Convention.* Urbana: University of Illinois, Institute of Government and Public Affairs, 1973.

"Changes in County Government." *Illinois Political Reporter* 9 (December 1969): 3 and 5.

Chapin, F. Stuart. *Cultural Change.* New York: The Century Co., 1928.

Clabaugh, Charles W. *Township School Trustees and Treasurers in Illinois: 1819–1980.* n.p.: Township School Trustees of Cook County, 1980.

Clayton, John. *The Illinois Fact Book and Historical Almanac: 1673–1968.* Carbondale and Edwardsville: Southern Illinois University Press, 1970.

Cole, Arthur Charles. *The Constitutional Debates of 1847.* Collections of the Illinois State Historical Library, vol. 14. Springfield: Illinois State Historical Library, 1919.

Colten, Craig. *Where the South Ends: A Critique of Regionalizing in Cultural Geography,* no. 66. Syracuse: Syracuse University, Department of Geography, 1980.

"Commissions Will Study Illinois Counties, Townships." *National Municipal Review* 46 (October 1957): 474–75.

"Constitutions for the Seventies." *National Civic Review* 60 (January 1971): 4–5.

Cook County: Township and County Services. Chicago: League of Women Voters of Cook County, 1973.

Cornelius, Janet. *A History of Constitution Making in Illinois.* Urbana: University of Illinois, Institute of Government and Public Affairs, 1969.

Dolan, Drew A. "Local Government Fragmentation: Does It Drive Up the Cost of Government?" *Urban Affairs Quarterly* 26 (September 1990): 28–45.

Drury, James W. "Townships Lose Ground." *National Municipal Review* 44 (January 1955): 10– 13.

Fairlie, John A. "County and Town Government in Illinois." *Annals of the American Academy of Political and Social Science* 47 (May 1913): 62–78.

———. "Governmental Reorganization in Illinois." *American Political Science Review* 9 (May 1915): 252–57.

Fliegel, Frederick C. *Diffusion Research and Prospects in Rural Sociology: The Record and Prospects for the Future.* Westport, Connecticut: Greenwood Press, 1993.

Frazer, Timothy C. "The Language of Yankee Cultural Imperialism: Pioneer Ideology and 'General American'." In *Heartland English: Variation and Transition in the American Midwest,* edited by Timothy C. Frazer. Tuscaloosa: University of Alabama Press, 1993.

Garrison, Milton. "History of Township Organization in Missouri." *The Missouri Historical Review* (October 1935): 47–56.

Gertz, Elmer, and Joseph P. Pisciotte. *Charter for a New Age: An Inside View of the Sixth Illinois Constitutional Convention.* Urbana: University of Illinois, Institute for Government and Public Affairs, 1980.

Gould, Peter. *The Geographer at Work.* London: Routledge & Kegan Paul, 1985.

———. *Spatial Diffusion: The Spread of Ideas and Innovations in Geographic Space.* New York: Learning Resources in International Studies, 1975.

Greenough, Katharine Croan. *Know Your County.* Washington: National League of Women Voters, 1937.

Haines, Elijah M. *A Compilation of the Laws of Illinois Relating to Township Organization and Management of County Affairs,* new rev. ed. Chicago: E. B. Myers and Company, 1875.

Hicks, H. S. "The Costliness of Local Government." *Illinois Journal of Commerce* 16 (September 1934): 12–13 and 27.

———. *How Many Taxing Bodies, How Much Government Do Taxpayers Need?: An Analysis of Expenditures by Minor Governmental Units of Macon County, Illinois.* [Decatur]: Association of Commerce of Decatur and Illinois Chamber of Commerce, 1936.

———. "Is the Township an Obsolete Form of Government?" *Illinois Journal of Commerce* 14 (October 1932): 14 and 25.

———. "Railroad Taxes in Illinois." *Illinois Journal of Commerce* 16 (October 1934): 21–23.

———. *Waste in Local Government: An Analysis of Local Expenditures of the Minor Governmental Units of Winnebago County, Illinois.* [Rockford]: Rockford Chamber of Commerce, 1934.

———. *Waste in Local Government: An Analysis of Local Government Costs in Sangamon County, Illinois.* [Springfield]: Springfield Chamber of Commerce, 1935.

———. *Waste in Local Government: An Analysis of the Expenditures of the Minor Governmental Units of St. Clair County, Illinois.* n.p.: Illinois Chamber of Commerce and East St. Louis Chamber of Commerce, 1935).

———. *The Wasteful System in Local Government: An Analysis of the Expenditures of the Minor Governmental Units of Kane County, Illinois.* n.p.: Illinois Chamber of Commerce, Aurora Chamber of Commerce, and Elgin Association of Commerce, 1936.

———. *Wasting Tax Money by the Millions: A Study in Township Road Building Costs in 12 Illinois Counties,* 2d ed. [Chicago]: Illinois Chamber of Commerce, 1935.

How Many Taxing Bodies, How Much Government Do Taxpayers Need?: An Analysis of the Expenditures of Minor Governmental Units of Tazewell County, Illinois. [Pekin]: Illinois Chamber of Commerce and Pekin Association of Commerce, 1937.

Howard, Robert P. *Illinois: A History of the Prairie State.* Grand Rapids, Michigan: William B. Eerdmans Publishing Company, 1972.

Howards, Irving. "Commission Studies Township Setup." *National Municipal Review* 47 (April 1958): 184–85.

———. "Illinois Township Commission Reports." *National Civic Review* 49 (January 1960): 35– 36.

Hunter, M. H. *Costs of Township and County Government in Illinois*, bul. 45. Urbana: University of Illinois, Bureau of Business Research, 1933.

Hurd, Harvey B. *The Revised Statutes of the State of Illinois. A.D. 1874.* Springfield: Illinois Journal Company, 1874.

Illinois. Commission on Taxation and Expenditures. *Report and Recommendations of the Illinois Commission on Taxation and Expenditures.* Springfield, 24 January 1933.

Illinois. Constitutional Convention. *Debates and Proceedings of the Constitutional Convention of the State of Illinois, Convened at the City of Springfield, Tuesday, December 13, 1869.* Springfield, 1870.

Illinois. Constitutional Convention. *Journal of the Constitutional Convention of the State of Illinois, Convened at Springfield, January 7, 1862.* Springfield, 1862.

Illinois. Constitutional Convention. *Journal of the Convention Assembled at Springfield June 7, 1847.* Springfield, 1847.

Illinois. Constitutional Convention. *Proceedings of the Constitutional Convention of the State of Illinois Convened January 6, 1920.* 5 vols. Springfield, 1920–1922.

Illinois. Joint Legislative Committee. *A Report of the Joint Legislative Committee of the 47th General Assembly.* Springfield, 1913.

Illinois. Legislative Reference Bureau. *List of Suggestions for Constitutional Change Together with Texts of Constitutions of Illinois.* Springfield, 1919.

Illinois. Retrenchment Commission. *Report of the Retrenchment Commission.* Kankakee, 1935.

Illinois. Secretary of State. *Counties of Illinois: Their Origin and Evolution.* n.p.: 1906.

Illinois. Sixth Illinois Constitutional Convention. "Journal of the Constitutional Convention, State of Illinois, 1969–1970." Springfield, 1969–1970.

———. *Record of Proceedings.* 7 vols. Springfield: Office of the Secretary of State, 1972.

Illinois. Township Government Laws Commission. *Report on the Township Government Laws Commission.* Springfield, 15 March 1978.

Johnson, Robin, and Norman Walzer. *Financial Picture of Townships in Illinois.* Macomb: Western Illinois University, Institute for Rural Affairs, 1996.

Jones, Howard P. "What Changes Should Be Made in Our Local Units of Government?." In *Rural Government: Proceedings of the Fourteenth American Country Life Conference, Ithaca, New York, August 17–20, 1931.* Chicago: The University of Chicago Press, 1932.

Kelly, Sheila. *County and Township Gazetteer.* Springfield: Illinois State Archives, 1988.

Kenney, David. *Basic Illinois Government: A Systematic Explanation.* Carbondale and Edwardsville: Southern Illinois University Press, 1974.

———. *Making a Modern Constitution: The Illinois Experience.* n.p.: Jackson County [Illinois] Historical Society, 1991.

Lind, Michael. "A Horde of Lilliputian Governments." *The New Leader,* 5 May 1997, 6–7.

Lovett, Lori. *Notes on the Location of Illinois County Seats.* Springfield: Illinois State Archives, 1988.

Lowden, Frank O. "The Problems of Rural Government." In *Rural Government: Proceedings of the Fourteenth American Country Life Conference, Ithaca, New York, August 17–20, 1931.* Chicago: The University of Chicago Press, 1932.

———. "Scrapping Useless Governments." *National Municipal Review* 22 (March 1933): 122– 24 and 136.

Manny, T. B. "Government Structure, Powers and Problems in the New Rural Municipality." In *Rural Government: Proceedings of the Fourteenth American Country Life Conference, Ithaca, New York, August 17–20, 1931.* Chicago: The University of Chicago Press, 1932.

———. "The Unit Requirement Idea for Local Government in Rural Areas." In *Standards of Living: Proceedings of the Thirteenth American Country Life Conference, Madison, Wisconsin, October 7–10, 1930.* Chicago: The University of Chicago Press, 1931.

Meek, J. T. "What the Referendum Brought Out." *Illinois Journal of Commerce* 16 (November 1934): 5–6, 16, and 29.

Melton, L. D. "The Elimination of Township Government in Oklahoma." *National Municipal Review* 27 (August 1938): 405–7.

Meyer, Douglas K. *Making the Heartland Quilt: A Geographical History of Settlement and Migration in Early-Nineteenth-Century Illinois.* Carbondale and Edwardsville: Southern Illinois University Press, 2000.

Miller, David R. *1970 Illinois Constitution: Annotated for Legislators,* 4th ed. Springfield: Illinois Legislative Research Unit, 1996.

Miller, George H. "Township Government." In *Illinois Local Government,* edited by James F. Keane and Gary Koch. Carbondale and Edwardsville: Southern Illinois University Press, 1990.

———. "Townships Defended: 'Last Bastion of Pure Democracy in an Otherwise Bureaucratized Republic.'" *Illinois Issues,* March 1993, 9.

Newell, Mason H. "Township Government in Illinois." *Transactions of the Illinois State Historical Society* 9 (1904): 467–504.

Nowlan, James D. *The Itinerant: A Heartland Story.* Winnetka, Illinois: Conversation Press, 2000.

Pisciotte, Joseph P. "How Illinois Did It." *National Civic Review* 58 (July 1969): 291–96.

"Proposals for the Reorganization of Local Government in Illinois: Part 1—The State Outside of Cook County." Chicago: Griffenhagen & Associates, 30 November 1932.

Richardson, Michael. "Abolish Townships: Anachronistic and Unnecessary Today in Illinois' Massive Political Pyramid." *Illinois Issues,* March 1993, 8.

Robertson, Mary Helen. "Constitutional Revision in Illinois: The League of Women Voters' Role." *National Civic Review* 60 (September 1971): 438–43 and 480.

Rogers, Everett M. *Diffusion of Innovations,* 3rd ed. New York: The Free Press, 1983.

Schwartz, Thomas F. "'Illinois & her Politicians': The Observations of Hezekiah Morse Wead, Delegate to the 1847 Illinois Constitutional Convention." *Journal of Illinois History* 3 (Spring 2000): 31–50.

Scott, Franklin William. *Newspapers and Periodicals of Illinois: 1814–1879.* Collections of the Illinois State Historical Library, vol. 6. Springfield: Illinois State Historical Library, 1910.

Snider, Clyde F. "Indiana Counties and Townships." *Indiana Magazine of History* 33 (June 1937): 120–52.

———. "The Twilight of the Township." *National Municipal Review* 41 (September 1952): 390– 96.

Sokolow, Alvin D. "A Change in the Role of Rural Townships?" *National Civic Review* 50 (April 1961): 209–10.

Spencer, Richard C. "Iowa Townships Still Here?" *National Municipal Review* 41 (September 1952): 397–99.

Sublett, Michael D. *Paper Counties: The Illinois Experience, 1825–1867.* New York: Peter Lang Publishing, Inc., 1990.

Township and Other Related Laws of Illinois. Charlottesville, Virginia: Matthew Bender & Company, Inc., 2000.

Township Organization for Illinois or Local Self-Government for the People. Alton, Illinois: Courier Steam Book and Job Printing House, 1859.

U.S. Works Progress Administration. Writers' Program. *County and Township Government.* [1940].

Van Sickle, Frederick Mercer. *Grassroots Democracy: A Celebration of the Township Officials of Illinois 75th Anniversary, 1907–1982.* Astoria, Illinois: Township Officials of Illinois, 1982.

Verlie, Emil Joseph, ed. *Illinois Constitutions,* Collections of the Illinois State Historical Library, vol. 13. Springfield: Illinois State Historical Library, 1919.

Wager, Paul W. "Can Local Self-Government Be Preserved in Our Rural Areas?" In *Rural Government: Proceedings of the Fourteenth American Country Life Conference, Ithaca, New York, August 17–20, 1931.* Chicago: The University of Chicago Press, 1932.

———. "Townships on Way Out." *National Municipal Review* 46 (October 1957): 456–60 and 475.

Walzer, Norman, and Nancy Baird. *Administration and Finance of Illinois Townships.* Macomb: Western Illinois University, Institute for Rural Affairs, [1990].

Watson, Johanna M. *Electing a Constitution: The Illinois Citizen and the 1970 Constitution.* Urbana: University of Illinois, Institute of Government and Public Affairs, 1980.

Wilson, Thomas Dwight. "Patterns of Change in Illinois County Government, 1960 to 1974." Ph.D. diss., University of Illinois, 1975.

Index

C

D

E

U

V

W

Y

Z